T0135598

Classification and Structure Theory of Lie Algebras of Smooth Sections

Klassifikation und Strukturtheorie von Lie-Algebren glatter Schnitte

Der Naturwissenschaftlichen Fakultät der
Friedrich-Alexander-Universität Erlangen-Nürnberg

zur

Erlangung des Doktorgrades Dr. rer. nat.

vorgelegt von

Hasan Gündoğan
aus Krefeld

Als Dissertation genehmigt von der Naturwissenschaftlichen Fakultät der
Universität Erlangen-Nürnberg

Tag der mündlichen Prüfung: 28. Oktober 2011

Vorsitzender der
Promotionskommission: Prof. Dr. Rainer Fink
Erstberichterstatter: Prof. Dr. Karl-Hermann Neeb
Zweitberichterstatterin: Prof. Dr. Catherine Meusburger

Bibliografische Information der Deutschen Nationalbibliothek

Die Deutsche Nationalbibliothek verzeichnet diese Publikation in der
Deutschen Nationalbibliografie; detaillierte bibliografische Daten sind
im Internet über http://dnb.d-nb.de abrufbar.

ISBN 978-3-8325-3024-2

Logos Verlag Berlin GmbH
Comeniushof, Gubener Str. 47,
10243 Berlin
Tel.: +49 (0)30 42 85 10 90
Fax: +49 (0)30 42 85 10 92
INTERNET: http://www.logos-verlag.de

anneme ve babama

Danksagung

Es gibt viele Menschen, denen ich danken möchte, weil es ohne sie diese Doktorarbeit nicht gäbe. Zuerst danke ich meinem Doktorvater Prof. Dr. Karl-Hermann Neeb, einem der beeindruckendsten Wissenschaftler und Menschen, denen ich begegnet bin. Neben seinem enormen mathematischen Wissen, auch in scheinbar entlegenen Gebieten, und seinen großen wissenschaftlichen Leistungen, fasziniert mich jedes Mal aufs Neue seine schier unendliche Geduld mit seinen Doktoranden und Studenten, ferner seine Fähigkeit und Bereitschaft, in persönlichen Gesprächen Sachverhalte beliebig genau und verständlich zu erklären, und schließlich seine bodenständige und sympathische Art.

Als nächstes danke ich all meinen ehemaligen und aktuellen Kollegen, mit denen ich zusammenarbeiten und Freundschaften aufbauen konnte. Hervorheben möchte ich zunächst meinen Doktorbruder Michael Klotz, mit dem ich neben mathematischen Gesprächen auch Unterhaltungen über buchstäblich „Gott und die Welt" führte und dabei viele neue Einblicke gewann. Auch meinen Erlanger Doktorbrüdern schulde ich viel: Christoph Zellner, den ich seit über zehn Jahren kenne, rettete mir das eine oder andere Mal in mathematischen Fragen das Leben, und Stefan Wagner, mit dem ich vor neun Jahren gleichzeitig in Darmstadt das Mathematik-Studium begann, initiierte immer wieder mathematische Diskussionen an der Tafel, so dass wir uns gegenseitig helfen konnten. Ein weiterer Dank gilt Gerlinde Gehring, die als Sekretärin meinen Kollegen und mir in der Darmstädter Zeit den Rücken freihielt.

Der Studienstiftung des deutschen Volkes danke ich für die Finanzierung meiner Promotion, für die Gelegenheiten viele interessante Studenten anderer Universitäten kennenzulernen, und für die vielen schönen Erlebnisse auf den Sommerakademien. Dank der Förderung durch die Studienstiftung war ich während meiner Promotion zeitlich und örtlich flexibel und konnte mich voll auf meine Forschung konzentrieren; das war ein Privileg, das nicht zu sehr hervorgehoben werden kann. In diesem Zusammnehang danke ich auch meiner Vertrauensdozentin Prof. Dr. Barbara Drossel von der Technischen Universität Darmstadt für ihre vorbildliche Betreuung.

Ich danke auch meiner Familie, insbesondere meiner Mutter, Döne Gündoğan, und meinem viel zu früh verstorbenen Vater, İmam Gündoğan: dafür, dass sie den Mut hatten, ihre Heimat Richtung Deutschland zu verlassen, um ihren Kindern ein besseres Leben zu ermöglichen; dafür, dass sie beide Tag und Nacht schufteten, um meine Schwestern und mich großzuziehen; dafür, dass sie uns in diese Welt setzten.

Zu guter Letzt danke ich meiner Freundin Claudia Ledwig für ihre unendliche Geduld mit mir, für ihre mentale Unterstützung während der letzten dreieinhalb Jahre und für die Tatsache, dass es sie gibt.

Zusammenfassung

Ausgehend von einem endlich-dimensionalen Lie-Algebren-Bündel $\pi : \mathfrak{K} \to M$ mit Faser $\mathfrak{k}$ definieren wir, für $k \in \mathbb{N}_0 \cup \{\infty\}$, auf dem Raum $\Gamma^k(\mathfrak{K})$ der C^k-Schnitte dieses Bündels eine Lie-Klammer über die Regel $[X, Y]_m := [X_m, Y_m]$ für $X, Y \in \Gamma(\mathfrak{K})$ und $m \in M$. Die Struktur- und Klassifikationstheorie dieses Typs von Lie-Algebren ist das Thema dieser Dissertation.

Anfangs werden elementare Eigenschaften diskutiert: das Zentrum, die Kommutator-Algebra, die Derivationen, das Zentroid und die Automorphismen von $\Gamma^k(\mathfrak{K})$. Hierzu werden dem Lie-Algebren-Bündel $\mathfrak{K}$ mit Faser $\mathfrak{k}$ auf natürliche Weise die Vektorbündel $\mathfrak{z}(\mathfrak{K})$, $[\mathfrak{K}, \mathfrak{K}]$, $\mathrm{Der}(\mathfrak{K})$, $\mathrm{Cent}(\mathfrak{K})$ und $\mathrm{Aut}(\mathfrak{K})$ mit jeweiliger Faser $\mathfrak{z}(\mathfrak{k})$, $[\mathfrak{k}, \mathfrak{k}]$, $\mathrm{Der}(\mathfrak{k})$, $\mathrm{Cent}(\mathfrak{k})$ und $\mathrm{Aut}(\mathfrak{k})$ zugeordnet, so dass sich natürliche Isomorphien $\mathfrak{z}\left(\Gamma^k(\mathfrak{K})\right) \cong \Gamma^k(\mathfrak{z}(\mathfrak{K}))$ und $\left[\Gamma^k(\mathfrak{K}), \Gamma^k(\mathfrak{K})\right] \cong \Gamma^k([\mathfrak{K}, \mathfrak{K}])$ ergeben. Für die weitere Betrachtung, in der unter anderem der Satz von Peetre, dass lokale Operatoren schon Differentialoperatoren sind, eine große Rolle spielt, ist die Voraussetzung vonnöten, dass die Faser-Algebra $\mathfrak{k}$ perfekt ist oder ein triviales Zentrum hat. Unter dieser Voraussetzung erhält man die Isomorphie $\mathrm{Der}\left(\Gamma^k(\mathfrak{K})\right) \cong \Gamma^k\left(\mathrm{Der}(\mathfrak{K})\right)$ für endliches k und eine kurze exakte Folge (von Vektorräumen)

$$0 \longrightarrow \Gamma(\mathrm{Der}\,(\mathfrak{K})) \longrightarrow \mathrm{Der}\,(\Gamma\,(\mathfrak{K})) \longrightarrow \Gamma(TM \otimes \mathrm{Cent}(\mathfrak{K})) \longrightarrow 0$$

für $\Gamma(\mathfrak{K}) = \Gamma^\infty(\mathfrak{K})$. Die Isomorphie $\mathrm{Cent}\left(\Gamma^k(\mathfrak{K})\right) \cong \Gamma^k(\mathrm{Cent}(\mathfrak{K}))$ gilt dann wieder für alle $k \in \mathbb{N}_0 \cup \{\infty\}$. Schließlich wird gezeigt, dass zwei Lie-Algebren von C^k-Schnitten mit jeweils zentral einfacher Faser genau dann isomorph sind, wenn die dazu gehörenden Lie-Algebren-Bündel isomorph sind, speziell ergibt sich eine kurze exakte Folge (von Gruppen)

$$1 \longrightarrow \Gamma^k\left(\mathrm{Aut}(\mathfrak{K})\right) \cong \mathrm{Aut}_{C^k(M, \mathbb{K})}\left(\Gamma^k(\mathfrak{K})\right) \longrightarrow \mathrm{Aut}\left(\Gamma^k(\mathfrak{K})\right) \longrightarrow \mathrm{Diff}^k(M)_\mathfrak{K} \longrightarrow 1.$$

Als nächstes wenden wir uns der Klassifikation von Lie-Algebren glatter Schnitte $\Gamma(\mathfrak{K})$ zu, die von einem Lie-Algebren-Bündel $\mathfrak{K} \to M$ mit zentral einfacher Faser $\mathfrak{k}$ kommen. Diese ist durch die Homotopieklassen $[M; B\,\mathrm{Aut}(\mathfrak{k})]_\sim$ gegeben, wobei $B\,\mathrm{Aut}(\mathfrak{k})$ der klassifizierende Raum der Lie-Gruppe $\mathrm{Aut}(\mathfrak{k})$ ist. Im Fall von $M = \mathbb{S}^1$ gibt es eine natürliche Bijektion zwischen $[M; B\,\mathrm{Aut}(\mathfrak{k})]_\sim$ und der Menge der Konjugationsklassen der endlichen Gruppe $\pi_0(\mathrm{Aut}(\mathfrak{k})) = \mathrm{Aut}(\mathfrak{k})/\mathrm{Aut}(\mathfrak{k})_0$, deren Bestimmung im Weiteren ausgeführt wird. Ein großer Teil dieser Arbeit beschäftigt sich mit der Frage, ob die zugehörige kurze exakte Folge von (Lie-)Gruppen

$$1 \longrightarrow \mathrm{Aut}(\mathfrak{k})_0 = \mathrm{Inn}(\mathfrak{k}) \longrightarrow \mathrm{Aut}(\mathfrak{k}) \longrightarrow \pi_0(\mathrm{Aut}(\mathfrak{k})) = \mathrm{Out}(\mathfrak{k}) \longrightarrow 1$$

für komplex und reell einfache Lie Algebren $\mathfrak{k}$ spaltet, die schließlich positiv beantwortet werden kann. Für geschlossene Mannigfaltigkeiten M der Dimension 2 zeigen wir, dass es eine natürliche Injektion von $[M; B\,\mathrm{Aut}(\mathfrak{k})]_\sim$ nach $\pi_1(\mathrm{Aut}(\mathfrak{k}))$ gibt, die den Kern $2 \cdot \pi_1(\mathrm{Aut}(\mathfrak{k}))$ hat, falls M nicht orientierbar ist, und sogar bijektiv ist, falls M orientierbar ist.

Anschließend definieren wir Verallgemeinerungen der Cartan-Killing-Form, nämlich universelle invariante symmetrische Bilinearformen, auf Lie-Algebren der Form $C \otimes \mathfrak{k}$, wobei $\mathfrak{k}$ eine endlich-dimensionale perfekte Lie-Algebra ist und C eine pseudo-unitale kommutative asso-ziative Algebra, eine Algebra, für die es zu je zwei Elementen $a, b \in C$ ein drittes Element $c \in C$ mit $ac = a$ und $bc = b$ gibt. Mit diesen Formen lässt sich dann eine universelle invariante symmetrische Bilinearform auf $\Gamma(\mathfrak{K})$ definieren, wobei diese auch stetig ist, wenn man $\Gamma(\mathfrak{K})$ mit der natürlichen Fréchet-Topologie versieht.

Auch universelle zentrale Erweiterungen werden thematisiert, und wie vorher werden diese erst für Tensorprodukt-Lie-Algebren $C \otimes \mathfrak{k}$ behandelt, wobei $\mathfrak{k}$ hier sogar halbeinfach sein muss. Für die Betrachtung unter Berücksichtigung einer Topologie auf C schränken wir auf den Fall ein, wo C eine strikte LF-Algebra ist und eine definierende Folge von unitalen Fréchet-Algebren besitzt. Schließlich kann man so, zumindest wenn die Basismannigfaltigkeit M kompakt ist, einen universellen 2-Kozykel auf $\Gamma(\mathfrak{K})$ finden.

Abgeschlossen wird diese Arbeit durch eine kurze Präsentation der Realisierung von affi-nen Kac-Moody-Algebren als erweiterte Schleifenalgebren (mit und ohne Twist) und einer Dis-kussion, inwiefern die Resultate der vorangegangen Kapitel über Lie-Algebren glatter Schnitte Verallgemeinerungen von Eigenschaften affiner Kac-Moody-Algebren sind.

Contents

Contents

1. Introduction

1.1. A Short History of Finite-Dimensional Lie Theory

The origins of modern Group Theory can be, at least, traced back to the French mathematician Évariste Galois (1811–1832), who used permutation groups to show that, in modern language, a polynomial equation is solvable by radicals if and only if its *Galois group* is solvable. Since then, mathematicians used the properties of symmetry groups to understand and solve equations. One of them was the Norwegian Marius Sophus Lie (1842–1899) who developed the theory of *continuous groups*, today known as *Lie groups*, in his famous books "Theorie der Transformationsgruppen" and "Vorlesungen über continuierliche Gruppen". These groups occured to him in the treatment of differential equations. Lie had the ingenious idea to examine the "infinitesimal" structure of Lie groups, i.e., in modern words, to study what effect the group multiplication has on the tangent space of the Lie group and encode this information in an additional structure, the *commutator bracket*, to obtain the corresponding *Lie algebra* of the group. Lie algebras can be defined independently of groups as vector spaces with a bilinear, alternating commutator bracket satisfying the Jacobi identity, named after the German Carl Gustav Jacobi (1804–1851). The fact that every abstractly defined finite-dimensional Lie algebra corresponds to a Lie group is known as "Lie's Third Theorem".

Other important figures in this mathematical field were the Germans Friedrich Engel (1861–1941), who was Lie's closest collaborator, and Wilhelm Killing (1847–1923), who gave an almost complete classification of the finite-dimensional complex simple Lie algebras; the classification was rigorously finished by the French mathematician Élie Cartan (1869–1951) and is done by constructing and classifying the corresponding *Cartan matrices*. The German mathematician and theoretical physicist Hermann Weyl (1885–1955) made great progress in the theory of compact Lie groups and it was not until then, that Lie theory became a basic concept of Quantum Mechanics.

The works of the Italian Eugenio Elia Levi (1883–1917) and the Russian Anatoly Malcev (1909–1967) showed that finite-dimensional complex Lie algebras decompose semidirectly into solvable and semisimple Lie algebras. Lie himself could characterize solvable Lie algebras as subalgebras of the algebra of strictly upper triangular matrices and semisimple Lie algebras are sums of the simple one, described by Cartan and Killing. Finally, the Russian Igor Ado (1910–1983) showed that any finite-dimensional complex Lie algebra can be described by matrices.

While the finite-dimensional Lie algebras are understood quite well by the above mentioned results, the theory of infinite-dimensional Lie algebras has still many open questions. For instance, one cannot expect a structural result like the one obtained by Levi and Malcev, but the existence of a *universal central extension* for a class of infinite-dimensional Lie algebras is an appropriate generalization.

1.2. Infinite-Dimensional Lie Algebras

The first infinite-dimensional Lie algebra naturally appearing in the geometry of manifolds is the Lie algebra of smooth vector fields $\mathcal{V}(M) = \Gamma(TM)$, the sections of the tangent bundle of a manifold M. Pursell and Shanks showed in [SP54] that isomorphic Lie algebras of smooth vector fields come from diffeomorphic manifolds.

Another class of infinite-dimensional Lie algebras is the one of *mapping algebras* $C^\infty(M, \mathfrak{g})$ for a manifold M and a finite-dimensional Lie algebra $\mathfrak{g}$, where the Lie bracket is defined pointwise: $[f, g](m) := [f(m), g(m)]$. More generally, if A is a commutative associative algebra and $\mathfrak{g}$ is a finite-dimensional Lie algebra, then the *current algebra* $A \otimes \mathfrak{g}$ is a Lie algebra with Lie bracket defined by $[a \otimes x, b \otimes y] := ab \otimes [x, y]$. Current algebras have universal central extensions, a result known as Kassel's Theorem (cf. [Kas84]). In [Mai02], Maier could extend this result to *topological current algebras*, i.e., to $A \otimes \mathfrak{g}$ for a commutative associative algebra A with a Fréchet topology. For details on the appearance of current algebras in mathematical physics we may refer to the monograph [Mic89].

A third class was independently developed by Victor Kac in [Kac68] and Robert Moody in [Moo68]: the nowadays called *Kac-Moody algebras*, who appear by reconstructing the Lie algebra from *generalized Cartan matrices*, i.e., matrices with all the properties induced by being a Cartan matrix of a complex simple Lie algebra, except for the finite dimension. For a survey on Kac-Moody algebras, see [Kac90]. The *affine* algebras are the infinite-dimensional Kac-Moody algebras which are "closest" to the finite-dimensional Lie algebras: the rank of an $n \times n$-Cartan matrix is n, the rank of an $n \times n$-generalized Cartan matrix belonging to an affine Kac-Moody algebra is $n - 1$. Moreover, they can also be realized as a two-dimensional extension of the *algebraic loop algebras* $\mathcal{L} \otimes \mathfrak{g}$, where $\mathcal{L}$ is the commutative associative algebra of complex-valued Laurent polynomials $\sum_{n \in \mathbb{Z}} a_n t^n$ and $\mathfrak{g}$ is a finite-dimensional complex simple Lie algebra.[1] Automorphisms of finite order and real forms of affine Kac-Moody algebras have been studied and completely classified by the work of many authors (cf. [Kob86], [BP87], [Lev88], [Rou88], [Rou89a], [Rou89b], [Bau89], [BR89], [Cor92a], [Cor92b], [Cor92c], [And92], [KW92], [BVBPBMR95], [Bat00], [JZ01], [BMR03], [BMR04]), but since this classification does not seem very instructive, there is another, more geometric approach by Heintze and Groß (cf. [HG09]): The Lie algebra $\mathcal{L} \otimes \mathfrak{g}$ is replaced by the *smooth loop algebra* $C^\infty(\mathbb{S}^1, \mathfrak{g})$ and the corresponding twisted versions. The discussion of finite order automorphisms and real forms then simplifies, and these new methods give even more complete answers and new insights. Affine Kac-Moody algebras are applied in conformal field theory (cf. [Fuc95]) and massive field theory (cf. [LeC94]), as well as in (super-)gravity theories (cf. [BM84]).

Lie algebras of sections are generalizations of mapping algebras: For $\mathbb{K} \in \{\mathbb{R}, \mathbb{C}\}$ let $\mathfrak{k}$ be a finite-dimensional $\mathbb{K}$-Lie algebra, M a smooth manifold and $\mathfrak{K}$ a *Lie algebra bundle*, i.e.,

- $\mathfrak{K}$ is a finite-dimensional smooth manifold with a surjective submersion $\pi : \mathfrak{K} \to M$,

- there is an open cover $(U_i)_{i \in I}$ of M with smooth maps $(\varphi_i : \pi^{-1}(U_i) \to \mathfrak{k})_{i \in I}$,

- the maps $(\pi, \varphi_i) : \pi^{-1}(U_i) \to U_i \times \mathfrak{k}$ for $i \in I$ are diffeomorphisms,

- if $m \in U_i \cap U_j$ for some $i, j \in I$ and $\mathfrak{K}_m := \pi^{-1}(m)$, then $\varphi_{i|\mathfrak{K}_m} \circ \varphi_{j|\mathfrak{K}_m}^{-1} \in \operatorname{Aut}(\mathfrak{k})$,

---

[1] More precisely, this is true for the *untwisted* affine Kac-Moody algebras. For the twisted ones, it is necessary to build in a $\mathfrak{g}$-automorphism of finite order into the tensor product $\mathcal{L} \otimes \mathfrak{g}$.

then there is a unique Lie algebra structure on each fiber $\mathfrak{K}_m$ such that the maps $\varphi_{i|\mathfrak{K}_m}$ are isomorphisms for all $m \in U_i$, $i \in I$. The space of sections $\Gamma^k(\mathfrak{K}) = \left\{ X \in \mathrm{C}^k(M, \mathfrak{K}) \,\middle|\, \pi \circ X = \mathrm{id}_M \right\}$ is, with the pointwise defined Lie bracket $[X, Y]_m := [X_m, Y_m]$, a Lie algebra, called *Lie algebra of C^k-sections*. For the trivial Lie algebra bundle $\mathfrak{K} = \mathbb{S}^1 \times \mathfrak{k}$, we obtain the untwisted smooth loop algebra $\mathrm{C}^\infty(\mathbb{S}^1, \mathfrak{k}) \cong \Gamma^\infty(\mathfrak{K})$.

The first to discuss Lie algebras of section intensively was Lecomte in [Lec79] and [Lec80], where he discussed, for centerfree fiber $\mathfrak{k}$, the center, the commutator algebra, the algebra of derivations and the centroid of the Lie algebra $\Gamma^k(\mathfrak{K})$, as well as the isomorphisms of two different section algebras. The case of Lie algebras of smooth sections, i.e., algebras $\Gamma(\mathfrak{K}) = \Gamma^\infty(\mathfrak{K})$ is of particular interest and subject to recent research (cf., e.g., [Nee04a], [NW09], [JW10]).

1.3. Requirements and Notation

The reading of this thesis requires the knowledge of Algebra, Topology, Differential Geometry and Functional Analysis on graduate level; as respective references we recommend the textbooks [Lan02], [Bre93], [Hel78] and [SW99]. Some general Lie theory will be presented in Section 2.1 and in Appendix A, for more on that subject we suggest the book [HN11]. Addenda to the facts on bundles worked out in Section 2.2 can be obtained from [Hus66].

Before starting with the main part of this dissertation, we want to clarify some conventions in notation and content.

- The set of non-negative integers $\{0, 1, 2, \ldots\}$ is denoted by $\mathbb{N}_0$, the set of positive integers $\{1, 2, \ldots\}$ is denoted by $\mathbb{N}$.

- All vector spaces and (multi-)linear maps are considered over the field $\mathbb{K} \in \{\mathbb{R}, \mathbb{C}\}$ of real or complex numbers, respectively. In addition, all topological vector spaces are meant to be Hausdorff.

- If V is a real vector space, then $V_\mathbb{C} = V \otimes_\mathbb{R} \mathbb{C} = V \oplus iV$ is a complex vector space. In this case $V_\mathbb{C}$ is called a complexification of V, and V is a real form of $V_\mathbb{C}$. If W is a complex vector space, then we can consider it as a real vector space by restricting the scalar multiplication $\mathbb{C} \times W \to W$ to $\mathbb{R} \times W \to W$ and we denote this real vector space by $W^\mathbb{R}$.

- A C^k-map for $k \in \mathbb{N}_0 \cup \{\infty\}$ is a k-times continuously differentiable function. A function is also called "smooth" instead of C^∞.

- Manifolds M will always be smooth, finite-dimensional over $\mathbb{R}$, hausdorff, paracompact and connected, if not stated otherwise.

- Depending on the context, the symbol $\mathbf{1}$ denotes a trivial group, the identity map of a vector space, the endomorphism of a vector bundle where each $\mathbf{1}_x$ is the identity of the x-fiber, or a constant map $X \to \{1\}$.

- Depending on the context, the symbol $\mathbf{0}$ denotes a trivial vector (sub)space, the trivial map of a vector space, the endomorphism of a vector bundle where each $\mathbf{0}_x$ is the trivial map of the x-fiber, or a constant map $X \to \{0\}$.

1.4. Outline of this Dissertation

We now give an outline of the most important results to be found in this thesis.

Chapter 2: Basic Notions and First Results

In the first section of this chapter, basic and important facts on Lie algebras are presented. After the obvious, we will see that perfect and centerfree Lie algebras $\mathfrak{g}$ have commutative centroids $\mathrm{Cent}(\mathfrak{g})$ and that, in the finite-dimensional case, the centroid decomposes directly as an associative algebra into a nilpotent part $\mathrm{N}(\mathfrak{g})$ and a semisimple part $\mathrm{S}(\mathfrak{g})$. We will also discuss a slighlty more general situation when a decomposition into indecomposable ideals is unique. We end this section with structural facts on semisimple Lie algebras.

The second section begins with the definitions of general fiber bundles and vector bundles with their morphisms. After defining corresponding frame bundles and giving an alternative description as bundles associated to principal bundles, we discuss quotient principal bundles, reduction of the structure group, pullbacks and the description of sections as equivariant maps.

Finally, the third section deals with Lie algebra bundles $\pi : \mathfrak{K} \to M$ with fiber $\mathfrak{k}$ and their section algebras $\Gamma^k(\mathfrak{K})$ for $\mathfrak{k} \in \mathbb{N}_0 \cup \{\infty\}$. We define related bundles $\mathfrak{z}(\mathfrak{K})$, $[\mathfrak{K}, \mathfrak{K}]$, $\mathrm{Der}(\mathfrak{K})$, $\mathrm{Cent}(\mathfrak{K})$ and $\mathrm{Aut}(\mathfrak{K})$, the center bundle, commutator bundle, derivation bundle, centroid bundle and automorphism bundle, respectively. It is easy to link the center and the commutator algebra of $\Gamma^k(\mathfrak{K})$ to the corresponding bundles:

Proposition 1.4.1. *We have* $\mathfrak{z}\left(\Gamma^k(\mathfrak{K})\right) \cong \Gamma^k(\mathfrak{z}(\mathfrak{K}))$ *and* $\left[\Gamma^k(\mathfrak{K}), \Gamma^k(\mathfrak{K})\right] \cong \Gamma^k([\mathfrak{K}, \mathfrak{K}])$ *by natural isomorphisms.*

From now on, we have to suppose that the fiber $\mathfrak{k}$ is a perfect or centerfree Lie algebra. For the calculation of the derivation algebra of $\Gamma^k(\mathfrak{K})$, we make use of the Peetre Theorem saying that local operators are differential operators and discuss the cases $k \in \mathbb{N}_0$ and $k = \infty$ separately: In the former case we obtain the following.

Theorem 1.4.2. *We have* $\mathrm{Der}\left(\Gamma^k(\mathfrak{K})\right) \cong \Gamma^k\left(\mathrm{Der}(\mathfrak{K})\right)$ *as Lie algebras.*

In order to describe the derivation algebra of $\Gamma^\infty(\mathfrak{K}) = \Gamma(\mathfrak{K})$, we introduce the notion of an x-derivation $\delta : \Gamma(\mathfrak{K}) \to \mathfrak{K}_x$. The set of all x-derivations for all $x \in M$ is denoted by $\mathcal{D}(\mathfrak{K})$. We deduce the following.

Theorem 1.4.3. *Every x-derivation of $\mathfrak{K}$ is a differential operator of order at most 1 and $\mathcal{D}(\mathfrak{K})$ is a vector bundle over M such that* $\mathrm{Der}\left(\Gamma(\mathfrak{K})\right)$ *can be naturally identified with $\Gamma(\mathcal{D}(\mathfrak{K}))$ and there exists a short exact sequence of vector bundles as follows:*

$$0 \longrightarrow \mathrm{Der}(\mathfrak{K}) \longrightarrow \mathcal{D}(\mathfrak{K}) \longrightarrow TM \otimes \mathrm{Cent}(\mathfrak{K}) \longrightarrow 0.$$

For a global description of $\mathrm{Der}(\Gamma(\mathfrak{K}))$ we define Lie connections $\nabla : \Gamma(TM) \to \mathrm{Der}(\Gamma(\mathfrak{K}))$, extend them to a connections $\nabla : \Gamma(TM \otimes \mathrm{Cent}(\mathfrak{K})) \to \mathrm{Der}(\Gamma(\mathfrak{K}))$ and prove the following fact.

Theorem 1.4.4. *A linear map $D : \Gamma(\mathfrak{K}) \to \Gamma(\mathfrak{K})$ is a derivation of $\Gamma(\mathfrak{K})$ if and only if there are sections $\mathcal{Y} \in \Gamma(TM \otimes \mathrm{Cent}(\mathfrak{K}))$ and $D_0 \in \Gamma(\mathrm{Der}(\mathfrak{K}))$ such that $D = \nabla_{\mathcal{Y}} + i(D_0)$. In this case the decomposition is unique.*

After applying the Peetre Theorem again, we obtain an easy description of the centroid of the Lie algebra $\Gamma^k(\mathfrak{K})$ for $k \in \mathbb{N}_0 \cup \{\infty\}$.

Theorem 1.4.5. *We have* $\mathrm{Cent}\left(\Gamma^k(\mathfrak{K})\right) \cong \Gamma^k\left(\mathrm{Cent}(\mathfrak{K})\right)$ *as associative algebras.*

This information helps us to see that $\Gamma^k(\mathfrak{K})$ is indecomposable if $\mathfrak{k}$ is so for perfect or center-free $\mathfrak{k}$. In the following we show two statements on decompositions, where $\mathfrak{k}$ is not necessarily perfect or centerfree. The first statement uses non-commutative Čech cohomology:

Theorem 1.4.6. *If* $\mathfrak{k}$ *is a finite-dimensional Lie algebra which has a unique decomposition* $\mathfrak{k} = \bigoplus_{i=1}^{n} \mathfrak{k}_i$ *into indecomposable ideals, and the Lie algebra bundle* $\pi : \mathfrak{K} \to M$ *with fiber* $\mathfrak{k}$ *is associated to the* $\mathrm{Aut}(\mathfrak{k})$-*principal bundle* $\rho : \mathrm{P}(\mathfrak{K}) \to M$, *then the following statements are equivalent:*

1. *There is a decomposition* $\Gamma^k(\mathfrak{K}) = \bigoplus_{i=1}^{n} \Gamma^k(\mathfrak{K}^i)$ *into ideals, for subbundles* $\pi_{|\mathfrak{K}^i} : \mathfrak{K}^i \to M$ *with indecomposable fibers* $\mathfrak{k}_i$.

2. $[\mathrm{P}(\mathfrak{K})] \in \check{\mathrm{H}}^1(M, \underline{\mathrm{Aut}(\mathfrak{k})})$ *is in the image of* $[\iota] : \check{\mathrm{H}}^1(M, \underline{\mathrm{H}(\mathfrak{k})}) \to \check{\mathrm{H}}^1(M, \underline{\mathrm{Aut}(\mathfrak{k})})$. *Here,* $\mathrm{H}(\mathfrak{k})$ *is then open normal subgroup of* $\mathrm{Aut}(\mathfrak{k})$ *stabilizing the decomposition* $\mathfrak{k} = \bigoplus_{i=1}^{n} \mathfrak{k}_i$.

Theorem 1.4.7. *Let* $\mathfrak{k}$ *be a finite-dimensional Lie algebra which has a unique decomposition* $\mathfrak{k} = \bigoplus_{i=1}^{n} \mathfrak{k}_i$ *and* $\pi : \mathfrak{K} \to M$ *be a Lie algebra bundle with fiber* $\mathfrak{k}$ *associated to the* $\mathrm{Aut}(\mathfrak{k})$-*principal bundle* $\rho : \mathrm{P}(\mathfrak{K}) \to M$. *Then there is a finite covering* $\widehat{q} : \widehat{M} \to M$ *such that, for the pullback Lie algebra bundle* $\widehat{q}^*\mathfrak{K} =: \mathfrak{G}$, *there is a decomposition*

$$\Gamma^k(\mathfrak{G}) = \bigoplus_{i=1}^{n} \Gamma^k\left(\mathfrak{G}^i\right).$$

into indecomposable ideals, where each $\mathfrak{G}^i \to \widehat{M}$ *is a subbundle of* $\mathfrak{G}$ *with fiber* $\mathfrak{k}_i$.

After stating and proving a Pursell-Shanks type theorem for mapping algebras, we prove the following theorem on isomorphisms of Lie algebras of C^k-sections and some corollaries, one of them is given below.

Theorem 1.4.8. *Let* $\pi : \mathfrak{K} \to M$ *and* $\varpi : \mathfrak{G} \to N$ *be two bundles of Lie algebras with perfect or centerfree fibers* $\mathfrak{k}$ *and* $\mathfrak{g}$, *respectively,* $\dim M = m$, $\dim N = n$, $\dim \mathfrak{k} = d$, $\dim \mathfrak{g} = e$ *and* $\mathrm{S}(\mathfrak{k}) = \mathbb{K} \cdot 1$, $\mathrm{S}(\mathfrak{g}) = \mathbb{K} \cdot 1$. *Suppose there exists an isomorphism of Lie algebras* $\mu : \Gamma^k(\mathfrak{K}) \to \Gamma^\ell(\mathfrak{G})$ *for some* $k, \ell \in \mathbb{N} \cup \{\infty\}$. *Then* $k = \ell$ *and*

(a) *if* $k \in \mathbb{N}$, *then* μ *is induced by a* C^k-*isomorphism of vector bundles* $\kappa : \mathfrak{K} \to \mathfrak{G}$, *i.e., if* $\kappa' : M \to N$ *is the map with* $\kappa' \circ \pi = \varpi \circ \kappa$, *then* $\mu(X) = \kappa \circ X \circ (\kappa')^{-1}$ *for all* $X \in \Gamma^k(\mathfrak{K})$. *In particular, the manifolds* M, N *are* C^k-*diffeomorphic and the Lie algebras* $\mathfrak{k}, \mathfrak{g}$ *are isomorphic.*

(b) *if* $k = \infty$, *then the manifolds* M, N *are smoothly diffeomorphic and the Lie algebras* $\mathfrak{k}, \mathfrak{g}$ *are isomorphic. After identifying the manifolds and the Lie algebras,* μ *turns into a linear differential operator of order at most* $d - 1$ *taking the following local form on a bundle chart* (U, φ) *of* $\pi : \mathfrak{K} \to M$ *with corresponding chart* (U, ξ) *of* M:

$$A^\varphi \overset{\mu^\varphi}{\longmapsto} \sum_{|\alpha| < d} \frac{1}{\alpha!} N^\alpha \cdot \left(\mu_0 \cdot \partial_\xi^\alpha A^\varphi\right),$$

where $\mu_0 : U \to \mathrm{Aut}(\mathfrak{g})$ *and* $N_1, \ldots, N_m \in \mathrm{C}^\infty(U, \mathrm{N}(\mathfrak{g}))$ *are smooth functions.*

Corollary 1.4.9. *If* $\pi : \mathfrak{K} \to M$ *is a Lie algebra bundle with central simple fiber* $\mathfrak{k}$, *then there is the following sequence of groups:*

$$1 \longrightarrow \Gamma^k \left(\mathrm{Aut}(\mathfrak{K}) \right) \cong \mathrm{Aut}_{C^k(M,\mathbb{K})} \left(\Gamma^k(\mathfrak{K}) \right) \hookrightarrow \mathrm{Aut} \left(\Gamma^k(\mathfrak{K}) \right) \longrightarrow \mathrm{Diff}^k(M)_{\mathfrak{K}} \longrightarrow 1.$$

Here, $\mathrm{Diff}^k(M)_{\mathfrak{K}}$ *denotes the set of* C^k-*diffeomorphisms of* M *that may be induced by an automorphism of* $\Gamma^k(\mathfrak{K})$.

Chapter 3: Classification by Homotopy Theory

This chapter starts with a brief step-by-step explanation why there is a bijective correspondence of the equivalence classes of Lie algebra bundles and the orbits of the group action $\sigma : \pi_0(\mathrm{Aut}(\mathfrak{g})) \times \pi_{n-1}(\mathrm{Aut}(\mathfrak{g})) \to \pi_{n-1}(\mathrm{Aut}(\mathfrak{g}))$, $([d],[f]) \mapsto [d \cdot f \cdot d^{-1}]$, using methods of differential topology and homotopy theory, yielding the following.

Theorem 1.4.10. *Lie algebra bundles* $\pi : \mathfrak{K} \to \mathbb{S}^n$ *with fiber* $\mathfrak{g}$, *for* $n \in \mathbb{N}$, *are classified by* $\pi_{n-1}(\mathrm{Aut}(\mathfrak{g}))/^\sigma \pi_0(\mathrm{Aut}(\mathfrak{g}))$ *and so are Lie algebras* $\Gamma(\mathfrak{K})$ *of smooth sections over* $\mathbb{S}^n$ *with fiber* $\mathfrak{g}$, *if* $\mathfrak{g}$ *is central simple. In particular, equivalence classes of Lie algebras* $\Gamma(\mathfrak{K})$ *over* $\mathbb{S}^1$ *with central simple* $\mathfrak{g}$ *are parametrized by the set of conjugacy classes* $\mathrm{Conj}(\pi_0(\mathrm{Aut}(\mathfrak{g})))$.

Motivated by this theorem, we calculate the component group $\pi_0(\mathrm{Aut}(\mathfrak{g})) = \mathrm{Out}(\mathfrak{g})$ and the number of its conjugacy classes for simple Lie algebras $\mathfrak{g}$ in the second section. Furthermore, we prove that the corresponding short exact sequence of groups

$$1 \longrightarrow \mathrm{Aut}(\mathfrak{g})_0 = \mathrm{Inn}(\mathfrak{g}) \longrightarrow \mathrm{Aut}(\mathfrak{g}) \longrightarrow \pi_0(\mathrm{Aut}(\mathfrak{g})) = \mathrm{Out}(\mathfrak{g}) \longrightarrow 1 \qquad (1.1)$$

is split. The latter is a classical fact for complex simple $\mathfrak{g}$, and $\pi_0(\mathrm{Aut}(\mathfrak{g}))$ is realized in $\mathrm{Aut}(\mathfrak{g})$ as the symmetry group of the corresponding Dynkin diagram. By the close connection of real simple compact Lie algebras to their complexifications, it is no surprise, and it has been shown before (cf., e.g., [HM06]), that the sequence (1.1) is split for real simple compact $\mathfrak{g}$. However, the splitness is true for all simple $\mathfrak{g}$. This discussion has also been published separately in [Gün10].

The chapter is ended by the following application of a theorem on cohomology for a classification of Lie algebras of smooth sections over closed manifolds of dimension 2.

Theorem 1.4.11. *Let* M *be a closed manifold of dimension 2 and* $\pi : \mathfrak{K} \to M$ *be a Lie algebra bundle with fiber* $\mathfrak{k}$ *such that* $\mathrm{Aut}(\mathfrak{k})$ *is connected.*

1. *If* M *is non-orientable, e.g., diffeomorphic to* $\mathbb{RP}^2$ *or the Klein bottle, then the equivalence classes of* $\Gamma(\mathfrak{K})$ *are parametrized by the quotient* $\pi_1(\mathrm{Aut}(\mathfrak{k}))/2 \cdot \pi_1(\mathrm{Aut}(\mathfrak{k}))$. *In particular, if* $\mathrm{Aut}(\mathfrak{k})$ *is 1-connected, then* $\mathfrak{K}$ *is trivial, so* $\Gamma(\mathfrak{K}) \cong C^\infty(M, \mathfrak{k})$.

2. *If* M *is orientable, e.g., diffeomorphic to* $\mathbb{S}^2$ *or* $\mathbb{T}^2$, *then the equivalence classes of* $\Gamma(\mathfrak{K})$ *are parametrized by* $\pi_1(\mathrm{Aut}(\mathfrak{k}))$. *In particular, if* $\mathrm{Aut}(\mathfrak{k})$ *is 1-connected, then* $\mathfrak{K}$ *is trivial, so* $\Gamma(\mathfrak{K}) \cong C^\infty(M, \mathfrak{k})$.

Chapter 4: Universal Invariant Symmetric Bilinear Forms

We discuss universal invariant symmetric bilinear forms for Lie algebras $\mathfrak{g}$ in the first section of this chapter. Such forms always exist and are basically unique by an abstract construction, and they map into a quotient $V_{\mathbb{K}}(\mathfrak{g})$ of the tensor product $\mathfrak{g} \otimes_{\mathbb{K}} \mathfrak{g}$. We show that, if $\mathfrak{g}$ is perfect, then universal invariant symmetric bilinear forms map into a quotient $V_A(\mathfrak{g}) = V(\mathfrak{g})$ of $\mathfrak{g} \otimes_A \mathfrak{g}$ for the commutative algebra $A = \mathrm{Cent}(\mathfrak{g})$. In this case the universal invariant symmetric bilinear form is denoted by $\kappa_{\mathfrak{g}} : \mathfrak{g} \times \mathfrak{g} \to V(\mathfrak{g})$. We also define, for a given Lie algebra bundle $\pi : \mathfrak{K} \to M$ with fiber $\mathfrak{k}$, the vector bundle $\nu : V(\mathfrak{K}) \to M$ with fiber $V(\mathfrak{k})$ and the natural invariant symmetric bilinear form $\kappa : \Gamma(\mathfrak{K}) \times \Gamma(\mathfrak{K}) \to V(\mathfrak{K})$, $\kappa(X, Y)(m) := \kappa_{\mathfrak{K}_m}(X_m, Y_m)$. Before showing that κ is universal, we discuss in the following two sections a class of commutative associative algebras, the commutative pseudo-unital algebras C, which possess, for each two elements $a, b \in C$ a third element $c \in C$ with $ac = a$ and $bc = b$, and show the following.

Proposition 1.4.12. *Let C be a commutative pseudo-unital algebra and $\mathfrak{k}$ be a finite-dimensional perfect Lie algebra. Then $\kappa_{C,\mathfrak{k}} : (C \otimes \mathfrak{k})^2 \to C \otimes V(\mathfrak{k})$ defined by $\kappa_{C,\mathfrak{k}}(a \otimes x, b \otimes y) := ab \otimes \kappa_{\mathfrak{k}}(x, y)$ is a universal invariant symmetric bilinear form.*

We show the universality of κ in the next section.

Theorem 1.4.13. *Let $\pi : \mathfrak{K} \to M$ be a finite-dimensional Lie algebra bundle with perfect fiber $\mathfrak{k}$. Then the invariant symmetric bilinear form $\kappa : \Gamma(\mathfrak{K}) \times \Gamma(\mathfrak{K}) \to \Gamma(V(\mathfrak{K}))$ is universal.*

After that we define continuous universal invariant symmetric bilinear forms for Fréchet-Lie algebras and finally show that κ is also universal in this category.

Chapter 5: Universal Central Extensions

In order to find universal central extensions of section algebras $\Gamma(\mathfrak{K})$, we first analyze, for a commutative pseudo-unital algebra C, a semisimple Lie algebra $\mathfrak{k}$ and a complete locally convex space X, the second cohomology group $\mathrm{H}^2(C \otimes \mathfrak{k}, X)$. Here, $C^+ = C \oplus \mathbb{K}$ is a unital algebra and $i : C \otimes \mathfrak{k} \to C^+ \otimes \mathfrak{k}$ is the natural inclusion.

Theorem 1.4.14. *The following is a group isomorphism:*

$$\mathrm{H}^2(i) : \mathrm{H}^2(C^+ \otimes \mathfrak{k}, X) \longrightarrow \mathrm{H}^2(C \otimes \mathfrak{k}, X)$$
$$[\omega] \longmapsto [\omega \circ i].$$

If C is a CPUSLF-algebra, i.e., a commutative pseudo-unital strict LF-algebra, then this fact is also true in the continuous category.

Theorem 1.4.15. *The following is a group isomorphism:*

$$\mathrm{H}^2_{\mathrm{ct}}(i) : \mathrm{H}^2_{\mathrm{ct}}(C^+ \otimes \mathfrak{k}, X) \longrightarrow \mathrm{H}^2_{\mathrm{ct}}(C \otimes \mathfrak{k}, X)$$
$$[\omega] \longmapsto [\omega \circ i].$$

Universal central extensions of Lie algebras $A \otimes \mathfrak{k}$ for complete locally convex algebras A, that are unital and commutative, can be described with the help of universal differential modules $(\Omega^1_{\mathcal{K}}(A), d_A)$ for a full subcategory $\mathcal{K}$ of the category of locally convex spaces (cf. [Mai02]). So we define this notion and give a proof of the following.

Theorem 1.4.16. *If C is a CPUSLF-algebra, then the continuous 2-cocycle*

$$\omega_C^+ : (C \otimes \mathfrak{k})^2 \longrightarrow \left(\Omega_{\mathcal{K}}^1(C^+)/\overline{d_{C^+}C^+} \right) \otimes \mathrm{V}(\mathfrak{k})$$

$$(f \otimes x, g \otimes y) \longmapsto [j(f) \cdot d_{C^+}(j(g))] \otimes \kappa_{\mathfrak{k}}(x, y).$$

for the natural injection $j : C \to C^+$ is weakly universal.

In the third section of this chapter, we use the universal invariant symmetric bilinear form κ from the preceding chapter and a Lie connection ∇ of $\mathfrak{K}$ to define a continuous 2-cocycle

$$\omega_{\nabla,c} : \Gamma_c(\mathfrak{K}) \times \Gamma_c(\mathfrak{K}) \longrightarrow \overline{\Omega}_c^1(M, \mathrm{V}(\mathfrak{K}))$$

$$(\xi, \eta) \longmapsto [\kappa(\xi, \nabla\eta)] \,,$$

where the subscript c stands for "compactly supported", and state the following result due to Wockel and Janssens (cf. Proposition I.1 of [JW10]).

Theorem 1.4.17. *If $\pi : \mathfrak{K} \to M$ is a finite-dimensional Lie algebra bundle with semisimple fiber $\mathfrak{k}$, then $\omega_{\nabla,c} \in Z_{ct}^2 \left(\Gamma_c(\mathfrak{K}), \overline{\Omega}_c^1(M, \mathrm{V}(\mathfrak{K})) \right)$ is weakly universal, i.e., for every locally convex space W, the following linear map is bijective:*

$$\mathrm{Hom}_{ct} \left(\overline{\Omega}_c^1(M, \mathrm{V}(\mathfrak{K})), W \right) \longrightarrow \mathrm{H}_{ct}^2(\Gamma_c(\mathfrak{K}), W)$$

$$\theta \longmapsto [\theta \circ \omega_{\nabla,c}] \,.$$

In particular, the 2-cocycle $\omega_\nabla : \Gamma(\mathfrak{K}) \times \Gamma(\mathfrak{K}) \to \overline{\Omega}^1(M, \mathrm{V}(\mathfrak{K}))$, $(\xi, \eta) \mapsto [\kappa(\xi, \nabla\eta)]$ is universal for compact M.

Chapter 6: Section Algebras as Kac-Moody Algebras

The first section of this short last chapter is dedicated to the realization of affine Kac-Moody algebras as extensions of loop algebras. We state facts on invariant symmetric bilinear forms and universal central extensions $\widetilde{\mathcal{L}}(\mathfrak{g})$ of algebraic loop algebras $\mathcal{L}(\mathfrak{g})$ for a complex simple Lie algebra $\mathfrak{g}$. In the second section we propose a possible generalization of affine Kac-Moody algebras $\widehat{\Gamma}(\mathfrak{K})$, starting with a Lie algebra bundle $\pi : \mathfrak{K} \to M$ with central simple fiber $\mathfrak{k}$, taking the universal central extension $\widetilde{\Gamma}(\mathfrak{K})$ of the section algebra $\Gamma(\mathfrak{K})$ and finally adding a derivation. We apply the results of the previous chapters to compare the properties of these objects to those of the usual affine Kac-Moody algebras.

Appendix A: More on Lie Algebras

In this appendix, we shortly present the classification of complex simple Lie algebras by Killing and Cartan using Cartan matrices and Dynkin diagrams and show how affine Kac-Moody algebras are defined with the help of generalized Cartan matrices. Futhermore, we give a short survey on universal central extensions of (topological) Lie algebras.

Appendix B: Notions of Differential Topology

The second appendix treats topological facts: We define the natural smooth topology on spaces of smooth functions, discuss strict LF-spaces and state an open mapping theorem for them, and finally introduce non-abelian Čech cohomology.

2. Basic Notions and First Results

In this chapter we will present some basic definitions and general properties of Lie algebras and fiber bundles, as well as some of the first results on Lie algebras of smooth sections.

2.1. Generalities on Lie Algebras

We will recapitulate some definitions, notions and basic facts on Lie algebras over the field $\mathbb{K} \in \{\mathbb{R}, \mathbb{C}\}$. The non-proven statements in this section can be found in standard books on Lie algebras, e.g., in [HN11].

Definition 2.1.1. A *Lie algebra* (over $\mathbb{K}$) $\mathfrak{g}$ is a $\mathbb{K}$-vector space endowed with a *Lie bracket* $[\cdot, \cdot]$, i.e., an alternating bilinear map $\mathfrak{g} \times \mathfrak{g} \to \mathfrak{g}$ fulfilling the Jacobi identity

$$[[x, y], z] + [[z, x], y] + [[y, z], x] = 0$$

for all $x, y, z \in \mathfrak{g}$. A vector subspace $\mathfrak{a} \subseteq \mathfrak{g}$ is called a *subalgebra*, symbolically $\mathfrak{a} \leq \mathfrak{g}$, if $[x, y] \in \mathfrak{a}$ for all $x, y \in \mathfrak{a}$, and an *ideal*, symbolically $\mathfrak{a} \trianglelefteq \mathfrak{g}$, if $[x, y] \in \mathfrak{a}$ for all $x \in \mathfrak{a}$, $y \in \mathfrak{g}$.

We now fix a Lie algebra $\mathfrak{g}$.

Definition 2.1.2. The ideal $[\mathfrak{g}, \mathfrak{g}] := \operatorname{span}_{\mathbb{K}} \{[x, y] \mid x, y \in \mathfrak{g}\} \trianglelefteq \mathfrak{g}$ is called the *commutator algebra* of $\mathfrak{g}$ and $\mathfrak{g}$ is called *perfect* if $\mathfrak{g} = [\mathfrak{g}, \mathfrak{g}]$. The ideal $\mathfrak{z}(\mathfrak{g}) := \{x \in \mathfrak{g} \mid [x, y] = 0 \text{ for all } y \in \mathfrak{g}\} \trianglelefteq \mathfrak{g}$ is called the *center* of $\mathfrak{g}$. The Lie algebra $\mathfrak{g}$ is *abelian* if $\mathfrak{z}(\mathfrak{g}) = \mathfrak{g}$ or, equivalently, $[\mathfrak{g}, \mathfrak{g}] = \mathbf{0}$. The Lie algebra $\mathfrak{g}$ is called *centerfree*, if $\mathfrak{z}(\mathfrak{g}) = \mathbf{0}$.

Definition 2.1.3. If $\mathfrak{g}$ is a real Lie algebra, then there is a natural Lie bracket on its *complexification* $\mathfrak{g}_{\mathbb{C}} := \mathfrak{g} \otimes_{\mathbb{R}} \mathbb{C} = \mathfrak{g} + i\mathfrak{g}$ defined as follows:

$$[x + iy, w + iz] := [x, w] - [y, z] + i([y, w] + [x, z]).$$

Lemma 2.1.4.

1. *Let V be a vector space. Then its vector space endomorphisms form an associative algebra where the multiplication of the algebra is $\circ$, the composition of functions. This associative algebra is called the* endomorphism algebra *of V, denoted by $\operatorname{End}(V)$.*

2. *If A is an associative algebra, then bilinear map $A \times A \to A$, $(a, b) \mapsto ab - ba$ is a Lie bracket, turning A into a Lie algebra denoted by A_L.*

3. *$(R, \cdot)$ being an algebra, the space*

$$\operatorname{Der}(R) := \{f \in \operatorname{End}(R) \mid f(r \cdot s) = f(r) \cdot s + r \cdot f(s) \text{ for all } r, s \in R\}$$

is a Lie subalgebra of $\operatorname{End}(R)_L$, called the Lie algebra of derivations *of R.*

4. Let $(R, \cdot)$ be an algebra and $I \trianglelefteq R$ an ideal, then the quotient vector space R/I equipped with the well-defined induced multiplication

$$R/I \times R/I \longrightarrow R/I$$
$$(r + I, s + I) \longmapsto (r \cdot s) + I$$

is an algebra, called quotient algebra of R modulo I, denoted by R/I.

Definition 2.1.5. The concept of a topological (Lie) algebra is important in infinite dimensions.

1. A *topological (Lie) algebra* is a (Lie) algebra, where the underlying vector space is a topological vector space such that the multiplication is continuous.

2. A *locally convex (Lie) algebra* is a locally convex topological (Lie) algebra.

3. A *Banach(-Lie) algebra* is a Banach space that is a topological (Lie) algebra

4. A *Fréchet(-Lie) algebra* is Fréchet space that is a topological (Lie) algebra

5. A norm $\|\cdot\|$ on an algebra $(R, \cdot)$ is called *submultiplicative*, if $\|r \cdot s\| \leq \|r\| \, \|s\|$ for all $r, s \in R$.

6. A topological Lie algebra $\mathfrak{g}$ is called *topologically perfect* if the commutator algebra $[\mathfrak{g}, \mathfrak{g}]$ is dense in $\mathfrak{g}$.

Remark 2.1.6. Since every bilinear map $V \times V \to V$ for a finite-dimensional vector space V is continuous and all the norms on V are equivalent, we can, for each finite-dimensional Lie algebra $(\mathfrak{g}, [\cdot, \cdot])$, choose a submultiplicative norm $\|\cdot\|$ such that $\mathfrak{g}$ is a Banach(-Lie) algebra.

Definition 2.1.7. If $(A, \cdot)$ is a not necessarily unital, associative algebra and a locally convex vector space with a family of generating seminorms $(p_i)_{i \in I}$, that is saturated[1], then the theory of locally convex algebras (cf. Section 4.4 of [Bal00]) says that the continuity of the multiplication of the algebra is equivalent to the following statement:

$$(\forall i \in I)(\exists j \in I)(\exists D > 0)(\forall a, b \in A) : p_i(a \cdot b) \leq D \cdot p_j(a) \cdot p_j(b). \tag{2.1}$$

Once given a locally convex associative algebra with continuous multiplication, we can assume that it has a *well-behaved* family of seminorms $(p_i)_{i \in I}$, i.e., it defines the topology on A, it is saturated and (2.1) is fulfilled for $D = 1$.

Definition 2.1.8. Let $(C, \cdot)$ be a commutative, but not necessarily unital, associative algebra and $(\mathfrak{h}, [\cdot, \cdot])$ a Lie algebra. Then the *tensor product Lie algebra* is the vector tensor product $C \otimes \mathfrak{h}$ with the Lie bracket defined by

$$[c_i \otimes x_k, c_j \otimes x_\ell] := c_i c_j \otimes [x_k, x_\ell] \,,$$

where $(c_i)_{i \in I}$ is a basis of C and $(x_k)_{k \in K}$ is a basis[2] of $\mathfrak{h}$.

---

[1] For $i, j \in I$ there is $k \in I$ such that $\max(p_i, p_j) \leq p_k$.

[2] It is not absolutely necessary to use a basis here, but we do it as a preparation of the following remark.

Remark 2.1.9. Let $(C, \cdot)$ be a commutative, but not necessarily unital, associative locally convex algebra with a well-behaved family of seminorms $(p_i)_{i \in I}$. Let $(\mathfrak{h}, [\cdot, \cdot])$ be a finite-dimensional Lie algebra with basis $(v_1, \ldots, v_d)$ and structure constants $H_{\ell,m}^t$, i.e., $[v_\ell, v_m] = \sum_t H_{\ell,m}^t v_t$. Then we define seminorms $(q_i)_{i \in I}$ on $C \otimes \mathfrak{h}$ by $q_i := \max_k q_{i,k}$, where the family $(q_{i,k})_{i \in I, 1 \leq k \leq d}$ is defined by $q_{i,k}\left(\sum_t c_t \otimes v_t\right) := p_i(c_k)$.

These seminorms turn $C \otimes \mathfrak{h}$ into a locally convex Lie algebra, i.e., the Lie bracket is continuous, by the following calculations:

$$q_{i,k}\left[\sum_\ell c_\ell \otimes v_\ell, \sum_m d_m \otimes v_m\right] = q_{i,k}\left(\sum_{\ell,m} c_\ell d_m \otimes [v_\ell, v_m]\right) = q_{i,k}\left(\sum_{\ell,m,t} H_{\ell,m}^t c_\ell d_m \otimes v_t\right)$$

$$= p_i\left(\sum_{\ell,m} H_{\ell,m}^k c_\ell d_m\right) \underset{\text{well-behaved}}{\leq} \sum_{\ell,m} \left|H_{\ell,m}^k\right| \cdot p_j(c_\ell) \cdot p_j(d_m)$$

$$= \sum_{\ell,m} \left|H_{\ell,m}^k\right| \cdot q_{j,\ell}\left(\sum_t c_t \otimes v_t\right) \cdot q_{j,m}\left(\sum_t d_t \otimes v_t\right) \tag{2.2}$$

and

$$q_i\left[\sum_\ell c_\ell \otimes v_\ell, \sum_m d_m \otimes v_m\right] = \max_k q_{i,k}\left[\sum_\ell c_\ell \otimes v_\ell, \sum_m d_m \otimes v_m\right]$$

$$\leq \max_k \sum_{\ell,m} \left|H_{\ell,m}^k\right| \cdot q_{j,\ell}\left(\sum_t c_t \otimes v_t\right) \cdot q_{j,m}\left(\sum_t d_t \otimes v_t\right) \tag{2.3}$$

$$\leq \left(\sum_{\ell,m} \max_k \left|H_{\ell,m}^k\right|\right) \cdot q_j\left(\sum_t c_t \otimes v_t\right) \cdot q_j\left(\sum_t d_t \otimes v_t\right).$$

Definition 2.1.10. The set of *Lie algebra morphisms* from $\mathfrak{g}$ to $\mathfrak{h}$ is

$$\mathrm{Hom}(\mathfrak{g}, \mathfrak{h}) := \{f : \mathfrak{g} \to \mathfrak{h} \text{ linear map} \mid f([x,y]) = [f(x), f(y)] \text{ for all } x, y \in \mathfrak{g}\}.$$

The set of *Lie algebra isomorphimsn* from $\mathfrak{g}$ to $\mathfrak{h}$ is

$$\mathrm{Iso}(\mathfrak{g}, \mathfrak{h}) := \{f : \mathfrak{g} \to \mathfrak{h} \text{ linear isomorphism} \mid f([x,y]) = [f(x), f(y)] \text{ for all } x, y \in \mathfrak{g}\}.$$

For $\mathfrak{g} = \mathfrak{h}$ the group $\mathrm{Aut}(\mathfrak{g}) := \mathrm{Iso}(\mathfrak{g}, \mathfrak{g})$ is called group of *Lie algebra automorphisms* of $\mathfrak{g}$.

Definition 2.1.11. Let $\mathfrak{g}$ be a Lie algebra and consider it as a $\mathfrak{g}$-module with respect to the adjoint representation $\mathrm{ad} : \mathfrak{g} \to \mathrm{Der}(\mathfrak{g})$, $\mathrm{ad}(x)(y) := [x, y]$.

1. $\mathfrak{g}$ is a *simple* Lie algebra, if it is a non-trivial simple $\mathfrak{g}$-module.

2. $\mathfrak{g}$ is a *semisimple* Lie algebra, if it is the direct sum of *non-trivial* simple $\mathfrak{g}$-modules, i.e., of simple Lie algebras.

3. $\mathfrak{g}$ is a *reductive* Lie algebra, if it is the direct sum of simple $\mathfrak{g}$-modules, i.e., of ideals of $\mathfrak{g}$.

Lemma 2.1.12.

1. *If $\mathfrak{g}$ is semisimple, then it is reductive, perfect, centerfree and $\mathrm{Der}(\mathfrak{g}) = \mathrm{ad}(\mathfrak{g})$. In particular, $\mathrm{ad}: \mathfrak{g} \to \mathrm{Der}(\mathfrak{g})$ is an isomorphism of Lie algebras.*

2. *If $\mathfrak{g}$ is reductive, then $\mathfrak{g} = [\mathfrak{g}, \mathfrak{g}] \oplus \mathfrak{z}(\mathfrak{g})$ and $[\mathfrak{g}, \mathfrak{g}]$ is semisimple. In particular, a reductive Lie algebra is semisimple if and only if it is centerfree if and only if it is perfect.*

3. *A real Lie algebra $\mathfrak{g}$ is semisimple if and only if its complexification $\mathfrak{g}_{\mathbb{C}}$ is so.*

Example 2.1.13. Let $\mathfrak{g}$ be a two-dimensional $\mathbb{K}$-vector space with basis (x_1, x_2). The unique skew-symmetric bilinear map $[\cdot, \cdot]: \mathfrak{g} \times \mathfrak{g} \to \mathfrak{g}$ determined by the relation $[x_1, x_2] := x_1$ is a Lie bracket. The Lie algebra $(\mathfrak{g}, [\cdot, \cdot])$ is centerfree, but is not perfect, thus not reductive.

Another example of a non-reductive Lie algebra can be found in Turkowski's classification of real Lie algebras of dimension up to 8 that admit a non-trivial Levi decomposition (cf. [Tur88]): Let $\mathfrak{h}$ be a six-dimensional real vector space with generators $v_1, \ldots, v_6$ and equipped with the Lie bracket with the only non-zero structure constants

$$-H_{21}^2 = H_{12}^2 = 2, \quad -H_{31}^3 = H_{13}^3 = -2, \quad -H_{32}^1 = H_{23}^1 = 1,$$
$$-H_{41}^4 = H_{14}^4 = 1, \quad -H_{52}^4 = H_{25}^4 = 1, \quad -H_{43}^5 = H_{34}^5 = 1,$$
$$-H_{51}^5 = H_{15}^5 = -1, \quad -H_{54}^6 = H_{45}^6 = 1,$$

i.e., $[v_i, v_j] = \sum_{k=1}^{6} H_{ij}^k v_k$ for all $i, j \in \{1, \ldots, 6\}$. Then $\mathfrak{h}$ is perfect with center $\mathbb{R} \cdot v_6$.

Definition 2.1.14. If $\mathfrak{g}$ is a Lie algebra of finite dimension, then $\mathrm{Aut}(\mathfrak{g}) \subseteq \mathrm{GL}(\mathfrak{g})$ is a topological group. If $\mathfrak{g}$ is also semisimple, then we call its identity component $\mathrm{Inn}(\mathfrak{g}) := \mathrm{Aut}(\mathfrak{g})_0$ the subalgebra of *inner automorphisms*. Elements of $\mathrm{Aut}(\mathfrak{g}) \setminus \mathrm{Inn}(\mathfrak{g})$ are called *outer automorphisms*.

Definition 2.1.15.

1. The *multiplier algebra* of an associative algebra A is the associate subalgebra

$$\mathrm{Mult}(A) := \{ f \in \mathrm{End}(A) \,|\, f(a \cdot b) = a \cdot f(b) \text{ for all } a, b \in A \}.$$

2. The *centroid* of a Lie algebra $\mathfrak{g}$ is the vector space

$$\begin{aligned}
\mathrm{Cent}(\mathfrak{g}) : &= \{ f \in \mathrm{End}(\mathfrak{g}) \,|\, [f, \mathrm{ad}_x] = 0 \text{ for all } x \in \mathfrak{g} \} \\
&= \{ f \in \mathrm{End}(\mathfrak{g}) \,|\, f \circ \mathrm{ad}_x = \mathrm{ad}_x \circ f \text{ for all } x \in \mathfrak{g} \} \\
&= \{ f \in \mathrm{End}(\mathfrak{g}) \,|\, f\,[x, y] = [x, f(y)] \text{ for all } x, y \in \mathfrak{g} \}.
\end{aligned}$$

The subsets of nilpotent and semisimple centroid elements are denoted by $\mathrm{N}(\mathfrak{g})$ and $\mathrm{S}(\mathfrak{g})$, respectively.

Remark 2.1.16. Let $\mathfrak{g}$ and $\mathfrak{h}$ be Lie algebras. The quotient Lie algebra $\mathfrak{g}/[\mathfrak{g}, \mathfrak{g}]$ and the Lie subalgebra $\mathfrak{z}(\mathfrak{h}) \subseteq \mathfrak{h}$ are abelian and so the Lie algebra morphisms $\mathfrak{g}/[\mathfrak{g}, \mathfrak{g}] \to \mathfrak{z}(\mathfrak{h})$ are the linear maps $\mathfrak{g}/[\mathfrak{g}, \mathfrak{g}] \to \mathfrak{z}(\mathfrak{h})$. In particular, we have $\mathrm{Hom}(\mathfrak{g}/[\mathfrak{g}, \mathfrak{g}], \mathfrak{z}(\mathfrak{h})) = \mathbf{0}$ if and only if $\mathfrak{g}$ is perfect or $\mathfrak{h}$ is centerfree.

Lemma 2.1.17.

1. $\operatorname{Cent}(\mathfrak{g})$ *is an associative subalgebra of* $\operatorname{End}(\mathfrak{g})$.

2. *We have the inclusions* $\operatorname{Cent}(\mathfrak{g}) \circ \operatorname{Der}(\mathfrak{g}) \subseteq \operatorname{Der}(\mathfrak{g})$ *and* $[\operatorname{Cent}(\mathfrak{g}), \operatorname{Der}(\mathfrak{g})] \subseteq \operatorname{Cent}(\mathfrak{g})$ *or, to say it in other words,* $\operatorname{Der}(\mathfrak{g})$ *is a* $\operatorname{Cent}(\mathfrak{g})$-*module and* $\operatorname{Der}(\mathfrak{g})$ *acts by derivations on* $\operatorname{Cent}(\mathfrak{g})$.

3. *There is a linear inclusion* $[\operatorname{Cent}(\mathfrak{g}), \operatorname{Cent}(\mathfrak{g})] \to \operatorname{Hom}(\mathfrak{g}/\left[\mathfrak{g}, \mathfrak{g}\right], \mathfrak{z}(\mathfrak{g})) \subseteq \operatorname{Cent}(\mathfrak{g})$. *In particular, if* $\mathfrak{g}$ *is perfect or centerfree, then* $\operatorname{Cent}(\mathfrak{g})$ *is commutative.*

4. *If* $\mathfrak{g}$ *is of finite dimension and* $\operatorname{Cent}(\mathfrak{g})$ *is commutative, then* $\operatorname{N}(\mathfrak{g}), \operatorname{S}(\mathfrak{g})$ *are associative subalgebras of* $\operatorname{Cent}(\mathfrak{g})$ *with* $\operatorname{Cent}(\mathfrak{g}) = \operatorname{N}(\mathfrak{g}) \oplus \operatorname{S}(\mathfrak{g})$.

5. *If* $\mathfrak{g}$ *is of finite dimension and semisimple, then* $\operatorname{N}(\mathfrak{g}) = \mathbf{0}$, *thus* $\operatorname{Cent}(\mathfrak{g}) = \operatorname{S}(\mathfrak{g})$.

Proof.

1. $\operatorname{Cent}(\mathfrak{g})$ is a vector subspace of $\operatorname{End}(\mathfrak{g})$. If $f, g \in \operatorname{Cent}(\mathfrak{g})$, then for all $x, y \in \mathfrak{g}$ we have:

$$(fg)\left[x, y\right] = f\left(g\left[x, y\right]\right) = f\left(\left[x, g(y)\right]\right) = \left[x, f(g(y))\right] = \left[x, (fg)(y)\right].$$

So, fg is also in $\operatorname{Cent}(\mathfrak{g})$, hence $\operatorname{Cent}(\mathfrak{g})$ is an associative subalgebra of $\operatorname{End}(\mathfrak{g})$.

2. Let $f \in \operatorname{Cent}(\mathfrak{g})$, $g \in \operatorname{Der}(\mathfrak{g})$ and $x, y \in \mathfrak{g}$. Then:

$$\begin{aligned}
\left[f, g\right]\left[x, y\right] &= f(g\left[x, y\right]) - g(f\left[x, y\right]) = f(\left[gx, y\right] + \left[x, gy\right]) - g(\left[x, fy\right]) \\
&= f(\left[gx, y\right]) + f(\left[x, gy\right]) - \left[gx, fy\right] - \left[x, gfy\right] \\
&= \left[gx, fy\right] + \left[x, fgy\right] - \left[gx, fy\right] - \left[x, gfy\right] \\
&= \left[x, fgy\right] - \left[x, gfy\right] = \left[x, fgy - gfy\right] = \left[x, \left[f, g\right](y)\right].
\end{aligned}$$

This implies $[\operatorname{Cent}(\mathfrak{g}), \operatorname{Der}(\mathfrak{g})] \subseteq \operatorname{Cent}(\mathfrak{g})$. We may also calculate:

$$(fg)\left[x, y\right] = f(\left[gx, y\right] + \left[x, gy\right]) = \left[(fg)x, y\right] + \left[x, (fg)y\right].$$

We thus get $\operatorname{Cent}(\mathfrak{g}) \circ \operatorname{Der}(\mathfrak{g}) \subseteq \operatorname{Der}(\mathfrak{g})$.

3. Let $f, g \in \operatorname{Cent}(\mathfrak{g})$ and $x, y \in \mathfrak{g}$. We calculate:

$$\begin{aligned}
\left[\left[f, g\right](x), y\right] &= \left[f(gx) - g(fx), y\right] = f\left[gx, y\right] - g\left[fx, y\right] = fg\left[x, y\right] - gf\left[x, y\right] \\
&= \left[fx, gy\right] - \left[fx, gy\right] = 0.
\end{aligned}$$

Therefore $[f, g](\mathfrak{g}) \subseteq \mathfrak{z}(\mathfrak{g})$. If $z \in [\mathfrak{g}, \mathfrak{g}]$ is arbitrarily chosen, then there exist finitely many elements $x_1, y_1, x_2, y_2, \ldots, x_n, y_n \in \mathfrak{g}$ such that $z = \sum_{i=1}^{n} [x_i, y_i]$ and thus we obtain

$$[f, g](z) = \sum_{i=1}^{n} fg\left[x_i, y_i\right] - gf\left[x_i, y_i\right] = 0$$

implying $[f, g]([\mathfrak{g}, \mathfrak{g}]) = 0$, so $[f, g]$ factors through a vector space morphism $\mathfrak{g}/\left[\mathfrak{g}, \mathfrak{g}\right] \to \mathfrak{z}(\mathfrak{g})$.

4. Now let $\text{Cent}(\mathfrak{g})$ be commutative and $\mathfrak{g}$ of finite dimension. Obviously, the subsets $\text{N}(\mathfrak{g}), \text{S}(\mathfrak{g})$ are closed under the multiplication by scalars in $\mathbb{K}$.

   Let $f, g \in \text{N}(\mathfrak{g})$ and $n, m \in \mathbb{N}$ such that $f^n = g^m = 0$. Obviously, $f \circ g$ is also nilpotent. Since $\text{Cent}(\mathfrak{g})$ is commutative, we have $[f, g] = 0$ and so we can apply the Binomial Theorem, yielding

$$(f+g)^{n+m+1} = \sum_{k=0}^{n+m-1} \binom{n+m-1}{k} f^k g^{n+m-1-k}$$

$$= \sum_{k=0}^{n-1} \binom{n+m-1}{n-1-k} f^{n-1-k} g^{m+k} + \sum_{k=0}^{m-1} \binom{n+m-1}{n+k} f^{n+k} g^{m-k-1}$$

$$= 0 + 0 = 0.$$

   Thus $f + g$ is also nilpotent. So $\text{N}(\mathfrak{g})$ is an associative subalgebra of $\text{Cent}(\mathfrak{g})$.

   Let $f, g \in \text{S}(\mathfrak{g})$. We want to show that $f + g$ and $f \circ g$ are also semisimple.

   If $\mathbb{K} = \mathbb{R}$, then we use the discussion in Definition 4.3.6 of [HN11] to say that $h \in \text{End}(\mathfrak{g})$ would be semisimple if and only if its complexification $h_{\mathbb{C}} := \text{id}_{\mathbb{C} \otimes \mathfrak{g}} \otimes h$ was semisimple. So we only consider the case $\mathbb{K} = \mathbb{C}$, where "semisimple" means the same as "diagonalizable". Since $\text{Cent}(\mathfrak{g})$ is commutative, we have $[f, g] = 0$ and so we can apply a fact about simultaneous diagonalization (cf., e.g., Exercise 1.1.1 (d) of [HN11]): Diagonalizable endomorphisms of a vector space of finite dimension are simultaneously diagonalizable if and only if they commute. So f and g are diagonal with respect to a fixed basis, thus also $f + g$ and $f \circ g$, so they are semisimple, too. This shows that $\text{S}(\mathfrak{g})$ is an associative subspace of $\text{Cent}(\mathfrak{g})$.

   The Jordan Decomposition Theorem 4.3.3 of [HN11] yields that each element $f \in \text{Cent}(\mathfrak{g})$ decomposes into the commuting sum of a nilpotent vector space endomorphism f_n and a semisimple one f_s and, because of the fact that f commutes with any ad_x for $x \in \mathfrak{g}$, this is also satisfied for f_n and f_s. Since $\text{Cent}(\mathfrak{g})$ is commutative, any two endomorphisms n, s with $f = n + s$, where $n \in \text{N}(\mathfrak{g})$ and $s \in \text{S}(\mathfrak{g})$, commute with f and so the theorem also implies that $n = f_n$ and $s = f_s$. This shows $\text{Cent}(\mathfrak{g}) = \text{N}(\mathfrak{g}) \oplus \text{S}(\mathfrak{g})$.

5. Let $d := \dim(\mathfrak{g})$. If $f \in \text{N}(\mathfrak{g})$, then $f^d(\mathfrak{g}) = 0$. On the other hand, $f(\mathfrak{g})$ is an ideal of $\mathfrak{g}$ because for $x, y \in \mathfrak{g}$ we have $[y, f(x)] = f[x, y] \in f(\mathfrak{g})$. If we denote $f(\mathfrak{g})$ by $\mathfrak{h} =: \mathfrak{h}^0$ and set $\mathfrak{h}^{n+1} := [\mathfrak{h}, \mathfrak{h}^n]$ for $n \in \mathbb{N}_0$, then we have $f(\mathfrak{g})^n = \mathfrak{h}^n \subseteq f^n(\mathfrak{g})$ for all $n \in \mathbb{N}_0$, implying $f(\mathfrak{g})^d = 0$, so $f(\mathfrak{g})$ is a nilpotent ideal of $\mathfrak{g}$. But $\mathfrak{g}$ is semisimple, so this ideal is trivial and $f = 0$. $\qquad\square$

Definition 2.1.18. A Lie algebra $\mathfrak{g}$ is called *decomposable* if it is the direct sum of two proper ideals. If there is no such decomposition, $\mathfrak{g}$ is called *indecomposable*.

Remark 2.1.19. A simple Lie algebra is automatically indecomposable, since $\mathbf{0}$ is its only proper ideal. If an indecomposable Lie algebra $\mathfrak{g}$ is reductive, then it is one-dimensional (and hence abelian) or simple because $\mathfrak{g} = \mathfrak{z}(\mathfrak{g}) \oplus [\mathfrak{g}, \mathfrak{g}]$ implies $\mathfrak{g} = \mathfrak{z}(\mathfrak{g})$ or $\mathfrak{g} = [\mathfrak{g}, \mathfrak{g}]$ and in the latter case a decomposition of the semisimple Lie algebra $[\mathfrak{g}, \mathfrak{g}]$ into simple Lie algebras may only have one summand. In general, indecomposable Lie algebras are not reductive, e.g., the Lie algebra $\mathfrak{g}$ in Example 2.1.13.

Remark 2.1.20. If $\mathfrak{g}$ is a perfect Lie algebra and $\mathfrak{i}, \mathfrak{j} \trianglelefteq \mathfrak{g}$ are ideals such that $\mathfrak{g} = \mathfrak{i} \oplus \mathfrak{j}$, then both ideals are $\mathrm{Cent}(\mathfrak{g})$-modules:

Indeed, by $\mathfrak{g} = [\mathfrak{g}, \mathfrak{g}] = [\mathfrak{i} \oplus \mathfrak{j}, \mathfrak{g}] = [\mathfrak{i}, \mathfrak{g}] \oplus [\mathfrak{j}, \mathfrak{g}]$, we know that $\mathfrak{i} = [\mathfrak{i}, \mathfrak{g}]$, so for and each $x \in \mathfrak{i}$ there are $x_1, \ldots, x_n \in \mathfrak{i}$, $y_1, \ldots, y_n \in \mathfrak{g}$ with $x = \sum_{i=1}^{n} [x_i, y_i]$. For $\varphi \in \mathrm{Cent}(\mathfrak{g})$ we have:

$$\varphi(x) = \sum_{i=1}^{n} \varphi[x_i, y_i] = \sum_{i=1}^{n} [x_i, \varphi(y_i)] \in [\mathfrak{i}, \mathfrak{g}] = \mathfrak{i}.$$

So if $\mathfrak{g}$ is a perfect Lie algebra with decomposition $\mathfrak{g} = \bigoplus_{i=1}^{n} \mathfrak{g}_i$ into ideals, then it is also a decomposition into $\mathrm{Cent}(\mathfrak{g})$-submodules.

The next remark on commutative centroids is the main idea of the proof of Proposition 2.1.26.

Remark 2.1.21. If $\mathfrak{g}$ is a Lie algebra with commutative centroid $\mathrm{Cent}(\mathfrak{g})$, e.g., $\mathfrak{g}$ perfect or centerfree, and $\mathfrak{g} = \bigoplus_{i=1}^{n} \mathfrak{g}_i$ is a decomposition into indecomposable non-zero ideals, then the decomposition is unique except for the order, i.e., $\{\mathfrak{g}_i : 1 \leq i \leq n\}$ is the set of indecomposable ideals of $\mathfrak{g}$.

Indeed, let $\mathfrak{g} = \mathfrak{i}_1 \oplus \mathfrak{i}_2$ be a decomposition into a direct sum of non-zero ideals and P^1 be the projection onto $\mathfrak{i}_1$ parallel to $\mathfrak{i}_2$ and for $i \in \{1, \ldots, n\}$ let P_i be the projection onto $\mathfrak{g}_i$ parallel to $\bigoplus_{j \neq i} \mathfrak{g}_j$. As projections associated to a direct sum decomposition, P^1 and the P_i's are in $\mathrm{Cent}(\mathfrak{g})$, which is commutative, so $P^1(\mathfrak{g}_i) = P_i(\mathfrak{i}_1) \subseteq \mathfrak{g}_i$. Due to the indecomposability of the $\mathfrak{g}_i$'s, this implies, for all $i \in \{1, \ldots, n\}$, either $P^1_{|\mathfrak{g}_i} = 0$ or $P^1_{|\mathfrak{g}_i} = 1$. This means that

$$\mathfrak{i}_1 = \bigoplus_{\ell=1}^{t} \mathfrak{g}_{j_\ell}$$

for some $1 \leq j_1 < \ldots < j_t \leq n$.

Theorem 2.1.24 below is due to Médina and Revoy, cf. [MR93], and gives us information about the centroids of Lie algebras which are not necessarily perfect or centerfree. Before stating it, we need a further notion.

Definition 2.1.22. Let $\mathfrak{g}$ be a Lie algebra. Then define the space

$$\mathrm{J}(\mathfrak{g}) := \{\, \varphi \in \mathrm{Hom}(\mathfrak{g}, \mathfrak{z}(\mathfrak{g})) |\, \mathrm{ad}_x \circ \varphi = 0 = \varphi \circ \mathrm{ad}_x \text{ for all } x \in \mathfrak{g} \} \cong \mathrm{Hom}(\mathfrak{g}/[\mathfrak{g}, \mathfrak{g}], \mathfrak{z}(\mathfrak{g})).$$

Remark 2.1.23. Let $\mathfrak{g}$ be a Lie algebra such that $\mathfrak{z}(\mathfrak{g}) \subseteq [\mathfrak{g}, \mathfrak{g}]$. For all $f \in \mathrm{Cent}(\mathfrak{g})$, $g \in \mathrm{J}(\mathfrak{g})$, $x, y \in \mathfrak{g}$ we have:

$$fg([x, y]) = f(0) = 0$$
$$[fg(x), y] = f[g(x), y] = f(0) = 0.$$
$$gf[x, y] = g[f(x), y] = 0$$
$$[gf(x), y] = [g(f(x)), y] = 0$$
$$g^2(x) = g(g(x)) \in g(\mathfrak{z}(\mathfrak{g})) \subseteq g([\mathfrak{g}, \mathfrak{g}]) = 0.$$

Thus $\mathrm{J}(\mathfrak{g})$ is a two-sided ideal of $\mathrm{Cent}(\mathfrak{g})$ with $\mathrm{J}(\mathfrak{g})^2 = 0$.

Theorem 2.1.24. *Let $\mathfrak{g}$ be a Lie algebra of finite dimension such that $\mathfrak{z}(\mathfrak{g}) \subseteq [\mathfrak{g}, \mathfrak{g}]$.*

1. *There are indecomposable idempotents $p_1, \ldots, p_n \in \mathrm{Cent}(\mathfrak{g})$ which are pairwise orthogonal, i.e. $p_i \circ p_j = \delta_{ij} p_i$ for $i, j \in \{1, \ldots, n\}$, satisfying $\sum_{i=1}^{n} p_i = 1$.*

2. *Setting $\mathfrak{g}_i := p_i(\mathfrak{g})$ for $i \in \{1, \ldots, n\}$ we have $\mathfrak{g} = \bigoplus_{i=1}^{n} \mathfrak{g}_i$, a decomposition into indecomposable non-zero ideals.*

3. *Each $\mathrm{Cent}(\mathfrak{g}_i)$ is isomorphic to $p_i \circ \mathrm{Cent}(\mathfrak{g}) \circ p_i$, its quotient $\mathrm{Cent}(\mathfrak{g}_i)/\mathrm{J}(\mathfrak{g}_i)$ modulo the maximal nilpotent ideal $\mathrm{J}(\mathfrak{g}_i)$ is a field.*

4. *Setting $C_{ij} := p_i \circ \mathrm{Cent}(\mathfrak{g}) \circ p_j$ for $i, j \in \{1, \ldots, n\}$ we have $\mathrm{Cent}(\mathfrak{g}) = \bigoplus_{i,j=1}^{n} C_{ij}$, a decomposition into ideals, each C_{ij} is a $\mathrm{Cent}(\mathfrak{g}_i)$-$\mathrm{Cent}(\mathfrak{g}_j)$-bimodule and, if $i \neq j \in \{1, \ldots, n\}$, then the vector space $\mathrm{Hom}(\mathfrak{g}_j/[\mathfrak{g}_j, \mathfrak{g}_j], \mathfrak{z}(\mathfrak{g}_i))$ is isomorphic to C_{ij}. Furthermore, there is a decomposition $\mathrm{J}(\mathfrak{g}) = \bigoplus_{i=1}^{n} \mathrm{J}(\mathfrak{g}_i) \oplus \bigoplus_{i \neq j=1}^{n} C_{ij}$ into ideals.*

We also need the following result, Proposition 22.1 of [Lam01], to show a more general result about the uniqueness of the decomposition of a finite-dimensional Lie algebra into indecomposable non-zero ideals.

Theorem 2.1.25. *Let R be a ring with unity element 1. Suppose there exists a decomposition of $1 \in R$ into a sum of indecomposable idempotents, say $1 = c_1 + \ldots + c_r$, such that $c_i c_j = \delta_{ij} c_i$ and $c_i \in Z(R)$ for all $i, j \in \{1, \ldots, n\}$. Then the decomposition is unique except for the order.*

Proposition 2.1.26. *Let $\mathfrak{g}$ be a finite-dimensional Lie algebra such that $\mathfrak{z}(\mathfrak{g}) \subseteq [\mathfrak{g}, \mathfrak{g}]$ with a decomposition into indecomposable non-zero ideals $\bigoplus_{i=1}^{n} \mathfrak{g}_i$ such that $\mathrm{Hom}(\mathfrak{g}_i/[\mathfrak{g}_i, \mathfrak{g}_i], \mathfrak{z}(\mathfrak{g}_j)) = 0$ if $i \neq j \in \{1, \ldots, n\}$. Then this decomposition is unique except for the order.*

Proof. By Theorem 2.1.24, we have $R := \mathrm{Cent}(\mathfrak{g}) \cong \bigoplus_{i=1}^{n} \mathrm{Cent}(\mathfrak{g}_i)$, $\mathrm{J}(\mathfrak{g}) = \bigoplus_{i=1}^{n} \mathrm{J}(\mathfrak{g}_i)$ and the quotient $\mathrm{Cent}(\mathfrak{g})/\mathrm{J}(\mathfrak{g}) \cong \bigoplus_{i=1}^{n} \mathrm{Cent}(\mathfrak{g}_i)/\mathrm{J}(\mathfrak{g}_i)$ is commutative. Now suppose there are two decompositions of $\mathfrak{g}$ into indecomposable non-zero ideals, say $\mathfrak{g} = \bigoplus_{i=1}^{n} \mathfrak{g}_i = \bigoplus_{j=1}^{m} \mathfrak{h}_j$ with corresponding systems of orthogonal projections $p_1, \ldots p_n$ and $q_1, \ldots, q_m$. Let $f \in \mathrm{Cent}(\mathfrak{g})$ and $f = f_1 + \ldots + f_n$ its decomposition into elements of the $\mathrm{Cent}(\mathfrak{g}_i) \subseteq \mathrm{Cent}(\mathfrak{g})$. Then, for fixed $i \in \{1, \ldots, n\}$, we have

$$p_i \circ f = \sum_{k=1}^{n} p_i \circ f_k = p_i \circ f_i = f_i = f_i \circ p_i = \sum_{k=1}^{n} f_k \circ p_i = f \circ p_i.$$

So the p_i's are in the center of R and, by a dual argument, all the q_j's are so. By Theorem 2.1.25, this implies $\{p_1, \ldots, p_n\} = \{q_1, \ldots, q_m\}$, thus $\{\mathfrak{g}_1, \ldots, \mathfrak{g}_n\} = \{\mathfrak{h}_1, \ldots, \mathfrak{h}_m\}$. $\square$

Remark 2.1.27. In general, the decomposition of a finite-dimensional Lie algebra $\mathfrak{g}$ into indecomposable non-zero ideals is not unique: Let $\mathfrak{g} = \mathfrak{g}_1 \oplus \mathfrak{g}_2$ be a decomposition into indecomposable non-zero ideals with corresponding indecomposable projections p_1, p_2 with non-trivial space $\mathrm{Hom}(\mathfrak{g}_1/[\mathfrak{g}_1, \mathfrak{g}_1], \mathfrak{z}(\mathfrak{g}_2))$. If $0 \neq \varphi : \mathfrak{g} \to \mathfrak{g}$ is a morphism mapping $\mathfrak{g}_1$ to $\mathfrak{z}(\mathfrak{g}_2)$ and the spaces $[\mathfrak{g}_1, \mathfrak{g}_1]$ and $\mathfrak{g}_2$ to 0, then $p'_1 := (1 + \varphi) \circ p_1 \circ (1 + \varphi)^{-1} = p_1 + \varphi$ is also an indecomposable idempotent of $\mathrm{Cent}(\mathfrak{g})$. Thus $\mathfrak{g} = \mathrm{im}\, p'_1 + \mathrm{im}(1 - p'_1)$ is another decomposition of $\mathfrak{g}$ into indecomposable non-zero ideals.

Remark 2.1.28. If $\mathfrak{g}_1, \ldots, \mathfrak{g}_n$ are the unique indecomposable non-zero ideals of $\mathfrak{g}$ and $f \in \mathrm{Aut}(\mathfrak{g})$ is a Lie algebra automorphism, then of course $f(\mathfrak{g}_1), \ldots, f(\mathfrak{g}_n)$ are the unique indecomposable non-zero ideals of $f(\mathfrak{g})$ and thus there is a permutation $\sigma \in \mathcal{S}_n$ such that $f(\mathfrak{g}_i) = \mathfrak{g}_{\sigma(i)}$ for all $i \in \{1, \ldots, n\}$.

Lemma 2.1.29. *Let $\mathfrak{g}$ be of finite dimension with commutative centroid.*

1. *If $\mathbb{K} = \mathbb{C}$, then $\mathfrak{g}$ is indecomposable if and only if $\mathrm{S}(\mathfrak{g}) = \mathbb{C} \cdot \mathbf{1}$.*

2. *If $\mathbb{K} = \mathbb{R}$, then $\mathfrak{g}$ is indecomposable if and only if*
   - *$\mathrm{S}(\mathfrak{g}) = \mathbb{R} \cdot \mathbf{1} + \mathbb{R} \cdot J$ for a complex structure J on $\mathfrak{g}$, i.e $J^2 = -\mathbf{1}$, or*
   - *$\mathrm{S}(\mathfrak{g}) = \mathbb{R} \cdot \mathbf{1}$ and $\mathfrak{g}$ does not admit a complex structure.*

Proof.

1. For any $f \in \mathrm{Cent}(\mathfrak{g})$ the eigenspaces of f are ideals of $\mathfrak{g}$: If $y \neq 0$ is an eigenvector of f for the eigenvalue λ and $x \in \mathfrak{g}$ then $f[x, y] = [x, fy] = [x, \lambda y] = \lambda[x, y]$, i.e., $[x, y]$ is also an eigenvector of f for the eigenvalue λ. If $\mathbb{K} = \mathbb{C}$, then the semisimple endomorphisms in $\mathrm{Cent}(\mathfrak{g})$ are exactly the diagonalizable endomorphisms in $\mathrm{Cent}(\mathfrak{g})$. Thus $\mathfrak{g}$ is indecomposable if and only if each endomorphism in $\mathrm{S}(\mathfrak{g})$ has exactly one eigenvalue, which is equivalent to $\mathrm{S}(\mathfrak{g}) = \mathbb{C} \cdot \mathbf{1}$.

2. If $\mathbb{K} = \mathbb{R}$ and $\mathfrak{g}$ is decomposable into the direct sum of non-trivial ideals $\mathfrak{g}_1$ and $\mathfrak{g}_2$ with corresponding orthogonal projections p_1 and p_2, then $p_1 p_2 = 0$ although $p_1 \neq 0 \neq p_2$, thus $\mathrm{Cent}(\mathfrak{g})$ has zero-divisors and is neither isomorphic to $\mathbb{R}$ nor to $\mathbb{C}$ as associative $\mathbb{R}$-algebras.

   If $\mathbb{K} = \mathbb{R}$ and $\mathfrak{g}$ is indecomposable, then any endomorphism in $\mathrm{S}(\mathfrak{g})$ admits exactly one real eigenvalue or exactly two non-real, complex conjugate eigenvalues. If every endomorphism in $\mathrm{S}(\mathfrak{g})$ admits exactly one real eigenvalue, then $\mathrm{S}(\mathfrak{g}) = \mathbb{R} \cdot \mathbf{1}$. Otherwise, if any $f \in \mathrm{S}(\mathfrak{g})$ admits the non-real complex eigenvalues $a + ib$ and $a - ib$, then $J := \frac{1}{b}(f - a\mathbf{1}) \in \mathrm{S}(\mathfrak{g})$ is a complex structure of $\mathfrak{g}$ because of the following calculation for the eigenvector $x \in \mathfrak{g}$ for the eigenvalue $a \pm ib$:

$$
J^2(x) = (\frac{1}{b^2}(f - a\mathbf{1})^2)(x) = \frac{1}{b^2}(f^2(x) - 2af(x) + a^2 x)
$$
$$
= \frac{1}{b^2}((a \pm ib)^2 x - 2a(a \pm ib)x + a^2 x) = \frac{1}{b^2}(a^2 \pm 2iab - b^2 - 2a^2 \mp 2iab + a^2)(x)
$$
$$
= -x.
$$

   The space of semisimple endomorphims in $\mathrm{Cent}(\mathfrak{g})$ in this case is the space of real linear combination of $\mathbf{1}$ and J, thus isomorphic to $\mathbb{C}$ as associative algebras over $\mathbb{R}$. Since the complex conjugation is the only $\mathbb{R}$-linear algebra automorphism on $\mathbb{C}$ different from the identity, the only complex structures on $\mathrm{S}(\mathfrak{g})$ are J and $-J$. $\qquad \square$

Definition 2.1.30. If $J \in \mathrm{Cent}(\mathfrak{g})$ with $J^2 = -\mathbf{1}$ is a complex structure of a real Lie algebra $\mathfrak{g}$, then the scalar multiplication $(r + is)x := rx + J(sx)$ for $r, s \in \mathbb{R}$, $x \in \mathfrak{g}$ turns $\mathfrak{g}$ into a complex Lie algebra for which we write $\mathfrak{g}^{\mathbb{C}}(J)$.

Definition 2.1.31. A simple $\mathbb{K}$-Lie algebra $\mathfrak{g}$ is called *central simple*, if $\mathrm{Cent}(\mathfrak{g}) = \mathbb{K} \cdot \mathbf{1}$.

A useful tool to analyze the structure of $\mathbb{K}$-Lie algebras is the Cartan-Killing form.

Definition 2.1.32.

1. Let $\mathfrak{g}$ be a $\mathbb{K}$-Lie algebra. The *Cartan-Killing form* of $\mathfrak{g}$ is the symmetric bilinear form

$$\kappa : \mathfrak{g} \times \mathfrak{g} \longrightarrow \mathbb{K}$$
$$(x, y) \longmapsto \operatorname{tr}\left(\operatorname{ad}(x) \circ \operatorname{ad}(y)\right).$$

Its *radical* is $\operatorname{rad}(\kappa) := \{\, y \in \mathfrak{g} \,|\, \kappa(x, y) = 0 \text{ for all } x \in \mathfrak{g} \,\}$.

2. A finite-dimensional real Lie algebra $\mathfrak{g}$ is *compact*, if one of the following equivalent conditions is fulfilled:

   a) There is a compact Lie group G such that $\mathbf{L}(G) \cong \mathfrak{g}$, i.e. $\mathfrak{g}$ is the Lie algebra associated to G.

   b) The topological subgroup $\operatorname{Inn}(\mathfrak{g}) := \left\langle e^{\operatorname{ad}(x)} : x \in \mathfrak{g} \right\rangle \leq \operatorname{Aut}(\mathfrak{g})$ has compact closure.

   c) The Cartan-Killing form of $\mathfrak{g}$ is negative definite.

   d) There is a euclidean scalar product $B : \mathfrak{g} \times \mathfrak{g} \to \mathbb{R}$ such that $B([x, y], z) = B(x, [y, z])$ for all $x, y, z \in \mathfrak{g}$.

Now we easily show some properties of the Cartan-Killing form.

Lemma 2.1.33. *Let $\mathfrak{g}$ be a $\mathbb{K}$-Lie algebra with Cartan-Killing form κ.*

1. For all $x, y, z \in \mathfrak{g}$ we have: $\kappa([x, y], z) = \kappa(x, [y, z])$. This property is the invariance *of κ.*

2. $\operatorname{rad}(\kappa) \trianglelefteq \mathfrak{g}$.

3. $\operatorname{Aut}(\mathfrak{g}) \subseteq \mathrm{O}(\mathfrak{g}, \kappa)$, i.e., for all $x, y \in \mathfrak{g}$ and $f \in \operatorname{Aut}(\mathfrak{g})$ we have: $\kappa(fx, y) = \kappa(x, f^{-1}y)$.

Proof. The following calculation shows the first statement:

$$\begin{aligned}
\kappa([x, y], z) &= \operatorname{tr}(\operatorname{ad}[x, y] \circ \operatorname{ad}(z)) = \operatorname{tr}([\operatorname{ad}(x), \operatorname{ad}(y)] \circ \operatorname{ad}(z)) \\
&= \operatorname{tr}(\operatorname{ad}(x) \circ \operatorname{ad}(y) \circ \operatorname{ad}(z) - \operatorname{ad}(y) \circ \operatorname{ad}(x) \circ \operatorname{ad}(z)) \\
&= \operatorname{tr}(\operatorname{ad}(x) \circ \operatorname{ad}(y) \circ \operatorname{ad}(z)) - \operatorname{tr}\left(\operatorname{ad}(y) \circ (\operatorname{ad}(x) \circ \operatorname{ad}(z))\right) \\
&= \operatorname{tr}(\operatorname{ad}(x) \circ \operatorname{ad}(y) \circ \operatorname{ad}(z)) - \operatorname{tr}\left((\operatorname{ad}(x) \circ \operatorname{ad}(z)) \circ \operatorname{ad}(y))\right) \\
&= \operatorname{tr}(\operatorname{ad}(x) \circ \operatorname{ad}(y) \circ \operatorname{ad}(z) - \operatorname{ad}(x) \circ \operatorname{ad}(z) \circ \operatorname{ad}(y)) \\
&= \operatorname{tr}(\operatorname{ad}(x) \circ [\operatorname{ad}(y), \operatorname{ad}(z)]) = \operatorname{tr}(\operatorname{ad}(x) \circ \operatorname{ad}[y, z]) \\
&= \kappa(x, [y, z]).
\end{aligned}$$

The second statement is true since if $z \in \operatorname{rad}(\kappa)$ and $x, y \in \mathfrak{g}$, then $\kappa(x, [y, z]) = \kappa([x, y], z) = 0$, so $[y, z] \in \operatorname{rad}(\kappa)$. For the third statement, we first note that

$$\operatorname{ad}(fx)(y) = [fx, y] = f\left[x, f^{-1}y\right] = f \circ \operatorname{ad}(x) \circ f^{-1}(y),$$

so $\operatorname{ad}(fx) = f \circ \operatorname{ad}(x) \circ f^{-1}$. We obtain:

$$\begin{aligned}
\kappa(fx, y) &= \operatorname{tr}(\operatorname{ad}(fx) \circ \operatorname{ad}(y)) = \operatorname{tr}\left(f \circ \operatorname{ad}(x) \circ f^{-1} \circ \operatorname{ad}(y)\right) \\
&= \operatorname{tr}\left(f^{-1} \circ f \circ \operatorname{ad}(x) \circ f^{-1} \circ \operatorname{ad}(y) \circ f\right) = \operatorname{tr}\left(\operatorname{ad}(x) \circ \operatorname{ad}(f^{-1}y)\right) = \kappa(x, f^{-1}y). \quad \square
\end{aligned}$$

Every finite-dimensional $\mathbb{K}$-Lie algebra can be decomposed into two parts: its solvable radical and a semisimple Levi complement. Before stating this important result, we first define some notions.

Definition 2.1.34.

1. A $\mathbb{K}$-Lie algebra $\mathfrak{g}$ is *nilpotent*, if its *lower central series* $\mathfrak{g}^0, \mathfrak{g}^1, \mathfrak{g}^2, \ldots$, which is defined by $\mathfrak{g}^0 := \mathfrak{g}$ and $\mathfrak{g}^{n+1} := [\mathfrak{g}, \mathfrak{g}^n]$ for $n \in \mathbb{N}_0$, becomes zero eventually.

2. A $\mathbb{K}$-Lie algebra $\mathfrak{g}$ is *solvable*, if its *derived series* $\mathfrak{g}^{(0)}, \mathfrak{g}^{(1)}, \mathfrak{g}^{(2)}, \ldots$, which is defined by $\mathfrak{g}^{(0)} := \mathfrak{g}$ and $\mathfrak{g}^{(n+1)} := [\mathfrak{g}^{(n)}, \mathfrak{g}^{(n)}]$ for $n \in \mathbb{N}_0$, becomes zero eventually. Since the sum of solvable ideals is solvable, every finite-dimensional $\mathbb{K}$-Lie algebra possesses a unique largest solvable ideal, the *radical* $\mathrm{rad}(\mathfrak{g})$.

The concept of a semidirect sum of Lie algebras is the infinitesimal analog to the concept of a semidirect product of (Lie) groups.

Definition 2.1.35. Given two $\mathbb{K}$-Lie algebras $\mathfrak{r}, \mathfrak{h}$ and a Lie algebra morphism $\alpha : \mathfrak{h} \to \mathrm{Der}(\mathfrak{r})$, we define the following Lie bracket on the vector space direct sum $\mathfrak{r} \oplus \mathfrak{h}$:

$$[(x, y), (x', y')] := (\alpha(y)(x') - \alpha(y')(x) + [x, x'], [y, y'])$$

for $x, x' \in \mathfrak{r}$ and $y, y' \in \mathfrak{h}$. This Lie algebra is called the *semidirect sum* with respect to α of $\mathfrak{r}$ and $\mathfrak{h}$, denoted by $\mathfrak{r} \rtimes_\alpha \mathfrak{h}$. Easy calculations show that $\{(x, 0) \in \mathfrak{r} \rtimes_\alpha \mathfrak{h} \mid x \in \mathfrak{r}\}$ is an ideal of $\mathfrak{r} \rtimes_\alpha \mathfrak{h}$ isomorphic to $\mathfrak{r}$ and $\{(0, y) \in \mathfrak{r} \rtimes_\alpha \mathfrak{h} \mid y \in \mathfrak{h}\}$ is a subalgebra of $\mathfrak{r} \rtimes_\alpha \mathfrak{h}$ isomorphic to $\mathfrak{h}$.

Similar to the global analog, the semidirect product of (Lie) groups, there is a more explicit description of a semidirect sum of Lie algebras, if the algebras in question are subspaces of another Lie algebra.

Definition 2.1.36. Let $\mathfrak{g}$ be a $\mathbb{K}$-Lie algebra with ideal $\mathfrak{r}$ and subalgebra $\mathfrak{h}$ such that $\mathfrak{r} + \mathfrak{h} = \mathfrak{g}$ and $\mathfrak{r} \cap \mathfrak{h} = \mathbf{0}$. Then the map $\alpha : \mathfrak{h} \to \mathrm{Der}(\mathfrak{r})$, $\alpha(y)(x) := [y, x]$ is morphism of Lie algebras. The corresponding semidirect sum $\mathfrak{r} \rtimes_\alpha \mathfrak{h}$ is also denoted by $\mathfrak{r} \rtimes \mathfrak{h}$ and called *inner semidirect sum* of $\mathfrak{r}$ and $\mathfrak{h}$. It is naturally isomorphic to $\mathfrak{g}$ via

$$\mathfrak{r} \rtimes \mathfrak{h} \longrightarrow \mathfrak{g}$$
$$(x, y) \longmapsto x + y,$$

so we may write $\mathfrak{g} = \mathfrak{r} \rtimes \mathfrak{h}$.

We can fully understand the class of finite-dimensional Lie algebras, if we understand the solvable and the semisimple ones. Roughly, this is result of the following, deep theorems.

Theorem 2.1.37 (Levi). *For each finite-dimensional $\mathbb{K}$-Lie algebra $\mathfrak{g}$ there is a semisimple subalgebra $\mathfrak{s}$ such that $\mathfrak{g} = \mathrm{rad}(\mathfrak{g}) \rtimes \mathfrak{s}$. Such a subalgebra $\mathfrak{s}$ is called* Levi complement *and the decomposition $\mathfrak{g} = \mathrm{rad}(\mathfrak{g}) \rtimes \mathfrak{s}$ is called* Levi decomposition.

Theorem 2.1.38 (Malcev). *For two Levi complements $\mathfrak{s}$ and $\mathfrak{t}$ of a finite-dimensional $\mathbb{K}$-Lie algebra $\mathfrak{g}$, there is an $x \in [\mathfrak{g}, \mathrm{rad}(\mathfrak{g})]$ such that $e^{\mathrm{ad}(x)}\mathfrak{s} = \mathfrak{t}$. In addition, Levi complements are precisely the maximal semisimple subalgebras of $\mathfrak{g}$.*

Remark 2.1.39. Let $\mathfrak{g}$ be a finite-dimensional $\mathbb{K}$-Lie algebra and $\mathfrak{g} = \mathrm{rad}(\mathfrak{g}) \rtimes \mathfrak{s}$ a Levi decomposition. We write $\alpha : \mathfrak{s} \to \mathrm{Der}(\mathrm{rad}(\mathfrak{g}))$, $\alpha(s)(r) := [s, r]$ and define the ideal $\mathfrak{s}_1 := \ker(\alpha)$ which is a semisimple Lie algebra, as is the quotient $\mathfrak{s}_2 := \mathfrak{s}/\mathfrak{s}_1$. With the injective Lie algebra morphism $\overline{\alpha} : \mathfrak{s}_2 \to \mathrm{Der}(\mathrm{rad}(\mathfrak{g}))$ induced by α, we have a semidirect sum $\mathrm{rad}(\mathfrak{g}) \rtimes_{\overline{\alpha}} \mathfrak{s}_2 \subseteq \mathfrak{g}$ and a decomposition $\mathfrak{g} = (\mathrm{rad}(\mathfrak{g}) \rtimes_{\overline{\alpha}} \mathfrak{s}_2) \oplus \mathfrak{s}_1$ into ideals. So $\mathfrak{g}$ has a semisimple direct summand if and only if α is not injective.

By Lie's and Ado's Theorems (cf. Theorems 4.4.8 and 6.5.1 of [HN11]), the solvable Lie algebras are well understood. We obtain the following theorem.

Theorem 2.1.40. *Every finite-dimensional complex solvable Lie algebra is isomorphic to a subalgebra of $\mathfrak{b}(n, \mathbb{C}) \leq \mathfrak{gl}(n, \mathbb{C})$, the Lie algebra of upper triangular matrices.*

For a better understanding of finite-dimensional real semisimple Lie algebras, we use the following theorem characterizing semisimple Lie algebras using the Cartan-Killing form.

Theorem 2.1.41 (Cartan). *A finite-dimensional $\mathbb{K}$-Lie algebra is semisimple if and only if its Cartan-Killing form is non-degenerate, i.e., has a trivial radical.*

The idea of the Cartan decomposition is decomposing a semisimple real Lie algebras into a compact and a non-compact part.

Theorem 2.1.42.

1. *Let $\mathfrak{g}$ be a semisimple real Lie algebra of finite dimension with Cartan-Killing form κ. Then it has a* Cartan decomposition *$\mathfrak{g} = \mathfrak{k} \oplus_\kappa^\tau \mathfrak{p}$, i.e., there is $\tau \in \mathrm{Aut}(\mathfrak{g})$ with $\tau^2 = \mathrm{id}_\mathfrak{g}$, called* Cartan involution, *such that $\kappa_{|\mathfrak{k} \times \mathfrak{k}}$ is negative-definite, where $\mathfrak{k} := \mathfrak{g}^\tau = \{\, x \in \mathfrak{g} \mid \tau(x) = x \,\}$, and $\kappa_{|\mathfrak{p} \times \mathfrak{p}}$ is positive-definite, where $\mathfrak{p} := \mathfrak{g}^{-\tau} = \{\, x \in \mathfrak{g} \mid -\tau(x) = x \,\}$. Furthermore, $\mathfrak{k}$ and $\mathfrak{p}$ are orthogonal with respect to κ, one has the relations $[\mathfrak{k}, \mathfrak{k}] \subseteq \mathfrak{k}$, $[\mathfrak{p}, \mathfrak{k}] \subseteq \mathfrak{p}$, $[\mathfrak{p}, \mathfrak{p}] \subseteq \mathfrak{k}$ and $\mathfrak{k}$ is compact.*

2. *If $\mathfrak{g} = \mathfrak{k} \oplus_\kappa^\tau \mathfrak{p}$ and $\mathfrak{g} = \mathfrak{c} \oplus_\kappa^\sigma \mathfrak{q}$ are two Cartan decompositions of a semisimple real Lie algebra $\mathfrak{g}$, then there is an inner automorphism $\gamma \in \mathrm{Inn}(\mathfrak{g})$ with $\sigma = \gamma \circ \tau \circ \gamma^{-1}$, $\gamma(\mathfrak{k}) = \mathfrak{c}$ and $\gamma(\mathfrak{p}) = \mathfrak{q}$.*

3. *There exists a bijective correspondence between the semisimple $\mathbb{R}$-Lie algebras with Cartan involution and the semisimple compact $\mathbb{R}$-Lie algebras with involution by*

$$\left(\mathfrak{g} = \mathfrak{g}^\tau \oplus \mathfrak{g}^{-\tau} \leq \mathfrak{g}_\mathbb{C} \,\middle|\, \tau \right) \longmapsto \left(\mathfrak{u} = \mathfrak{g}^\tau \oplus i\mathfrak{g}^{-\tau} \leq \mathfrak{g}_\mathbb{C} \,\middle|\, \mathrm{id}_{\mathfrak{g}^\tau} \oplus -\mathrm{id}_{i\mathfrak{g}^{-\tau}} \right)$$

with inverse

$$\left(\mathfrak{u} = \mathfrak{u}^\sigma \oplus \mathfrak{u}^{-\sigma} \leq \mathfrak{u}_\mathbb{C} \,\middle|\, \sigma \right) \longmapsto \left(\mathfrak{g} = \mathfrak{u}^\sigma \oplus i\mathfrak{u}^{-\sigma} \leq \mathfrak{u}_\mathbb{C} \,\middle|\, \mathrm{id}_{\mathfrak{u}^\sigma} \oplus -\mathrm{id}_{i\mathfrak{u}^{-\sigma}} \right)$$

and $\mathfrak{g}$ is simple if and only if $\mathfrak{u}$ is so.

Complex semisimple Lie algebras of finite dimension possess two essentially different real forms: compact real forms and split real forms. In order to define the latter type of real forms, we need to define Cartan subalgebras.

Definition 2.1.43. A subalgebra $\mathfrak{h} \leq \mathfrak{g}$ of a $\mathbb{K}$-Lie algebra is called *Cartan subalgebra*, if is nilpotent and self-normalizing, i.e., $\mathfrak{h} = \{x \in \mathfrak{g} \,|\, [x, y] \in \mathfrak{h} \text{ for all } y \in \mathfrak{h}\}$.

Here are some well-known remarks on Cartan subalgebras.

Remark 2.1.44.

1. Cartan subalgebras are maximal nilpotent, but not every maximal nilpotent subalgebra of a $\mathbb{K}$-Lie algebra is a Cartan subalgebra. For instance, the strictly upper triangular matrices in $\mathfrak{sl}(n, \mathbb{C})$ for $n \geq 2$ form a maximal nilpotent subalgebra $\mathfrak{a}$ which is normalized by diagonal matrices.

2. If $\mathfrak{h}$ is a subalgebra of the real Lie algebra $\mathfrak{g}$, then $\mathfrak{h}$ is a Cartan subalgebra of $\mathfrak{g}$ if and only if $\mathfrak{h}_\mathbb{C}$ is a Cartan subalgebra of $\mathfrak{g}_\mathbb{C}$.

3. Surjective morphisms of Lie algebras map Cartan subalgebras onto Cartan subalgebras.

4. Every finite-dimensional $\mathbb{K}$-Lie algebra $\mathfrak{g}$ possesses a Cartan subalgebra.

5. The Cartan subalgebras of a finite-dimensional complex Lie algebras $\mathfrak{g}$ are conjugate by an automorphism of $\mathfrak{g}$. In particular, the Cartan subalgebras of $\mathfrak{g}$ all have the same dimension, called *rank* of $\mathfrak{g}$.

6. Every Cartan subalgebra $\mathfrak{h}$ of a finite-dimensional semisimple $\mathbb{K}$-Lie algebra $\mathfrak{g}$ is abelian and the morphism $\mathrm{ad}_\mathfrak{g}(x) : \mathfrak{g} \to \mathrm{Der}(\mathfrak{g})$ is semisimple for all $x \in \mathfrak{h}$.

Now we can define the notion of split Lie algebras and split real forms.

Definition 2.1.45. A semisimple $\mathbb{K}$-Lie algebra $\mathfrak{g}$ is called *split*, if there exists a Cartan subalgebra $\mathfrak{h} \leq \mathfrak{g}$ such that $\mathrm{ad}_\mathfrak{g}(x)$ is triangularizable for all $x \in \mathfrak{h}$. [3] In this case, the $\mathrm{ad}_\mathfrak{g}(x)$ are automatically diagonalizable by their semisimplicity and, by the abelianness of $\mathfrak{h}$, the family $\{\mathrm{ad}_\mathfrak{g}(x) \,|\, x \in \mathfrak{h}\}$ is simultaneously diagonalizable. So we have a *splitting*, i.e., a vector space decomposition

$$\mathfrak{g} = \mathfrak{h} \oplus \bigoplus_{\lambda \in \Delta} \mathfrak{g}^\lambda,$$

for $\mathfrak{g}^\lambda := \{y \in \mathfrak{g} \,|\, [x, y] = \lambda(x) \cdot y \text{ for all } x \in \mathfrak{h}\}$, $\lambda \in \mathfrak{h}^*$. Let $\Delta := \{\lambda \in \mathfrak{h}^* \setminus \{0\} \,|\, \dim(\mathfrak{g}^\lambda) \neq 0\}$, called the *root system* of $\mathfrak{g}$ with respect to $\mathfrak{h}$. The elements of Δ are called *roots* and $\mathfrak{g}^\lambda$ the *root space* of λ for a root $\lambda \in \Delta$. Note that $\mathfrak{h} = \mathfrak{g}^0$.

Definition 2.1.46. A real semisimple split Lie algebra $\mathfrak{g}$ is called *split real form* of its complexification $\mathfrak{g}_\mathbb{C}$. A real semisimple compact Lie algebra $\mathfrak{u}$ is called *compact real form* of its complexification $\mathfrak{u}_\mathbb{C}$.

The following well-known facts of the next theorem can be found in Chapitre VI of [Ser66].

Theorem 2.1.47. *Let $\mathfrak{g}$ be a complex semisimple Lie algebra of finite dimension.*

---

[3] If $\mathfrak{g}$ is split, we implicitly fix an appropriate Cartan subalgebra $\mathfrak{h} \leq \mathfrak{g}$. There are real semisimple Lie algebras $\mathfrak{g}$ with two different Cartan subalgebras $\mathfrak{h}_1, \mathfrak{h}_2$ such that $\mathrm{ad}_\mathfrak{g}(x)$ is diagonalizable for all $x \in \mathfrak{h}_1$, but not for all $x \in \mathfrak{h}_2$.

1. $\mathfrak{g}$ *is split, i.e., for the unique Cartan subalgebra* $\mathfrak{h} \leq \mathfrak{g}$ *(except for conjugation by an automorphism of* $\mathfrak{g}$*) there is a splitting*

$$\mathfrak{g} = \mathfrak{h} \oplus \bigoplus_{\lambda \in \Delta} \mathfrak{g}^{\lambda}.$$

Moreover, the root spaces are all of dimension 1, fulfill the relation $\left[\mathfrak{g}^{\lambda}, \mathfrak{g}^{\nu}\right] = \mathfrak{g}^{\lambda+\nu}$ *for* $\lambda, \nu \in \Delta$, $\lambda + \nu \neq 0$ *and* $\mathfrak{h}^{\lambda} := \left[\mathfrak{g}^{\lambda}, \mathfrak{g}^{-\lambda}\right]$ *is also of dimension 1 for* $\lambda \in \Delta$.

2. *There is a subset* $\Pi \subseteq \Delta$, *called* root basis, *such that* $\Delta \subseteq \mathbb{N}_0[\Pi] \sqcup -\mathbb{N}_0[\Pi]$.

3. *By choosing, for* $\lambda \in \Pi$, *the unique element* $\check{\lambda} := h_{\lambda} \in \mathfrak{h}^{\lambda}$ *such that* $\lambda(h_{\lambda}) = 2$ *and an arbitrary element* $e_{\lambda} \in \mathfrak{g}^{\lambda} \setminus \{0\}$, *there exists a unique element* $f_{\lambda} \in \mathfrak{g}^{-\lambda}$ *such that* $[e_{\lambda}, f_{\lambda}] = h_{\lambda}$, $[h_{\lambda}, e_{\lambda}] = 2e_{\lambda}$ *and* $[h_{\lambda}, f_{\lambda}] = -2f_{\lambda}$, *i.e., the subalgebra* $\mathfrak{g}(\lambda)$ *generated by* $e_{\lambda}, f_{\lambda}, h_{\lambda}$ *is isomorphic to* $\mathfrak{sl}(2, \mathbb{C})$.

4. *The system of subalgebras* $\{\mathfrak{g}(\lambda) | \lambda \in \Pi\}$ *generates* $\mathfrak{g}$.

Corollary 2.1.48. *The* $\mathbb{R}$-*subalgebra* $\mathfrak{s} \leq \mathfrak{g}$ *generated by* $e_{\lambda}, f_{\lambda}, h_{\lambda}$ *for* $\lambda \in \Pi$ *is a split real form of* $\mathfrak{g}$.

Example 2.1.49. We give examples of split and compact real forms for the classical Lie algebras (cf. Example 4.1.1 and Example 4.1.6 of [Vin94]).

1. For $n \in \mathbb{N}$, split real forms of $\mathfrak{sl}(n+1, \mathbb{C})$, $\mathfrak{so}(2n+3, \mathbb{C})$, $\mathfrak{sp}(2n+4, \mathbb{C})$, $\mathfrak{so}(2n+6, \mathbb{C})$ are given by $\mathfrak{sl}(n+1, \mathbb{R})$, $\mathfrak{so}(n+1, n+2, \mathbb{R})$, $\mathfrak{sp}(2n+4, \mathbb{R})$, $\mathfrak{so}(n+3, n+3, \mathbb{R})$, respectively.

2. For $n \in \mathbb{N}$, compact real forms of $\mathfrak{sl}(n+1, \mathbb{C})$, $\mathfrak{so}(2n+3, \mathbb{C})$, $\mathfrak{sp}(2n+4, \mathbb{C})$, $\mathfrak{so}(2n+6, \mathbb{C})$ are given by $\mathfrak{su}(n+1, \mathbb{C})$, $\mathfrak{so}(2n+3, \mathbb{R})$, $\mathfrak{sp}(n+2, \mathbb{H})$, $\mathfrak{so}(2n+6, \mathbb{R})$, respectively.

2.2. Bundles

In this section we will repeat some basic definitions and properties of bundles, based on standard literature such as [Hus66].

Definition 2.2.1. Let $\mathbb{F}, M, F$ be manifolds such that there is surjective submersion $\pi : \mathbb{F} \to M$ and each fiber $\mathbb{F}_x := \pi^{-1}(x)$ for $x \in M$ is diffeomorphic to F. If there is an open cover $\mathfrak{U} = (U_i)_{i \in I}$ of M and a family of smooth maps $\Phi = (\varphi_i)_{i \in I}$, each $\varphi_i : \pi^{-1}(U_i) \to F$ inducing a diffeomorphism $(\pi, \varphi_i) : \pi^{-1}(U_i) \to U_i \times F$ (*local triviality*), then the sextuple $(\mathbb{F}, M, \pi, F, \mathfrak{U}, \Phi)$ is called a *fiber bundle* with *bundle space* $\mathbb{F}$ over the *base space* M with *bundle projection* π, *fiber* F and *bundle atlas* $(\mathfrak{U}, \Phi)$. A pair (U_i, φ_i) is called *bundle chart*. As a shorter formulation we will say "$\pi : \mathbb{F} \to M$ is a bundle with fiber F" without explicit mention of the bundle atlas.

There is a strong topological result, allowing us always to assume the existence of a finite bundle atlas. We prepare to formulate it.

Definition 2.2.2.

1. Let $(U_i)_{i \in I}$ be an open cover of a topological space X. A *refinement* of $(U_i)_{i \in I}$ is a triple $(J, f, (V_j)_{j \in J})$ of a set J, a function $f : J \to I$ and an open cover $(V_j)_{j \in J}$ of X such that $V_j \subseteq U_{f(j)}$ for all $j \in J$.

2. A *Palais refinement*[4] of a given open cover $(U_i)_{i \in I}$ is a refinement $(J, f, (V_j)_{j \in J})$, for which there exists a number $r \in \mathbb{N}$ and partition $(J_t)_{t=1}^r$ of J as follows: For all $t \in \{1, \ldots, r\}$ the set J_t is countable and if $i, j \in J_t$, then the condition $i \neq j$ already implies $V_i \cap V_j = \emptyset$.

Remark 2.2.3. If M is a finite-dimensional, paracompact and connected manifold, then Proposition I.1.I of [GHV72] says that for any open cover $(U_i)_{i \in I}$ of M, there is a Palais refinement $(J, f, (V_j)_{j \in J}, (J_t)_{t=1}^r)$.

If $(U_i, \varphi_i)_{i \in I}$ is a bundle atlas of an F-fiber bundle over M with atlas $(U_i, \chi_i)_{i \in I}$ of M and with Palais refinement $(J, f, (V_j)_{j \in J}, (J_t)_{t=1}^r)$, then we can define, for all $t \in \{1, \ldots, r\}$, the open set $W_t := \bigcup_{j \in J_t} V_j$, on which the chart maps $\psi_t : W_t \to F$ and $\xi_t : W_t \to \mathbb{R}^{\dim(M)}$ are well-defined by $\psi_t(w) := \varphi_i(w)$ and $\xi_t(w) := \chi_i(w)$, respectively, for $w \in U_i$.

Definition 2.2.4. A fiber bundle $\pi' : \mathbb{F}' \to M$ is a *subbundle* of a fiber bundle $\pi : \mathbb{F} \to M$, provided $\mathbb{F}' \subseteq \mathbb{F}$ and $\pi'(x) = \pi(x)$ for all $x \in \mathbb{F}'$.

For each fiber bundle there is a class of interesting functions, the so-called sections.

Definition 2.2.5. Let $\pi : \mathbb{F} \to M$ be a fiber bundle and $k \in \mathbb{N} \cup \{\infty\}$. A C^k-map

$$X : M \longrightarrow \mathbb{F}$$
$$x \longmapsto X(x) = X_x$$

is called C^k-*section* of the bundle, if $\pi \circ X = \mathrm{id}_M$. The set of all C^k-sections of the fiber bundle $\pi : \mathbb{F} \to M$ is denoted by $\Gamma^k(\mathbb{F})$ and we also write $\Gamma(\mathbb{F})$ instead of $\Gamma^\infty(\mathbb{F})$.

We now give a first definition of a vector bundle.

Definition 2.2.6. Let V be a finite-dimensional vector space and $\pi : \mathbb{V} \to M$ a fiber bundle with fiber V with vector space structures on each fiber $\mathbb{V}_x$ for $x \in M$, isomorphic to V by $(\varphi_i)_{|_{\mathbb{V}_x}} : \mathbb{V}_x \to V$ for each $i \in I$. Note that this induces a pointwise vector space structure on $\Gamma(\mathbb{V})$. Then the sextuple $(\mathbb{V}, M, \pi, V, \mathfrak{U}, \Phi)$ is called a *vector bundle*. As a shorter formulation we will say "$\pi : \mathbb{V} \to M$ is a vector bundle with fiber V".

There are categorial morphisms of vector bundles.

Definition 2.2.7. Let $\pi_1 : \mathbb{V}_1 \to M_1$, $\pi_2 : \mathbb{V}_2 \to M_2$ be vector bundles with fibers V_1, V_2, respectively.

1. A *vector bundle morphism* from $\pi_1 : \mathbb{V}_1 \to M_1$ to $\pi_2 : \mathbb{V}_2 \to M_2$ is a pair of smooths maps $\mu : \mathbb{V}_1 \to \mathbb{V}_2$, $\lambda : M_1 \to M_2$ such that the diagram

$$
\begin{array}{ccc}
\mathbb{V}_1 & \xrightarrow{\ \mu\ } & \mathbb{V}_2 \\
{\scriptstyle \pi_1}\downarrow & & \downarrow{\scriptstyle \pi_2} \\
M_1 & \xrightarrow{\ \lambda\ } & M_2
\end{array}
$$

commutes and $\mu : (\mathbb{V}_1)_x \to (\mathbb{V}_2)_{\lambda(x)}$ is a vector space morphism for all $x \in M_1$. The space of vector bundle morphisms from $\pi_1 : \mathbb{V}_1 \to M_1$ to $\pi_2 : \mathbb{V}_2 \to M_2$ is denoted by $\mathrm{Hom}(\mathbb{V}_1, \mathbb{V}_2)$. We also write $\mathrm{End}(\mathbb{V}_1) := \mathrm{Hom}(\mathbb{V}_1, \mathbb{V}_1)$, each element is a *vector bundle endomorphism*.

---

[4]Although this type of cover is called "Palais" in various francophone articles like [FLS74] and [Lec80], the origins of this naming remain unclear to the author.

2. A vector bundle morphism (μ, λ) is a *vector bundle isomorphism*, if there exists a vector bundle morphism $\iota \in \mathrm{Hom}(\mathbb{V}_2, \mathbb{V}_1)$, such that μ is inverse to ι. In this case the vector bundles $\pi_1 : \mathbb{V}_1 \to M_1$ and $\pi_2 : \mathbb{V}_2 \to M_2$ are called *isomorphic*.

3. In the case of $M_1 = M_2$, a vector bundle isomorphism (μ, λ) is an *vector bundle equivalence*, if $\lambda = \mathrm{id}_{M_1}$. In this case the vector bundles $\pi_1 : \mathbb{V}_1 \to M_1$ and $\pi_2 : \mathbb{V}_2 \to M_1$ are called *equivalent*.

4. The V_1-vector bundle $\pi_1 : \mathbb{V}_1 \to M_1$ is called *trivial*, if it is equivalent to the vector bundle $(M_1 \times V_1, M_1, \mathrm{pr}_1, V_1, \{M\}, \{\mathrm{id}_M\})$.

Example 2.2.8. The first natural examples of vector bundles on a manifold M are the trivial vector bundles $M \times V \to V$ for vector spaces V, the tangent bundle $TM \to M$, where the fiber $(TM)_x$ for each $x \in M$ is (naturally isomorphic to) the tangent space $T_x M$, and the cotangent bundle $T^*M \to M$, where the fiber $(T^*M)_x$ of each $x \in M$ is (naturally isomorphic to) the cotangent space $(T_x M)^* = \mathrm{Hom}(T_x M, \mathbb{R})$.

An alternative definition of a vector bundle needs the definition of a principal bundle and a bundle associated to a principal bundle.

Definition 2.2.9. Let G be a Lie group and $\pi : P \to M$ a fiber bundle with fiber G such that π is the canonical projection of a smooth right action

$$R : P \times G \longrightarrow P$$
$$(p, g) \longmapsto p \cdot g = R_g(p) = R^p(g).$$

If the following diagram commutes for all $i \in I$, $g \in G$

$$
\begin{array}{ccc}
\pi^{-1}(U_i) & \xrightarrow{\ R_g\ } & \pi^{-1}(U_i) \\
{\scriptstyle \varphi_i}\big\downarrow & & \big\downarrow{\scriptstyle \varphi_i} \\
G & \xrightarrow[\ \rho_g\ ]{} & G,
\end{array}
$$

then the septuple $(P, M, \pi, G, R, \mathfrak{U}, \Phi)$ is called a *principal bundle* with *structure group* G and *bundle action* R. As a shorter formulation we will say "$\pi : P \to M$ is a G-principal bundle" without explicit mention of the bundle action. A *covering* is a principal bundle with discrete structure group.

Remark 2.2.10. The smooth right action $R : P \times G \to P$ of a principal bundle is free and proper, i.e., if $R(p, g) = p$ for some $p \in P$, $g \in G$, then $g = 1$, and with R' being the map $P \times G \to P \times P$, $(p, g) \mapsto (R(p, g), p)$ the R'-preimages of compact sets are compact.

It can be shown that any free and proper smooth right action R of a Lie group G on a manifold P defines a principal bundle.

Example 2.2.11. If M is a connected manifold, then its universal covering $q_M : \widetilde{M} \to M$ is a $\pi_1(M)$-principal bundle. The corresponding right action $R : \widetilde{M} \times \pi_1(M) \to \widetilde{M}$ is the action by deck transformations.

Note that for principal bundles there are the same notions (morphism, equivalence, etc.) totally analogous to those in Definition 2.2.7 for vector bundles. There is a natural principal bundle corresponding to a given vector bundle.

Definition 2.2.12. Let $\pi : \mathbb{V} \to M$ be a V-vector bundle. The set $\mathrm{Fr}(\mathbb{V}) := \bigcup_{m \in M} \mathrm{Iso}(V, \mathbb{V}_m)$ carries the structure of a $\mathrm{GL}(V)$-principal bundle with respect to $\mathrm{Fr}(\mathbb{V}) \times \mathrm{GL}(V) \to \mathrm{Fr}(\mathbb{V})$, $(\psi, g) \mapsto \psi \circ g$ with base space M, called *frame bundle* of $\mathbb{V}$.

If $\pi : P \to M$ is a G-principal bundle and F is a manifold on which G acts from the left, then we can construct a new bundle with fiber F. We will use this general construction of associated bundles to give another definition of vector bundles.

Definition 2.2.13. Let $\pi : P \to M$ be a G-principal bundle and F a manifold with a smooth left action

$$\begin{aligned} L : G \times F &\longrightarrow F \\ (g, f) &\longmapsto g \cdot f. \end{aligned}$$

By defining a left action as follows

$$\begin{aligned} L' : G \times (P \times F) &\longrightarrow P \times F \\ (g, (p, f)) &\longmapsto g \cdot (p, f) := \left(p \cdot g^{-1}, g \cdot f \right). \end{aligned}$$

and considering the orbit space $P \times_G F := \mathbb{F} := (P \times F)/G$, we obtain a bundle $\pi' : \mathbb{F} \to M$ with fiber F, called a *bundle associated* to the G-principal bundle $\pi : P \to M$ with fiber F. Its bundle projection is well-defined by

$$\begin{aligned} \pi' : \mathbb{F} &\longrightarrow M \\ [(p, f)] := G \cdot (p, f) &\longmapsto \pi(p). \end{aligned}$$

The bundle atlas $(U_i, \varphi_i)_{i \in I}$ of the G-principal bundle $\pi : P \to M$ induces a well-defined bundle atlas $(U_i, \varphi'_i)_{i \in I}$ of $\pi' : \mathbb{F} \to M$ by

$$\begin{aligned} \varphi'_i : \pi'^{-1}(U_i) &\longrightarrow F \\ [(p, f)] &\longmapsto \varphi_i(p) \cdot f. \end{aligned}$$

Finally, each fiber $\mathbb{F}_x = \pi'^{-1}(x)$ for $x \in M$ is diffeomorphic to F:

$$\begin{aligned} \pi'^{-1}(x) &= \{ [(p, f)] \in (P \times F)/G \mid \pi(p) = x \} = \{ [(p, f)] \in (P \times F)/G \mid p \in P_x \} \\ &= (P_x \times F)/G \cong F. \end{aligned}$$

Definition 2.2.14. Let V be a finite-dimensional vector space, $\tau : G \times V \to V$ a smooth linear Lie group action and $\pi : P \to M$ a G-principal bundle. Then the bundle $\pi' : P \times_G V \to M$ is called a *vector bundle* associated to P.

Of course, both definitions of vector bundles coincide. We formulate this equivalence in the following statement, for details on the proof we refer to Section 5.2 of [Hus66].

Theorem 2.2.15.

1. *Given a finite-dimensional vector space V, a smooth Lie group action $\tau : G \times V \to V$ and a G-principal bundle $P \to M$, the bundle $\mathbb{V} = P \times_G V \to M$ is a vector bundle in the sense of Definition 2.2.6 by defining the following well-defined operations for $\lambda \in \mathbb{K}$, $v, w \in \mathbb{V}_x$, $x \in M$ and some bundle chart (U, φ) with $x \in U$:*

$$v + w := \varphi_{|\mathbb{V}_x}^{-1}(\varphi(v) + \varphi(w)),$$
$$\lambda \cdot v := \varphi_{|\mathbb{V}_x}^{-1}(\lambda \cdot \varphi(v)).$$

2. *Given a vector bundle $\pi : \mathbb{V} \to M$ with fiber V in the sense of Definition 2.2.6, there is an equivalence of vector bundles as follows:*

$$\mathrm{Fr}(\mathbb{V}) \times_{\mathrm{GL}(V)} V \longrightarrow \mathbb{V}$$
$$\left\{ \left(\psi \circ g^{-1}, g(v) \right) : g \in \mathrm{GL}(V) \right\} = [(\psi, v)] \longmapsto \psi(v)$$

for the frame bundle $\mathrm{Fr}(\mathbb{V})$ and the natural action $\lambda : \mathrm{GL}(V) \times V \to V$.

Definition 2.2.16. Let $\pi' : \mathbb{V} \to M$ be a vector bundle with fiber V associated to its frame bundle $\pi : \mathrm{Fr}(\mathbb{V}) \to M$ with bundle atlas $(U_i, \varphi_i')_{i \in I}$ and $(U_i, \varphi_i)_{i \in I}$, respectively. We define $\mathcal{I} := \{ (i, j) \in I \times I \mid U_i \cap U_j \neq \emptyset \}$. For $(i, j) \in \mathcal{I}$ and $x \in U_i \cap U_j$ we define $g_{ji}(x) \in \mathrm{GL}(V)$ by

$$(\pi, \varphi_j) \circ (\pi, \varphi_i)^{-1} : (U_i \cap U_j) \times G \longrightarrow (U_i \cap U_j) \times G$$
$$(x, g) \longmapsto (x, g_{ji}(x)g).$$

This element can also be defined by

$$\left(\pi', \varphi_j' \right) \circ \left(\pi', \varphi_i' \right)^{-1} : (U_i \cap U_j) \times V \longrightarrow (U_i \cap U_j) \times V$$
$$(x, v) \longmapsto (x, g_{ji}(x)(v)).$$

We obtain smooth functions $g_{ji} : U_i \cap U_j \to \mathrm{GL}(V)$ for $(i, j) \in \mathcal{I}$, called *transition maps*.

Another equivalent way to define vector bundles is given by Proposition I.5.5.1 of [Hus66].

Proposition 2.2.17. *Let V be a finite-dimensional vector space, G a closed subgroup of $\mathrm{GL}(V)$, M a manifold, $(U_i)_{i \in I}$ an open cover of M and $g_{ji} : U_i \cap U_j \to G$ for $(i, j) \in \mathcal{I}$ a familiy of maps satisfying the conditions*

$$g_{ii}(x) = 1 \text{ for all } x \in U_i,$$
$$g_{ji}(x) \cdot g_{ik}(x) \cdot g_{kj}(x) = 1 \text{ for all } x \in U_i \cap U_j \cap U_k,$$

where $\mathcal{I} := \{ (i, j) \in I \times I \mid U_i \cap U_j \neq \emptyset \}$. Then there is a vector bundle $\pi : \mathbb{V} \to M$ with fiber V, bundle atlas $(U_i, \varphi_i)_{i \in I}$ and transition maps $(g_{ji})_{(i,j) \in \mathcal{I}}$ and it is unique up to equivalence.

Example 2.2.18. Let M be a manifold of dimension d with tangent bundle $p : TM \to M$ and cotangent bundle $q : T^*M \to M$ (cf. Example 2.2.8). The latter bundle is associated to the frame bundle $\pi : \mathrm{Fr}(TM) \to M$, considered as a $\mathrm{GL}(\mathbb{R}^d)$-principal bundle, with respect to the action $\lambda : \mathrm{GL}(\mathbb{R}^d) \times (\mathbb{R}^d)^* \to (\mathbb{R}^d)^*$, $\lambda(A, \alpha)(v) := \alpha(A \cdot v)$.

If a vector bundle is given, it is sometimes inconvenient to consider it associated to its frame bundle, but one wants to consider it associated to a principal bundle with a "smaller" structure group. In the following, we will explain under which circumstances this is possible and how.

Remark 2.2.19. Let $\pi : \mathbb{V} \to M$ be a vector bundle with fiber V associated to its frame bundle $\rho : \mathrm{Fr}(\mathbb{V}) \to M$. If $H \leq \mathrm{GL}(V)$ is a subgroup and $(U_i, \varphi_i)_{i \in I}$ is bundle atlas of $\mathbb{V}$ such that the corresponding transition maps have values in H only, then $\pi : \mathbb{V} \to M$ can be considered associated to an H-principal bundle $\varrho : P \to M$ by applying a construction analogous of that of the frame bundle, replacing $\mathrm{GL}(V)$ by H.

Another way to change the structure group of a principal bundle is to take a quotient. This is based on the following.

Proposition 2.2.20. *Let $\rho : P \to M$ be a principal bundle with right action $\sigma : P \times G \to P$, $(p, g) \mapsto p.g$. If $N \trianglelefteq G$ is a closed normal subgroup, then for $P' := P/N := \{ p.N \,|\, p \in P \}$ and $H := G/N$ we have an H-principal bundle $\rho' : P' \to M$, $p.N \mapsto \rho(p)$ with right action $\sigma' : P' \times H \to P'$, $(p.N, [g]) \mapsto (p.g).N$.*

Proof. If (U, θ) is a bundle chart in $p \in P$, then U is an m-neighborhood in M, where $m := \rho(p)$, and $\theta : \rho^{-1}(U) \to G$ is a smooth G-equivariant map. Note that P' has a canonical manifold structure such that it is the base space of an N-principal bundle $P \to P'$, obtained by restricting the right action $\sigma : P \times G \to P$. The map $\theta' : (\rho')^{-1}(U) \to H = G/N$, $[p] = p.N \mapsto [\theta(p)]$ is well-defined because θ is equivariant and smooth because of the closedness of N. $\qquad\square$

Definition 2.2.21. If $\rho : P \to M$ is a G-principal bundle and $N \trianglelefteq G$ is a closed normal subgroup with quotient $H := G/N$, then the H-principal bundle $\rho' : P/N \to M$ is called the *quotient principal bundle* of P modulo N.

Remark 2.2.22. Let $\pi : \mathbb{V} \to M$ be a vector bundle associated to a principal bundle $\rho : P \to M$ with respect to the left action $\lambda : G \times V \to V$ and let $\sigma : P \times G \to P$, $(p, g) \mapsto p.g$ be the right action of the principal bundle. Let $N \trianglelefteq G$ be a closed normal subgroup with quotient $H := G/N$. If, in addition, we have $\lambda(n, v) = v$ for all $n \in N$, $v \in V$, then λ factors through an action $\lambda' : H \times V \to V$, $([g], v) \mapsto \lambda(g, v)$ and for the quotient principal bundle $P' := P/N$, there is an equivalence of vector bundles as follows:

$$\Phi : \mathbb{V} = P \times_G V \longrightarrow P' \times_H V$$

$$\left\{ \left[\sigma\left(p, g^{-1} \right), \lambda(g, v) \right] : g \in G \right\} = [p, v] \longmapsto [p.N, v] = \left\{ \left[\sigma'\left(p.N, [g]^{-1} \right), \lambda'([g], v) \right] : [g] \in H \right\}.$$

Sections of associated bundles, hence also of vector bundles, can be described as equivariant functions on the total space of the principal bundle. The next statement is Theorem 4.8.1 of [Hus66].

Proposition 2.2.23. *Let $P \times_G F$ be a bundle associated to a principal bundle $\rho : P \to M$. For*

$$\mathrm{C}^\infty(P, F)^G := \left\{ \alpha \in \mathrm{C}^\infty(P, F) \,|\, \alpha(p.g^{-1}) = g.\alpha(p) \text{ for all } p \in P, g \in G \right\}$$

there is a bijection

$$\begin{aligned} \Psi : \mathrm{C}^\infty(P, F)^G &\longrightarrow \Gamma(P \times_G F) \\ \Psi(\alpha)(\rho(p)) &:= [(p, \alpha(p))] . \end{aligned} \tag{2.4}$$

If $P \times_G F = \mathbb{V}$ *is a vector bundle, then* (2.4) *even defines a linear isomorphism.*

The next construction for general fiber bundles helps to identify sections more explicitly.

Definition 2.2.24. Let $\pi : \mathbb{F} \to M$ be an F-fiber bundle, M' a manifold and $q : M' \to M$ a smooth function. We define the set

$$q^*\mathbb{F} := \left\{ (m', f) \in M' \times \mathbb{F} \,\middle|\, q(m') = \pi(f) \right\}$$

and write $\pi' : q^*\mathbb{F} \to M'$ and $\widetilde{q} : q^*\mathbb{F} \to \mathbb{F}$ for the natural projections, so the diagram

$$
\begin{array}{ccc}
q^*\mathbb{F} & \overset{\widetilde{q}}{\longrightarrow} & \mathbb{F} \\
{\scriptstyle \pi'} \downarrow & & \downarrow {\scriptstyle \pi} \\
M' & \overset{q}{\longrightarrow} & M
\end{array}
$$

commutes. With the smooth subspace structure $q^*\mathbb{F} \subseteq M' \times \mathbb{F}$ we obtain a new F-fiber bundle $\pi' : q^*\mathbb{F} \to M'$, the *pullback bundle* of $\pi : \mathbb{F} \to M$ with respect to q.

Remark 2.2.25. Here are some important properties of pullback bundles.

1. Let $\pi : \mathbb{V} \to M$ be a vector bundle associated to a principal bundle $\rho : P \to M$ with respect to the left action $\lambda : G \times V \to V$ and let $\sigma : P \times G \to P$ be the right action of the principal bundle. If M' is a manifold and $q : M' \to M$ is a smooth function, then $\rho' : q^*P \to M'$ is a principal bundle with right action $\sigma' : q^*P \times G \to G$, $((m', p), g) \mapsto (m', \sigma(p, g))$ to which the vector bundle $\pi' : q^*\mathbb{V} \to M'$ is associated with respect to the left action λ.

2. If $\rho : P \to M$ is a G-principal bundle, then the pullback G-principal bundle $\rho' : \rho^*P \to P$ is trivial, since it has the section $s : P \to \rho^*P$, $p \mapsto (p, p)$, so the map $P \times G \to \rho^*P$, $(p, g) \mapsto s(p).g = (p, p.g)$ is an equivalence.

3. If $\pi : \mathbb{V} \to M$ is a V-vector bundle associated to a G-principal bundle $\rho : P \to M$, then the pullback V-vector bundle $\rho' : \rho^*\mathbb{V} \to P$ is trivial, since there is an equivalence of vector bundles as follows: $P \times V \to \rho^*\mathbb{V}$, $(p, v) \mapsto (p, [(p, v)])$, where $[(p, v)] \in P \times_G V = \mathbb{V}$. The isomorphism $C^\infty(P, V) \cong \Gamma(\rho^*\mathbb{V})$ is given by $s(p) = (p, [p, f(p)])$ for $p \in P$, $s \in \Gamma(\rho^*\mathbb{V})$, $f \in C^\infty(P, V)$.

4. Let $\rho : P \to M$ be a G-principal bundle and $N \trianglelefteq G$ be a closed normal subgroup with quotient $H := G/N$. If $\rho' : P/N \to M$ is the quotient H-principal bundle of P modulo N, and $q : M' \to M$ is a smooth map, then $q^*(P/N) \to M'$ is naturally equivalent to the quotient H-principal bundle of q^*P modulo N via

$$
\begin{aligned}
q^*(P/N) &\longrightarrow (q^*P)/N \\
(m', [p]) &\longmapsto [(m', p)] ,
\end{aligned}
$$

because the principal bundle right action on $q^*(P)$ is defined by $(m', p).g := (m', p.g)$.

Quotient principal bundles are flat, if the corresponding quotient group is discrete (cf. Remark B.3.3.2). We use the description of flat bundles from this remark for the following.

Proposition 2.2.26. *Let $\rho : P \to M$ be a G-principal bundle and $N \trianglelefteq G$ be a closed normal subgroup with finite quotient $H := G/N$. Then there is a finite covering $\widehat{q} : \widehat{M} \to M$ such that the pullback principal bundle $\widehat{q}^*(P/N) \to \widehat{M}$ is trivial.*

Proof. The quotient principal bundle $P/N \to M$ is flat, hence, for some $\chi \in \operatorname{Hom}(\pi_1(M), H)$, equivalent to $P_\chi = \widetilde{M} \times_{\pi_1(M)} H$, the bundle associated to the universal covering $q_M : \widetilde{M} \to M$ with respect to the action $\pi_1(M) \times H \to H$, $(\varphi, h) \mapsto \chi(\varphi)h$. Here, χ is unique up to conjugation by an element of H. Let $\widehat{M}$ be the orbit manifold of the action $\widetilde{M} \times \ker(\chi) \to \widetilde{M}$, the restriction of the $\pi_1(M)$-action on $\widetilde{M}$ by deck transformations, and $\widehat{q} : \widehat{M} \to M$ be the corresponding projection. Note that $\widehat{M}$ does not depend on the choice of $\chi' \in [\chi]$ because $\chi' = h.\chi$ for some $h \in H$ implies $\ker(\chi) = \ker(\chi')$. Let $\Lambda := \pi_1(M)/\ker(\chi)$. Then P_χ is equivalent to $P_\Lambda := \widehat{M} \times_\Lambda H$, the bundle associated to the Λ-principal bundle $\widehat{q} : \widehat{M} \to M$ with respect to the action $\Lambda \times H \to H$, $([\varphi], h) \mapsto \chi(\varphi)h$.

For $Q := \widehat{q}^*P$ the quotient principal bundle Q/N is naturally equivalent to $\widehat{q}^*P_\Lambda$ (cf. Remark 2.2.25.4), but the latter (and hence the former) bundle is trivial by Remark 2.2.25.2. $\square$

2.3. Lie Algebra Bundles and Lie Algebras of Sections

We now define Lie algebra bundles and their section algebras and prove first structural theorems.

Definition 2.3.1.

1. Let $\mathfrak{k}$ be a finite-dimensional Lie algebra and M a manifold. A *Lie algebra bundle* is a vector bundle $\pi : \mathfrak{K} \to M$ with fiber $\mathfrak{k}$ which has a bundle atlas $(U_i, \varphi_i)_{i \in I}$ such that the transition maps g_{ij} defined by

$$(\pi, \varphi_j) \circ (\pi, \varphi_i)^{-1} : (U_i \cap U_j) \times \mathfrak{k} \longrightarrow (U_i \cap U_j) \times \mathfrak{k}$$
$$(x, v) \longmapsto (x, g_{ji}(x)(v))$$

have values in $\operatorname{Aut}(\mathfrak{k})$ only. By Remark 2.2.19, it is associated to an $\operatorname{Aut}(\mathfrak{k})$-principal bundle $\rho : \operatorname{P}(\mathfrak{K}) \to M$.

2. We justify the name "Lie algebra bundle" by defining a Lie bracket on each fiber $\mathfrak{K}_x$ for $x \in M$ by a bundle chart (U, φ) in x and the Lie bracket $[\cdot, \cdot]$ on $\mathfrak{k}$:

$$[\cdot, \cdot] : \mathfrak{K}_x \times \mathfrak{K}_x \longrightarrow \mathfrak{K}_x$$
$$(v, w) \longmapsto [v, w]_\varphi = \varphi^{-1}_{|\mathfrak{K}_x} [\varphi(v), \varphi(w)].$$

In order to see that this is well-defined, we make use of the fact that all maps $g_{ij}(x)$ are automorphisms of $\mathfrak{k}$:

$$[v, w]_{\varphi_j} = \varphi^{-1}_{j|\mathfrak{K}_x} [\varphi_j(v), \varphi_j(w)] = \varphi^{-1}_{i|\mathfrak{K}_x} (g_{ij}(x) [\varphi_j(v), \varphi_j(w)])$$
$$= \varphi^{-1}_{i|\mathfrak{K}_x} [g_{ij}(x)(\varphi_j(v)), g_{ij}(x)(\varphi_j(w))] = \varphi^{-1}_{i|\mathfrak{K}_x} [\varphi_i(v), \varphi_i(w)]$$
$$= [v, w]_{\varphi_i}.$$

3. Using the Lie algebra structure on each fiber of $\mathfrak{K}$, we define a Lie bracket on the space of sections $\Gamma^k(\mathfrak{K})$ for $k \in \mathbb{N}_0 \cup \{\infty\}$ by

$$[X,Y](x) := [X(x), Y(x)] \quad \text{for } X, Y \in \Gamma^k(\mathfrak{K}), x \in M.$$

The Lie algebra $\Gamma^k(\mathfrak{K})$ is called *Lie algebra of* C^k-*sections.* We call $\Gamma(\mathfrak{K}) = \Gamma^\infty(\mathfrak{K})$ also *Lie algebra of smooth sections.*

For Lie algebra bundles there are the same notions (morphism, equivalence, etc.) completely analogous to those in Definition 2.2.7 for vector bundles, where each morphism is has to respect the Lie brackets.

Example 2.3.2. If $\mathfrak{k}$ is a finite-dimensionale Lie algebra and M a manifold, then the trivial Lie algebra bundle $\mathrm{pr}_1 : M \times \mathfrak{k} \to M$ possess a section algebra $\Gamma(M \times \mathfrak{k})$ naturally isomorphic to $C^\infty(M, \mathfrak{k})$ by $X \mapsto X^{\mathrm{pr}_2}$ for the bundle chart (M, pr_2), i.e., $X^{\mathrm{pr}_2} = \mathrm{pr}_2 \circ X$. So a Lie algebra of smooth sections is, algebraically, a generalization of a smooth mapping algebra $C^\infty(M, \mathfrak{k})$, which is a generalization of a *smooth loop Lie algebra* $C^\infty(\mathbb{S}^1, \mathfrak{k})$ with pointwise defined bracket.

Isomorphic Lie algebra bundles possess isomorphic section algebras. Later we will see that the opposite is also true for central simple fibers.

Proposition 2.3.3. *Let* $\pi : \mathfrak{K} \to M$ *and* $\varpi : \mathfrak{G} \to M$ *be Lie algebra bundles with fiber* $\mathfrak{k}$. *If* $\Lambda : \mathfrak{K} \to \mathfrak{G}$ *is an isomorphism of Lie algebra bundles, then* $\Gamma^k(\mathfrak{K}) \cong \Gamma^k(\mathfrak{G})$ *as Lie algebras.*

Proof. Let $\lambda : M \to M$ be the C^k-diffeomorphism induced by Λ, i.e., $\varpi \circ \Lambda = \lambda \circ \pi$. Define a map $\mu : \Gamma^k(\mathfrak{K}) \to \Gamma^k(\mathfrak{G})$ as follows: Let $X \in \Gamma^k(\mathfrak{K})$ and $m \in M$. Then $\mu(X)_m := \Lambda\left(X_{\lambda^{-1}(m)}\right) \in \mathfrak{G}_m$. Clearly, μ is well-defined and fiberwise an isomorphism of Lie algebras since Λ is an isomorphism of Lie algebra bundles. Hence μ is a morphism of Lie algebras with trivial kernel. The surjectivity of μ follows from the fact that for $Y \in \Gamma^k(\mathfrak{G})$ we can define a section $X \in \Gamma^k(\mathfrak{K})$ with $\mu(X) = Y$ by $X_m := \Lambda^{-1}\left(Y_{\lambda(m)}\right)$ for all $m \in M$. $\square$

Remark 2.3.4. We equip $\Gamma(\mathfrak{K})$ with the Fréchet space structure from Remark B.1.12 and obtain a Fréchet-Lie algebra $\Gamma(\mathfrak{K})_\infty$, i.e., the Lie bracket is continuous with respect to this topology:

We fix a finite bundle atlas $(U_i, \varphi_i)_{i \in F}$, smooth covers $(U_{i,n,\ell}, C_{i,n,\ell})_{n,\ell \in \mathbb{N}_0}$ of U_i for $i \in F$ (cf. Definition B.1.6), and a submultiplicative norm $\|\cdot\|$ on $\mathfrak{k}$. We write $\pi^M_{r,s} : T^r M \to T^s M$, where $r > s \in \mathbb{N}_0$, for the iterated tangent projections. The Fréchet space $C^\infty(U_i, \mathfrak{k})_\infty$ has the continuous seminorms

$$p_{i,n,\ell} : C^\infty(U_i, \mathfrak{k}) \to \mathbb{R}_+, \quad f \mapsto \sup_{x \in C_{i,n,\ell}} \|d^n f(x)\|.$$

Our first estimate is the following:

$$p_{i,0,\ell}([f,g]) = \sup_{x \in C_{i,0,\ell}} \|[f,g](x)\| \leq \sup_{x \in C_{i,0,\ell}} \|f(x)\| \cdot \|g(x)\| \leq p_{i,0,\ell}(f) \cdot p_{i,0,\ell}(g).$$

Now fix $i \in F$ and let $f, g \in C^\infty(U_i, \mathfrak{k})$, $x \in U_i$ and $v \in T_x M$. By the Chain Rule and the fact that any bilinear map $\beta : V \times V \to V$ for a finite-dimensional vector space V is smooth with differential $d_{(a,b)}\beta(v,w) = \beta(v,b) + \beta(a,w)$, we obtain:

$$d_x[f,g](v) = d_x([\cdot,\cdot] \circ (f,g))(v) = d_{(f(x),g(x))}[\cdot,\cdot](d_x(f,g)(v))$$
$$= [f(x), d_x g(v)] + [d_x f(v), g(x)].$$

We obtain the next estimate:

$$
\begin{aligned}
p_{i,1,\ell}\left([f,g]\right) &= \sup_{v \in C_{i,1,\ell}} \left\| d\left[f,g\right](v) \right\| \leq \sup_{v \in C_{i,1,\ell}} \left\| \left[f(\pi_{1,0}^M(v)), dg(v)\right] + \left[df(v), g(\pi_{1,0}^M(v))\right] \right\| \\
&\leq p_{i,1,\ell}\left(\left[f \circ \pi_{1,0}^M, dg\right]\right) + p_{i,1,\ell}\left(\left[df, g \circ \pi_{1,0}^M\right]\right) \\
&\leq p_{i,0,\ell}\left(f\right) \cdot p_{i,1,\ell}\left(g\right) + p_{i,1,\ell}\left(f\right) \cdot p_{i,0,\ell}\left(g\right).
\end{aligned}
\tag{2.5}
$$

By iterating the estimate (2.5), we obtain:

$$
p_{i,n,\ell}\left([f,g]\right) \leq \sum_{k=0}^{n} \binom{n}{k} \sup_{v \in C_{i,n,\ell}} \left\| \left[d^k f \circ \pi_{n,k}^M, d^{n-k} g \circ \pi_{n,n-k}^M\right] \right\| \leq \sum_{k=0}^{n} \binom{n}{k} p_{i,k,\ell}\left(f\right) \cdot p_{i,n-k\ell}\left(g\right).
$$

Since $\Gamma(\mathfrak{K})_\infty$ is a Fréchet space with respect to the seminorms $p_{i,n,\ell}$ for $i \in F$, $\ell, n \in \mathbb{N}_0$, this estimate proves the continuity of the Lie bracket on $\Gamma(\mathfrak{K})_\infty$. In particular, Lie algebras of smooth sections are topological generalizations of smooth loop Lie algebras $C^\infty\left(\mathbb{S}^1, \mathfrak{k}\right)_\infty$ with pointwise defined bracket.

As for usual vector bundles, there is a notion of a sum of Lie algebra bundles (over the same base space) and the corresponding section Lie algebras behave as one would expect.

Definition 2.3.5. Let $\pi : \mathfrak{K} \to M$ and $\varpi : \mathfrak{G} \to M$ be Lie algebra bundles with fiber $\mathfrak{k}$ and $\mathfrak{g}$, respectively. Their *sum* is the Whitney sum as vector bundles $(\pi \oplus \varpi) : \mathfrak{K} \oplus \mathfrak{G} \to M$ with fiber $\mathfrak{k} \oplus \mathfrak{g}$ (cf. Definition 3.2.6 of [Hus66]).

The proofs of the following two propositions are fairly straightforward.

Proposition 2.3.6. *If $\pi : \mathfrak{K} \to M$ and $\varpi : \mathfrak{G} \to M$ are Lie algebra bundles with fiber $\mathfrak{k}$ and $\mathfrak{g}$, respectively, then its sum $(\pi \oplus \varpi) : \mathfrak{K} \oplus \mathfrak{G} \to M$ is a Lie algebra bundle with fiber $\mathfrak{k} \oplus \mathfrak{g}$.*

Proposition 2.3.7. *If $\pi : \mathfrak{K} \to M$ and $\varpi : \mathfrak{G} \to M$ are Lie algebra bundles with fiber $\mathfrak{k}$ and $\mathfrak{g}$, respectively, with sum $(\pi \oplus \varpi) : \mathfrak{K} \oplus \mathfrak{G} \to M$, then there is an isomorphism of Lie algebras $\Gamma^k(\mathfrak{K}) \oplus \Gamma^k(\mathfrak{G}) \cong \Gamma^k(\mathfrak{K} \oplus \mathfrak{G})$.*

There are certain bundles associated to a Lie algebra bundle which will help us to analyze the structure of the Lie algebra of its C^k-sections.

Definition 2.3.8. Let $\pi : \mathfrak{K} \to M$ be a Lie algebra bundle with fiber $\mathfrak{k}$ associated to an $\mathrm{Aut}(\mathfrak{k})$-principal bundle $\rho : \mathrm{P}(\mathfrak{K}) \to M$, where $\mathrm{P}(\mathfrak{K})$ is the union of Lie algebra isomorphisms $\bigcup_{m \in M} \mathrm{Iso}(\mathfrak{k}, \mathfrak{K}_m)$

1. Let $I \trianglelefteq \mathfrak{k}$ be an $\mathrm{Aut}(\mathfrak{k})$-stable ideal of $\mathfrak{k}$. We define $\varpi : \mathfrak{K}[I] \to M$ to be the bundle associated to $\rho : \mathrm{P}(\mathfrak{K}) \to M$ with fiber I. It can be identified with a subbundle of $\pi : \mathfrak{K} \to M$ as follows: A typical element of $\mathfrak{K}[I]_x \subseteq \mathfrak{K}[I]$ takes the form

$$
[(\psi, x)] := \left\{ (\psi f^{-1}, f(x)) : f \in \mathrm{Aut}(\mathfrak{k}) \right\}
$$

for a Lie algebra isomorphism $\psi : \mathfrak{k} \to \mathfrak{K}_m$ and an element $x \in I$. The element $\psi(x) \in \mathfrak{K}_m$ is clearly well-defined and so we may embed $\mathfrak{K}[I]$ into $\mathfrak{K}$. In the cases of I being $[\mathfrak{k}, \mathfrak{k}]$ and $\mathfrak{z}(\mathfrak{k})$ we also write $[\mathfrak{K}, \mathfrak{K}]$ instead of $\mathfrak{K}[[\mathfrak{k}, \mathfrak{k}]]$, called the *commutator bundle* of $\mathfrak{K}$ and $\mathfrak{z}(\mathfrak{K})$ instead of $\mathfrak{K}[\mathfrak{z}(\mathfrak{k})]$, called the *center bundle* of $\mathfrak{K}$, respectively. Note that $[\mathfrak{K}, \mathfrak{K}]_m \cong [\mathfrak{K}_m, \mathfrak{K}_m]$ and $\mathfrak{z}(\mathfrak{K})_m \cong \mathfrak{z}(\mathfrak{K}_m)$ via $[(\psi, x)] \mapsto \psi(x)$ for $x \in [\mathfrak{k}, \mathfrak{k}]$ and $x \in \mathfrak{z}(\mathfrak{k})$, respectively.

2. Let $J \subseteq \operatorname{End}(\mathfrak{k})$ be a Lie subalgebra of $\operatorname{End}(\mathfrak{k})_L$, an associative subalgebra of $\operatorname{End}(\mathfrak{k})$ or a closed (Lie) subgroup of $\operatorname{Aut}(\mathfrak{k})$, which is invariant under the left action

$$\operatorname{Aut}(\mathfrak{k}) \times \operatorname{End}(\mathfrak{k}) \longrightarrow \operatorname{End}(\mathfrak{k})$$
$$(f, g) \longmapsto f \circ g \circ f^{-1}.$$

We define $\varpi : \mathfrak{K}(J) \to M$ to be the bundle associated to $\rho : \mathrm{P}(\mathfrak{K}) \to M$ with fiber J. It can be identified with a subbundle of $\Pi : \operatorname{End}(\mathfrak{K}) \to M$ as follows: A typical element of $\mathfrak{K}(J)_m \subseteq \mathfrak{K}(J)$ takes the form

$$[(\psi, j)] := \operatorname{Aut}(\mathfrak{k}) \cdot (\psi, j) = \left\{ (\psi f^{-1}, f j f^{-1}) : f \in \operatorname{Aut}(\mathfrak{k}) \right\}$$

for a Lie algebra isomorphism $\psi : \mathfrak{k} \to \mathfrak{K}_m$ and an endomorphism $j : \mathfrak{k} \to \mathfrak{k}$, $j \in J$. The map $\psi \circ j \circ \psi^{-1} : \mathfrak{K}_m \longrightarrow \mathfrak{K}_m$ is clearly well-defined and, by applying this construction to all classes $[(\psi, j)]$ for $\psi \in \operatorname{Iso}(\mathfrak{k}, \mathfrak{K}_m)$, where m runs through M and $j \in J$ is fixed, we obtain an element of $\operatorname{End}(\mathfrak{K})$. In the cases of J being $\operatorname{Der}(\mathfrak{k})$, $\operatorname{Cent}(\mathfrak{k})$ and $\operatorname{Aut}(\mathfrak{k})$, we also write $\operatorname{Der}(\mathfrak{K})$ instead of $\mathfrak{K}(\operatorname{Der}(\mathfrak{k}))$ and $\operatorname{Cent}(\mathfrak{K})$ instead of $\mathfrak{K}(\operatorname{Cent}(\mathfrak{k}))$ and $\operatorname{Aut}(\mathfrak{K})$ instead of $\mathfrak{K}(\operatorname{Aut}(\mathfrak{k}))$, respectively. Note that $\operatorname{Der}(\mathfrak{K})_m \cong \operatorname{Der}(\mathfrak{K}_m)$ and $\operatorname{Cent}(\mathfrak{K})_m \cong \operatorname{Cent}(\mathfrak{K}_m)$ and $\operatorname{Aut}(\mathfrak{K})_m \cong \operatorname{Aut}(\mathfrak{K}_m)$ via $[(\psi, j)] \mapsto \psi \circ j \circ \psi^{-1}$ for $j \in \operatorname{Der}(\mathfrak{k})$, $j \in \operatorname{Cent}(\mathfrak{k})$ and $j \in \operatorname{Aut}(\mathfrak{k})$, respectively.

3. Let $\pi : \mathfrak{K} \to M$ and $\varpi : \mathfrak{G} \to M$ be Lie algebra bundles with fiber $\mathfrak{k}$ and an isomorphism $\Lambda : \mathfrak{K} \to \mathfrak{G}$. The induced map $\lambda : M \to M$ is a diffeomorphism and lets us define the *λ-isomorphism bundle* $\operatorname{Iso}_\lambda(\mathfrak{K}, \mathfrak{G})$ to be the subbundle of $\operatorname{Hom}(\mathfrak{K}, \mathfrak{G}) \to M$ consisting of the fibers $\operatorname{Iso}(\mathfrak{K}_m, \mathfrak{G}_{\lambda(m)})$ for $m \in M$. If $\mathfrak{K} = \mathfrak{G}$ we also write $\operatorname{Aut}_\lambda(\mathfrak{K})$ for $\operatorname{Iso}_\lambda(\mathfrak{K}, \mathfrak{K})$ and note that $\operatorname{Aut}_{\operatorname{id}_M}(\mathfrak{K}) \cong \operatorname{Aut}(\mathfrak{K})$.

Remark 2.3.9. If $\pi : \mathfrak{K} \to M$ is a Lie algebra bundle with semisimple fiber $\mathfrak{k}$, associated to the $\operatorname{Aut}(\mathfrak{k})$-principal bundle $\rho : \mathrm{P}(\mathfrak{K}) \to M$, then $\mathfrak{K} \cong \operatorname{Der}(\mathfrak{K})$ via the following equivalence of Lie algebra bundles:

$$\operatorname{ad} : \mathfrak{K} \longrightarrow \operatorname{Der}(\mathfrak{K})$$
$$v \longmapsto \left[(\psi, \psi^{-1} \circ \operatorname{ad}_{\mathfrak{K}_{\pi(v)}}(v) \circ \psi) \right].$$

This equivalence does not depend on the choice of $\psi \in \operatorname{Iso}(\mathfrak{k}, \mathfrak{K}_{\pi(v)}) \in \mathrm{P}(\mathfrak{K})_{\pi(v)}$. Alternatively, we can define the equivalence via

$$\mathfrak{K} \supseteq \mathfrak{K}_m \longrightarrow \operatorname{Der}(\mathfrak{K}_m) \cong \operatorname{Der}(\mathfrak{K})_m \subseteq \operatorname{Der}(\mathfrak{K})$$
$$v \longmapsto \operatorname{ad}_{\mathfrak{K}_m}(v).$$

Remark 2.3.10. Let $\pi : \mathfrak{K} \to M$ be a Lie algebra bundle with fiber $\mathfrak{k}$.

1. There are natural inclusions $[\mathfrak{K}, \mathfrak{K}] \to \mathfrak{K}$ and $\mathfrak{z}(\mathfrak{K}) \to \mathfrak{K}$, thus the section spaces $\Gamma^k([\mathfrak{K}, \mathfrak{K}])$ and $\Gamma^k(\mathfrak{z}(\mathfrak{K}))$ can be embedded into $\Gamma^k(\mathfrak{K})$, too. Moreover, since, for all $x \in M$, the fibers $[\mathfrak{K}, \mathfrak{K}]_x$ and $\mathfrak{z}(\mathfrak{K})_x$ are ideals of $\mathfrak{K}_x$, the spaces $\Gamma^k([\mathfrak{K}, \mathfrak{K}])$ and $\Gamma^k(\mathfrak{z}(\mathfrak{K}))$ are ideals of $\Gamma^k(\mathfrak{K})$.

2. Each fiber of the vector bundle $\operatorname{Der}(\mathfrak{K})$ is a Lie algebra isomorphic to $\operatorname{Der}(\mathfrak{k})$ and each fiber of the vector bundle $\operatorname{Cent}(\mathfrak{K})$ is an associative algebra isomorphic to $\operatorname{Cent}(\mathfrak{k})$. So the corresponding section algebras $\Gamma^k(\operatorname{Der}(\mathfrak{K}))$ and $\Gamma^k(\operatorname{Cent}(\mathfrak{K}))$ have the pointwise structure of a Lie algebra and an associative algebra, respectively.

In order to discuss sections of bundles locally, we introduce the following notation.

Definition 2.3.11. Let (U, φ) be a bundle chart of a vector bundle $\pi : \mathbb{V} \to M$ with fiber V. The local form of a section $X \in \Gamma^k(\pi^{-1}(U))$, where $k \in \mathbb{N}_0 \cup \{\infty\}$, is $X^\varphi \in \mathrm{C}^k(U, V)$, defined by requiring the following diagram to be commutative:

$$
\begin{array}{ccc}
\pi^{-1}(U) & \xrightarrow{(\pi, \varphi)} & U \times V \\
{\scriptstyle X} \uparrow & \nearrow {\scriptstyle (\mathrm{id}_U, X^\varphi)} & \\
U. & &
\end{array}
$$

If N is a set and $\eta : \Gamma^k(\pi^{-1}(U)) \to N$ is a map, then the corresponding local form of η is denoted by $\eta^\varphi : \mathrm{C}^k(U, V) \to N$.

We will also deal with differential operators, for which we need a convenient local description.

Definition 2.3.12. If $\alpha = (\alpha_1, \ldots, \alpha_m) \in \mathbb{N}_0^m$ is a multi-index with $|\alpha| := \sum_{i=1}^m \alpha_i$, then we use the notation $f^\alpha := f_1^{\alpha_1} \circ \ldots \circ f_m^{\alpha_m}$ for functions $f_1, \ldots, f_m$.

Furthermore, we write $\partial_\xi^\alpha f := \partial_{\xi_1}^{\alpha_1} \cdots \partial_{\xi_m}^{\alpha_m} f$, where $\partial_\xi f(x) := \frac{d}{dt}\big|_{t=0} f\left(\xi^{-1}(\xi(x) + te)\right)$, for the canonical $\mathbb{R}^m$-basis $(e_1, \ldots, e_m)$ and a sufficiently often differentiable map f, defined on an x-neighborhood U with trivializing chart ξ.

Finally, let Z be a set such that, for all $z \in Z$, there are maps $T_z : X \to Y$, maps $S_z : W \to X$ and points $A_z \in X$. Then the map $W \to Y$, $z \mapsto T_z \circ S_z$ is denoted by $T \circ S$ and the map $Z \to Y$, $z \mapsto T_z(A_z)$ is denoted by $T \cdot A$.

Lie connections are certain connections on Lie algebra bundles, which always exist if the base manifold is paracompact.

Definition 2.3.13.

1. Let $\pi : \mathbb{V} \to M$ be a vector bundle. A *connection* of $\mathbb{V}$ is a $\mathrm{C}^\infty(M, \mathbb{K})$-linear map $\nabla : \Gamma(TM) \to \mathrm{End}(\Gamma(\mathbb{V}))$, $X \mapsto \nabla_X$, such that $\nabla_X(f \cdot A) = (X.f) \cdot A + f \cdot \nabla_X A$ for all $X \in \Gamma(TM)$, $f \in \mathrm{C}^\infty(M, \mathbb{K})$, $A \in \Gamma(\mathbb{V})$. The *curvature* of a connection ∇ is the alternating map

$$
\begin{aligned}
F(\nabla) : \Gamma(TM) \times \Gamma(TM) &\longrightarrow \mathrm{End}(\Gamma(\mathbb{V})) \\
(X, Y) &\longmapsto \nabla_X \nabla_Y - \nabla_Y \nabla_X - \nabla_{[X,Y]}.
\end{aligned}
$$

A connection with vanishing curvature is called *flat*.

2. Let $\pi : \mathfrak{K} \to M$ be a Lie algebra bundle with fiber $\mathfrak{k}$ and ∇ a connection of $\mathfrak{K}$. We call ∇ a *Lie connection* if for all $X \in \Gamma(TM)$ and all $A, B \in \Gamma(\mathfrak{K})$ we have

$$
\nabla_X [A, B] = [\nabla_X A, B] + [A, \nabla_X B],
$$

i.e., ∇ is a map $\Gamma(TM) \to \mathrm{Der}(\Gamma(\mathfrak{K}))$.

Remark 2.3.14. Given a Lie algebra bundle $\pi : \mathfrak{K} \to M$ with fiber $\mathfrak{k}$, there always exists a Lie connection $\nabla : \Gamma(TM) \to \mathrm{Der}(\Gamma(V(\mathfrak{K})))$.

To this end, we note that a connection $\nabla : \Gamma(TM) \to \mathrm{End}(\Gamma(\mathfrak{K}))$ is locally given by its *Christoffel symbols*: If (U, φ) is a bundle chart of $\mathfrak{K}$ and (U, ξ) a chart of M, then there are unique mappings $\Gamma_i^\varphi \in C^\infty(U, \mathrm{End}(\mathfrak{k}))$ for $i \in \{1, \ldots, \dim(M)\}$ such that

$$(\nabla_X A)^\varphi = \sum_{i=1}^{\dim(M)} X^i \left(\partial_{\xi_i} A^\varphi + \Gamma_i^\varphi \cdot A^\varphi\right)$$

for all sections A and all vector fields X with local form $\sum_{i=1}^m X^i \partial_{\xi_i}$. It is easy to show that ∇ is a Lie connection if and only if for all bundle charts (U, φ) of $\mathfrak{K}$ and all corresponding charts (U, ξ) of M the Christoffel symbols are in in $C^\infty(U, \mathrm{Der}(\mathfrak{k}))$. For the construction of a Lie connection, let $(U_j, \varphi_j)_{j \in J}$ be a locally finite bundle atlas of $\mathfrak{K}$ with corresponding atlas $(U_j, \xi_j)_{j \in J}$ of M. For each $j \in J$ we define a Lie connection ∇^{U_j} on $\pi^{-1}(U_j)$ by:

$$\left(\nabla_X^{U_j} A\right)^{\varphi_j} := \sum_{i=1}^{\dim(M)} X^i \partial_{(\xi_j)_i} A^\varphi$$

for $X \in \Gamma(TU)$ and $A \in \Gamma\left(\pi^{-1}(U_j)\right)$. We choose a smooth partition of unity $(\rho_j : M \to [0,1])_{j \in J}$ such that $\mathrm{supp}(\rho_j) \subseteq U_j$ for all $j \in J$, and define

$$\nabla := \sum_{j \in J} \rho_j \cdot \nabla^{U_j}.$$

Since ∇ is locally the finite sum of Lie connections, it is also a Lie connection.

From now on, we generalize the results from [Lec79] and [Lec80]. We have $k \in \mathbb{N}_0 \cup \{\infty\}$ and a finite-dimensional Lie algebra bundle $\pi : \mathfrak{K} \to M$ with fiber $\mathfrak{k}$. The dimensions of M and $\mathfrak{k}$ are denoted by m and d, respectively.

2.3.1. Center and Commutator

Proposition 2.3.15. $\mathfrak{z}\left(\Gamma^k(\mathfrak{K})\right) = \Gamma^k\left(\mathfrak{z}(\mathfrak{K})\right)$. *In particular, $\Gamma^k(\mathfrak{K})$ is centerfree if and only if $\mathfrak{k}$ is centerfree.*

Proof. If $X \in \Gamma^k\left(\mathfrak{z}(\mathfrak{K})\right)$ then, for all $x \in M$, we have $X_x \in \mathfrak{z}(\mathfrak{K}_x)$ and thus $[X, Y]_x = [X_x, Y_x] = 0$ for all $Y \in \Gamma^k(\mathfrak{K})$ and $x \in M$. This proves $\mathfrak{z}\left(\Gamma^k(\mathfrak{K})\right) \supseteq \Gamma^k\left(\mathfrak{z}(\mathfrak{K})\right)$. If $X \in \mathfrak{z}\left(\Gamma^k(\mathfrak{K})\right)$, $x \in M$ and $u \in \mathfrak{K}_x$, then choose a section $Y \in \Gamma^k(\mathfrak{K})$ such that $Y_x = u$, thus $[X_x, u] = [X, Y]_x = 0$ proving $X_x \in \mathfrak{z}(\mathfrak{K}_x)$. This proves $\mathfrak{z}\left(\Gamma^k(\mathfrak{K})\right) \subseteq \Gamma^k\left(\mathfrak{z}(\mathfrak{K})\right)$. $\qquad\square$

For the calculation of the commutator of $\Gamma^k(\mathfrak{K})$ we first have to prove a technical lemma.

Lemma 2.3.16. *There exists an $r \in \mathbb{N}$ such that, for each bundle chart (U, φ) and each section $X \in \Gamma^k([\mathfrak{K}, \mathfrak{K}])$ with compact support contained in U, there are sections $Y_1, Z_1, \ldots, Y_r, Z_r \in \Gamma^k(\mathfrak{K})$ with compact supports contained in U with*

$$X = \sum_{t=1}^r [Y_t, Z_t].$$

Proof. Let $([u_1, w_1], \ldots [u_r, w_r])$ be a basis of $[\mathfrak{k}, \mathfrak{k}]$. Let $F \subseteq M$ be compact such that

$$\operatorname{supp} X \subseteq F^\circ \subseteq F \subseteq U$$

and $\rho : M \to [0, 1]$ a smooth function with compact support contained in U such that $\rho_{|F} = 1$. Then there are functions $f_1, \ldots, f_r \in C^k(M, \mathbb{K})$ with supports contained in F such that we have, for each $x \in U$:

$$X^\varphi(x) = \sum_{t=1}^r f_t(x) [u_t, w_t].$$

For $t \in \{1, \ldots, r\}$ we define $Y_t, Z_t \in \Gamma^k(\mathfrak{K})$ by

$$Y_t(x) = Z_t(x) = 0 \text{ for } x \in M \backslash U$$
$$Y_t^\varphi(x) = f_t(x)u_t \text{ and } Z_t^\varphi(x) = \rho(x)w_t \text{ for } x \in U.$$

For $x \in M \backslash F$ we have $X(x) = 0 = \sum_{t=1}^r 0 = \sum_{t=1}^r [Y_t(x), Z_t(x)]$. For $x \in F$ we have

$$X^\varphi(x) = \sum_{t=1}^r [f_t(x)u_t, 1 \cdot w_t] = \sum_{t=1}^r [Y_t^\varphi(x), Z_t^\varphi(x)].$$

Thus $X = \sum_{t=1}^r [Y_t, Z_t]$. $\qquad \square$

Proposition 2.3.17. *For $k \in \mathbb{N} \cup \{\infty\}$ we have $\left[\Gamma^k(\mathfrak{K}), \Gamma^k(\mathfrak{K})\right] = \Gamma^k\left([\mathfrak{K}, \mathfrak{K}]\right)$. In particular, $\Gamma^k(\mathfrak{K})$ is perfect if and only if $\mathfrak{k}$ is perfect.*

Proof. There is finite atlas $(U_i, \xi_i)_{i \in F}$ of M and finite bundle atlas $(U_i, \varphi_i)_{i \in F}$ of $\mathfrak{K}$ (cf. Remark 2.2.3). Let $(\rho_i : M \to [0, 1])_{i \in F}$ be a partition of unity such that $\operatorname{supp}(\rho_i)$ is a compact subset of U_i for all $i \in F$. Any section $X \in \Gamma^k(\mathfrak{K})$ takes the form $X = \sum_{i \in F} X_i$ for $X_i := \rho_i X$. We fix $X \in \Gamma^k([\mathfrak{K}, \mathfrak{K}])$ and $i \in F$. Then, by Lemma 2.3.16, there are sections $Y_1^i, Z_1^i, \ldots, Y_s^i, Z_s^i \in \Gamma^k(\mathfrak{K})$ with compact supports contained in U_i, such that $X_i = \sum_{p=1}^s [Y_p^i, Z_p^i] \in \left[\Gamma^k(\mathfrak{K}), \Gamma^k(\mathfrak{K})\right]$, hence $X \in \left[\Gamma^k(\mathfrak{K}), \Gamma^k(\mathfrak{K})\right]$ and $\left[\Gamma^k(\mathfrak{K}), \Gamma^k(\mathfrak{K})\right] \supseteq \Gamma^k([\mathfrak{K}, \mathfrak{K}])$.

In order to show $\left[\Gamma^k(\mathfrak{K}), \Gamma^k(\mathfrak{K})\right] \subseteq \Gamma^k([\mathfrak{K}, \mathfrak{K}])$ we consider $X = \sum_{s=1}^p [Y_p, Z_p]$ for appropriate sections $Y_p, Z_p \in \Gamma^k(\mathfrak{K})$. For any $x \in M$ we have $X(x) = \sum_{s=1}^p [Y_p(x), Z_p(x)] \in [\mathfrak{K}, \mathfrak{K}]_x$. $\qquad \square$

The calculations of the derivation algebra and the centroid of a given section algebra need more preparation.

2.3.2. Derivations

Definition 2.3.18. Let $\pi_1 : \mathbb{V}_1 \to M$, $\pi_2 : \mathbb{V}_2 \to M$ be two vector bundles and $k_1, k_2 \in \mathbb{N} \cup \{\infty\}$. A linear operator $T : \Gamma^{k_1}(\mathbb{V}_1) \to \Gamma^{k_2}(\mathbb{V}_2)$ is called *local*, if for any $X \in \Gamma^{k_1}(\mathbb{V}_1)$ and any open set $U \subseteq M$ the condition $X_x = 0 \in (\mathbb{V}_1)_x$ for all $x \in U$ implies $(TX)_x = 0 \in (\mathbb{V}_2)_x$ for all $x \in U$.

The meaning of an operator to be local is shown in the following lemma.

Lemma 2.3.19. *If $Y, Z \in \Gamma^{k_1}(\mathbb{V}_1)$ are identical sections on an open set $U \subseteq M$ and we have a local operator $T : \Gamma^{k_1}(\mathbb{V}_1) \to \Gamma^{k_2}(\mathbb{V}_2)$, then $TY, TZ \in \Gamma^{k_1}(\mathbb{V}_2)$ are identical on U.*

Proof. The section $X := Y - Z \in \Gamma^{k_1}(\mathbb{V}_1)$ vanishs on U. The locality of T yields that $TX = TY - TZ$ also vanishs on U. So $TY_{|U} = TZ_{|U}$. $\qquad\square$

Example 2.3.20. Let $\nabla : \Gamma(TM) \to \text{End}(\Gamma(\mathbb{V}))$ be a connection on a vector bundle $\pi : \mathbb{V} \to M$ with fiber V and $X \in \Gamma(TM)$ a vector field. Then ∇_X is a local operator:

Let $Y \in \Gamma(\mathbb{V})$ be a section, which is zero on an open set $U \subseteq M$, let $x \in U$ be a point and let $\rho : M \to [0,1]$ be a smooth map such that $\rho_{|M\setminus U} = \mathbf{0}$ and $\rho_{|W} = \mathbf{1}$ for a smaller x-neighborhood $W \subseteq U$. We easily calculate:

$$(\nabla_X Y)(x) = \rho(x)(\nabla_X Y)(x) = (\nabla_X(\rho \cdot Y))(x) - ((X.\rho)Y)(x) = 0 - 0 = 0.$$

The Peetre Theorem, proven in [Nar85], p. 175-176, can be stated as follows.

Theorem 2.3.21 (Peetre Theorem, Version 1)**.** *Let M be a smooth m-dimensional manifold and $\pi_i : \mathbb{V}_i \to M$ a vector bundle with fiber V_i for $i = 1, 2$. If $T : \Gamma(\mathbb{V}_1) \to \Gamma(\mathbb{V}_2)$ is a local operator, then for any point $x \in M$ there exists an open neighborhood $U \subseteq M$, bundle charts (U, φ_1), (U, φ_2) of $\mathbb{V}_1$, $\mathbb{V}_2$, respectively, a chart (U, ξ) of M, a number $n \in \mathbb{N}$ and a familiy of functions $f_\alpha \in C^\infty(U, \text{Hom}(V_1, V_2))$, $\alpha \in \mathbb{N}^m$, $|\alpha| \le n$, such that for all $X \in \Gamma\left(\pi_1^{-1}(U)\right)$ we have*

$$(TX)^{\varphi_2} = \sum_{|\alpha| \le n} f_\alpha \cdot \left(\partial_\xi^\alpha X^{\varphi_1}\right). \qquad (2.6)$$

Definition 2.3.22. Formula (2.6) says that $T_{|U}$ is a *differential operator of order at most n*.

A modification of the proof of Theorem 2.3.21 gives the following (cf. [Lec79], page 52).

Theorem 2.3.23 (Peetre Theorem, Version 2)**.** *Let M be a smooth m-dimensional manifold and $\pi_i : \mathbb{V}_i \to M$ a vector bundle with fiber V_i and $k_i \in \mathbb{N}_0$ for $i = 1, 2$. If $T : \Gamma^{k_1}(\mathbb{V}_1) \to \Gamma^{k_2}(\mathbb{V}_2)$ is a local operator, then T is a differential operator of order at most $k_1 - k_2$. In particular, if $k_1 = k_2$ then T is a differential operator of order 0 and if $k_1 < k_2$ then $T = 0$. Furthermore, we obtain $T = 0$ if $T : \Gamma^{k_1}(\mathbb{V}_1) \to \Gamma(\mathbb{V}_2) \subseteq \Gamma^{k_1+1}(\mathbb{V}_2)$ is local for $k_1 \in \mathbb{N}$.*

Remark 2.3.24. In the situations of the above theorems, if $T : \Gamma^k(\mathbb{V}_1) \to \Gamma^k(\mathbb{V}_2)$ is a differential operator of order 0 for $k \in \mathbb{N}_0 \cup \{\infty\}$, then it can be identified with a C^k-section of the bundle $\text{Hom}(\mathbb{V}_1, \mathbb{V}_2)$ as follows: Fix an $x \in M$ and an open neighborhood $U \subseteq M$ of x such that $\pi_1^{-1}(U)$ is trivial. For vectors $v \in (\mathbb{V}_1)_x$ we choose sections $X : U \to \pi_1^{-1}(U)$ such that $X_x = v$ and define the linear map

$$\tau_x : (\mathbb{V}_1)_x \longrightarrow (\mathbb{V}_2)_x$$
$$v \longmapsto (TX)_x.$$

Note that $(TX)_x$ depends on X only via X_x because T is an operator of order 0. The required section is the map $M \to \text{Hom}(\mathbb{V}_1, \mathbb{V}_2)$, $x \mapsto \tau_x$.

Theorem 2.3.25. *If $\mathfrak{k}$ is perfect or centerfree and $k \in \mathbb{N}_0$, then $\text{Der}\left(\Gamma^k(\mathfrak{K})\right) \cong \Gamma^k(\text{Der}(\mathfrak{K}))$ as Lie algebras.*

Proof. Let D be a derivation of $\Gamma^k(\mathfrak{K})$. We want to show that D is automatically a local operator. Let $X \in \Gamma^k(\mathfrak{K})$ be zero on an open set $U \subseteq M$.

- Suppose $\mathfrak{z}(\mathfrak{k}) = 0$. Then for all $x \in U$ and $Y \in \Gamma^k(\mathfrak{K})$ with $\mathrm{supp}(Y) \subseteq U$ we have $[X, Y] = 0$ and therefore:

$$[(DX)_x, Y_x] = (D\,[X, Y])_x - [X_x, (DY)_x] = 0 - 0 = 0.$$

This implies $[(DX)_x, v] = 0$ for all $v \in \mathfrak{K}_x$, yielding $(DX)_x \in \mathfrak{z}(\mathfrak{K}_x) = 0$ for all $x \in U$.

- Suppose $[\mathfrak{k}, \mathfrak{k}] = \mathfrak{k}$. Since $\Gamma^k(\mathfrak{K})$ is also perfect, there exist sections $Y_1, Z_1, \ldots, Y_r, Z_r \in \Gamma^k(\mathfrak{K})$ such that $X = \sum_{t=1}^r [Y_t, Z_t]$. Now, if $x \in U$ then there is an x-neighborhood $W \subseteq U$ and a smooth map $\rho : M \to [0, 1]$ such that $\rho_{|W} = \mathbf{1}$ and $\rho_{M \setminus U} = \mathbf{0}$. We set $X' := (1 - \rho)^2 \cdot X$. Thus $X = X'$ on M and we obtain $X = \sum_{t=1}^r [(1 - \rho) \cdot Y_t, (1 - \rho) \cdot Z_t]$, yielding

$$(DX)_x = \sum_{t=1}^r \left[D((1 - \rho) \cdot Y_t)(x), (1 - \rho)(x) \cdot Z_t(x) \right] + \left[(1 - \rho)(x) \cdot Y_t(x), D((1 - \rho) \cdot Z_t)(x) \right]$$
$$= 0.$$

In both cases we have $(DX)_{|U} = \mathbf{0}$, thus D is local.

Applying Theorem 2.3.23 and Remark 2.3.24 gives a linear map

$$\Psi : \mathrm{Der}\left(\Gamma^k(\mathfrak{K})\right) \to \Gamma^k\left(\mathrm{End}(\mathfrak{K})\right),$$

well-defined by $\Psi(D)(X_x) := (DX)_x$ for $X \in \Gamma^k(\mathfrak{K})$, $x \in M$. Since we have

$$\Psi(D)\,[X_x, Y_x] = (D\,[X, Y])_x = ([DX, Y])_x + ([X, DY])_x = [\Psi(D)(X_x), Y_x] + [X_x, \Psi(D)(Y_x)]$$

for $X, Y \in \Gamma^k(\mathfrak{K})$ and $x \in M$, the image of Ψ is contained in $\Gamma^k(\mathrm{Der}(\mathfrak{K}))$. Furthermore, the map Ψ is compatible with the Lie brackets on $\mathrm{Der}\left(\Gamma^k(\mathfrak{K})\right)$ and $\Gamma^k(\mathrm{Der}(\mathfrak{K}))$:

$$\Psi\left[D, D'\right](X_x) = \left([D, D']\,X\right)_x = \left(D(D'X)\right)_x - \left(D'(DX)\right)_x = \left[\Psi(D), \Psi(D')\right](X_x).$$

Finally, the function $\Phi : \Gamma^k(\mathrm{Der}(\mathfrak{K})) \to \mathrm{Der}\left(\Gamma^k(\mathfrak{K})\right)$, defined by $(\Phi\,(\mathfrak{D})\,(X))_x := \mathfrak{D}_x(X_x)$ for a section $\mathfrak{D} \in \Gamma^k(\mathrm{Der}(\mathfrak{K}))$, $X \in \Gamma^k(\mathfrak{K})$, $x \in M$, is the inverse of Ψ. We conclude that Ψ is an isomorphism of Lie algebras. $\qquad \square$

In order to analyze $\Gamma(\mathfrak{K}) = \Gamma^\infty(\mathfrak{K})$ we define the notion of an x-derivation of $\mathfrak{K}$, a certain notation and a technical lemma.

Definition 2.3.26. Let $x \in M$.

1. An *x-derivation* or *derivation in x* of $\mathfrak{K}$ is a linear map $\delta : \Gamma(\mathfrak{K}) \to \mathfrak{K}_x$ such that the following condition is satisfied for all $X, Y \in \Gamma(\mathfrak{K})$:

$$\delta\,[X, Y] = [\delta(X), Y_x] + [X_x, \delta(Y)].$$

The vector space of all x-derivations of $\mathfrak{K}$ is denoted by $\mathcal{D}_x(\mathfrak{K})$, the union $\bigcup_{x \in M} \mathcal{D}_x(\mathfrak{K})$ is denoted by $\mathcal{D}(\mathfrak{K})$ and we have a projection $\overline{p} : \mathcal{D}(\mathfrak{K}) \to M$ defined by $\overline{p}(\mathcal{D}_x(\mathfrak{K})) = \{x\}$ for $x \in M$.

2. Note that $\mathrm{Der}(\mathfrak{K})$ can be embedded into $\mathcal{D}(\mathfrak{K})$ via the natural injection $i : \mathrm{Der}(\mathfrak{K}) \to \mathcal{D}(\mathfrak{K})$, where $D \in \mathrm{Der}(\mathfrak{K}_x)$ is mapped to the x-derivation $i(D) : \Gamma(\mathfrak{K}) \to \mathfrak{K}_x$, $X \mapsto D(X_x)$.

Definition 2.3.27. If $f : M \to V$ is a C^k-map between a manifold and a finite-dimensional vector space, $n \in \mathbb{N}_0$ with $n \leq k$ and $x \in M$, then we define the notation

$$j_x^n(f) = 0 \quad :\Longleftrightarrow \quad \begin{cases} f(x) = 0 & \text{if } n = 0 \\ j_x^{n-1}(f) = 0 \text{ and } T_x^n f = 0 & \text{if } n > 0. \end{cases}$$

The following lemma will be very useful in some future proofs.

Lemma 2.3.28. *If we have a vector bundle $\pi : \mathbb{V} \to M$ with fiber V, a section $A \in \Gamma(\mathbb{V})$, a point $x \in M$ and an integer $s \in \mathbb{N}_0$ such that $j_x^s(A) = 0$, then there is an integer $r \in \mathbb{N}_0$ and there are sections $A_1, \ldots, A_r \in \Gamma(\mathbb{V})$ and functions $a_1, \ldots, a_r \in C^\infty(M, \mathbb{R})$ with $a_1(x) = \ldots = a_r(x) = 0$ such that*

$$A = \sum_{i=1}^{r} a_i^{s+1} A_i. \tag{2.7}$$

Proof. Let n be the vector space dimension of V and m the dimension of M. The lemma will be proven in two steps.

1. We consider the case of M being a convex open neighborhood $U \subseteq \mathbb{R}^m$ of $x = 0$ and $\mathbb{V} = U \times \mathbb{K}^n$ and $A \in C^\infty(U, \mathbb{K}^n)$ with compact support $F \subseteq U$. Then, by the Taylor Formula, there is a family of functions $A_\alpha \in C^\infty(U, \mathbb{K}^n)$, $\alpha \in \mathbb{N}^m$, $|\alpha| = s + 1$, such that

$$A(y) = \sum_{|\alpha| = s+1} \frac{1}{\alpha!} \cdot y_1^{\alpha_1} \cdots y_m^{\alpha_m} \cdot A_\alpha(y)$$

for all $y = (y_1, \ldots, y_m) \in U$. Let $W \subseteq U$ be an open set with $F \subseteq W$ and $\rho : U \to [0,1]$ be a smooth function with compact support such that $\rho_{|W} = \mathbf{1}$. Then we have

$$A(y) = \rho(y)^{s+2} \cdot A(y) = \sum_{|\alpha| = s+1} (\rho(y)y_1)^{\alpha_1} \cdots (\rho(y)y_m)^{\alpha_m} \cdot \frac{1}{\alpha!} \cdot \rho(y) \cdot A_\alpha(y) \tag{2.8}$$

for all $y \in U$. By the Multinomial Theorem, we have

$$(t_1 + \ldots + t_m)^k = \sum_{|\alpha| = k} \binom{k}{\alpha!} t_1^{\alpha_1} \cdots t_m^{\alpha_m}$$

for $t_1, \ldots, t_m \in \mathbb{R}$ and, by the inversion formula presented in [MHBRC94], there are, for $\alpha \in \mathbb{N}_0^m$, scalars $\lambda_{ij} = \lambda_{ij}(\alpha) \in \mathbb{R}$ and $\mu_j = \mu_j(\alpha) \in \mathbb{R}$, where $i \in \{1, \ldots, m\}$ and $j \in \{1, \ldots, p\}$ for an integer $p = p(\alpha) \in \mathbb{N}_0$, such that

$$t_1^{\alpha_1} \cdots t_m^{\alpha_m} = \sum_{j=1}^{p} \mu_j (\lambda_{1j} t_1 + \ldots + \lambda_{mj} t_m)^{|\alpha|}.$$

Equation (2.8) then turns into:

$$A(y) = \sum_{|\alpha|=s+1} \sum_{j=1}^{p(\alpha)} \Big(\underbrace{\lambda_{1j}\rho(y)y_1 + \ldots + \lambda_{mj}\rho(y)y_m}_{=:a_\alpha(y)} \Big)^{s+1} \cdot \underbrace{\frac{\mu_j}{\alpha!} \cdot \rho(y) \cdot A_\alpha(y)}_{=:\widetilde{A}_\alpha(y)}$$

$$= \sum_{|\alpha|=s+1} a_\alpha(y)^{s+1} \cdot \widetilde{A}_\alpha(y).$$

Since $\widetilde{A}_\alpha$ is of compact support and $a_\alpha(0) = 0$ for all $\alpha \in \mathbb{N}_0^m$ with $|\alpha| = s+1$, we have shown (2.7) for this local case.

2. For the general case, we take an x-neighborhood $U \subseteq M$, a bundle chart (U, φ), a chart (U, ξ) of M such that $\xi(x) = 0$ and $\xi(U)$ is convex. Choose a smooth function $\rho : M \to [0, 1]$ such that $\mathrm{supp}(\rho)$ is a compact subset of U and $\rho_{|W} = 1$ for an x-neighborhood $W \subseteq U$. By the argument in the first step, we may write

$$(\rho \cdot A)^\varphi \circ \xi^{-1} = \sum_{i=1}^{r} \widetilde{a}_i^{s+1} \widetilde{A}_i$$

where $\widetilde{A}_1, \ldots \widetilde{A}_r \in \mathrm{C}^\infty(\xi(U), V)$, $\widetilde{a}_1, \ldots, \widetilde{a}_r \in \mathrm{C}^\infty(\xi(U), \mathbb{R})$ with $\widetilde{a}_1(x) = \ldots = \widetilde{a}_r(x) = 0$. For $i \in \{1, \ldots r\}$, we define $A_i \in \Gamma(\mathbb{V})$ by $A_i := \rho \cdot A_i'$, where $(A_i')^\varphi(y) := \widetilde{A}_i(\xi(y))$ for all $y \in U$. Now we define $a_i \in \mathrm{C}^\infty(M, \mathbb{R})$ with $a_1(x) = \ldots = a_r(x) = 0$ by $a_i := \rho \cdot (\widetilde{a}_i \circ \xi)$. This yields $A = \sum_{i=1}^{r} a_i^{s+1} A_i$ on M. $\qquad\square$

Theorem 2.3.29. *If $\mathfrak{k}$ is perfect or centerfree, then every x-derivation of $\mathfrak{K}$ is a differential operator of order at most 1 and the triple $(\mathcal{D}(\mathfrak{K}), \overline{p}, M)$ admits a vector bundle structure such that $\mathrm{Der}(\Gamma(\mathfrak{K}))$ can be naturally identified with $\Gamma(\mathcal{D}(\mathfrak{K}))$ and there exists a short exact sequence of vector bundles as follows:*

$$0 \longrightarrow \mathrm{Der}(\mathfrak{K}) \xrightarrow{\ i\ } \mathcal{D}(\mathfrak{K}) \xrightarrow{\ \sigma\ } TM \otimes \mathrm{Cent}(\mathfrak{K}) \longrightarrow 0.$$

Proof. Let $\delta \in \mathcal{D}_x(\mathfrak{K})$ be an x-derivation. By Lemma 2.3.28, every section $X \in \Gamma(\mathfrak{K})$ with $j_x^1(X) = 0$ takes the form

$$X = \sum_{i=1}^{r} f_i^2 X_i \tag{2.9}$$

for functions $f_i \in \mathrm{C}^\infty(M, \mathbb{K})$, $f_i(x) = 0$ and $X_i \in \Gamma(\mathfrak{K})$.

- Suppose $\mathfrak{z}(\mathfrak{k}) = 0$. If $Y \in \Gamma(\mathfrak{K})$ is a smooth section, then:

$$[\delta(X), Y(x)] = \sum_{i=1}^{r} \big[\delta\left(f_i^2 X_i\right), Y(x)\big] = \sum_{i=1}^{r} \Big(\delta\big[f_i^2 X_i, Y\big] - \underbrace{\big[f_i(x)^2 X_i(x), \delta(Y)\big]}_{=0} \Big)$$

$$= \sum_{i=1}^{r} \delta\left[f_i X_i, f_i Y\right] = \sum_{i=1}^{r} [\delta(f_i X_i), f_i(x) Y(x)] + [f_i(x) X_i(x), \delta(f_i Y)]$$

$$= \sum_{i=1}^{r} 0 + 0 = 0,$$

thus $\delta(X) \in \mathfrak{z}(\mathfrak{K}_x) = 0$.

- Suppose $[\mathfrak{k}, \mathfrak{k}] = \mathfrak{k}$. Since $\Gamma(\mathfrak{K})$ is also perfect, there exist, for each $i \in \{1, \ldots, r\}$, sections $Y_{i1}, Z_{i1}, \ldots, Y_{is}, Z_{is} \in \Gamma(\mathfrak{K})$ such that $X_i = \sum_{t=1}^{s} [Y_{it}, Z_{it}]$. Then:

$$\delta(X) = \sum_{i=1}^{r} \delta\left(f_i^2 X_i\right) = \sum_{i=1}^{r} \sum_{t=1}^{s} \delta\left[f_i Y_{it}, f_i Z_{it}\right]$$
$$= \sum_{i=1}^{r} \sum_{t=1}^{s} [\delta(f_i Y_{it}), f_i(x) Z_{it}(x)] + [f_i(x) Y_{it}(x), \delta(f_i Z_{it})] = \sum_{i=1}^{r} \sum_{t=1}^{s} 0 = 0.$$

For both cases, we conclude:

$$\delta(X) = 0 \text{ if } j_x^1(X) = 0. \tag{2.10}$$

Let (U, φ) be a bundle chart in x and (U, ξ) a corresponding chart of M. Since $\pi : \mathfrak{K} \to M$ is defined by an $\text{Aut}(\mathfrak{k})$-valued cocycle, φ restricted to any fiber is an isomorphism of Lie algebras, therefore the Lie bracket on $\Gamma\left(\pi^{-1}(U)\right)$ corresponds to the natural pointwise Lie bracket on $\text{C}^{\infty}(U, \mathfrak{k})$ meaning $[X, Y]^{\varphi} = [X^{\varphi}, Y^{\varphi}]$ for $X, Y \in \Gamma\left(\pi^{-1}(U)\right)$. Note, by relation (2.10), that δ has a local form without terms of order greater than 1:

$$X^{\varphi} \xmapsto{\delta^{\varphi}} D\left(X^{\varphi}(x)\right) + \sum_{i=1}^{m} S^i \left(\partial_{\xi_i} X^{\varphi}(x)\right) \tag{2.11}$$

for certain $D, S^1, \ldots, S^m \in \text{End}(\mathfrak{k})$.

Let us show now that a local form of the type (2.11) is satisfied for an x-derivation δ^{φ} if and only if $D \in \text{Der}(\mathfrak{k})$ and $S^1, \ldots, S^m \in \text{Cent}(\mathfrak{k})$. Equation (2.11) implies the following two equations:

$$\delta^{\varphi} [X^{\varphi}, Y^{\varphi}] = D\left([X^{\varphi}(x), Y^{\varphi}(x)]\right) + \sum_{i=1}^{m} S^i \left(\partial_{\xi_i} [X^{\varphi}, Y^{\varphi}] (x)\right)$$
$$= D\left([X^{\varphi}(x), Y^{\varphi}(x)]\right) + \sum_{i=1}^{m} S^i \left([\partial_{\xi_i} X^{\varphi}(x), Y^{\varphi}(x)] + [X^{\varphi}(x), \partial_{\xi_i} Y^{\varphi}(x)]\right). \tag{2.12}$$

$$[\delta^{\varphi} \left(X^{\varphi}\right), Y^{\varphi}(x)] + [X^{\varphi}(x), \delta^{\varphi} \left(Y^{\varphi}\right)] = [D\left(X^{\varphi}(x)\right), Y^{\varphi}(x)] + [X^{\varphi}(x), D\left(Y^{\varphi}(x)\right)]$$
$$+ \sum_{i=1}^{m} \left[S^i \left(\partial_{\xi_i} X^{\varphi}(x)\right), Y^{\varphi}(x)\right] + \left[X^{\varphi}(x), S^i \left(\partial_{\xi_i} Y^{\varphi}(x)\right)\right]. \tag{2.13}$$

By comparing the equations (2.12) and (2.13) for constant sections X, Y we obtain

$$D\left([X^{\varphi}(x), Y^{\varphi}(x)]\right) = [D\left(X^{\varphi}(x)\right), Y^{\varphi}(x)] + [X^{\varphi}(x), D\left(Y^{\varphi}(x)\right)],$$

thus D is a derivation of $\mathfrak{k}$. Knowing this, we compare the equations (2.12) and (2.13) for constant X and any Y and obtain, after the cancellation of the derivation part:

$$\sum_{i=1}^{m} S^i [X^{\varphi}(x), \partial_{\xi_i} Y^{\varphi}(x)] = \sum_{i=1}^{m} \left[X^{\varphi}(x), S^i \left(\partial_{\xi_i} Y^{\varphi}(x)\right)\right].$$

Since for any $i \in \{1, \ldots, m\}$ we have

$$\partial_{\xi_i} Y^\varphi(x) = \frac{d}{dt}\bigg|_{t=0} Y^\varphi\left(\xi^{-1}(\xi(x) + te_i)\right) = (T_x Y^\varphi)\left[\left(T_{\xi(x)}\xi^{-1}\right)(e_i)\right]$$

and $T_{\xi(x)}\xi^{-1}$ is a linear bijection, there is, for any $x \in U$ and $v \in \mathfrak{k}$, a smooth map $Y^\varphi : U \to \mathfrak{k}$ with $\partial_{\xi_i} Y^\varphi(x) = \delta_{ij} v$ for $j \in \{1, \ldots, m\}$. Therefore we may conclude that each S^i is contained in the centroid of $\mathfrak{k}$.

We obtain a bijective correspondence:

$$\bigcup_{x \in U} \mathcal{D}_x \left(U \times \mathfrak{k}\right) \longrightarrow U \times \left(\mathrm{Der}(\mathfrak{k}) \times \mathrm{Cent}(\mathfrak{k})^m\right) \tag{2.14}$$
$$\delta^\varphi \longmapsto \left(\overline{p}\left(\delta^\varphi\right), \left(D, \left(S^1, \ldots, S^m\right)\right)\right).$$

By extending this construction to all charts of a maximal bundle atlas of M, we construct trivializations of $\mathcal{D}(\mathfrak{K})$. In order to show that we obtain a bundle structure on $\mathcal{D}(\mathfrak{K})$ we will show that the transitions are linear: Let (U, φ) and (V, ψ) be bundle charts of $\pi : \mathfrak{K} \to M$ in x with corresponding charts (U, ξ) and (V, η) of M with $\xi(x) = \eta(x) = 0$ and let $D, \widetilde{D} \in \mathrm{Der}(\mathfrak{k})$, $S^1, \ldots, S^m, \widetilde{S}^1, \ldots, \widetilde{S}^m \in \mathrm{Cent}(\mathfrak{k})$ be the endomorphisms such that for all $X^\varphi, X^\psi \in C^\infty(U \cap V, \mathfrak{k})$ we have

$$\delta^\varphi\left(X^\varphi\right) = D\left(X^\varphi(x)\right) + \sum_{i=1}^m S^i\left(\partial_{\xi_i} X^\varphi(x)\right)$$

and

$$\delta^\psi\left(X^\psi\right) = \widetilde{D}\left(X^\psi(x)\right) + \sum_{i=1}^m \widetilde{S}^i\left(\partial_{\eta_i} X^\psi(x)\right).$$

By the definition of the local forms, we have

$$\left(\left(\varphi_{|\mathfrak{K}_x}\right)^{-1} \circ \delta^\varphi\right)\left(X^\varphi\right) = \delta(X) = \left(\left(\psi_{|\mathfrak{K}_x}\right)^{-1} \circ \delta^\psi\right)\left(X^\psi\right)$$

and

$$\left(\varphi_{|\mathfrak{K}_x}\right)^{-1}\left(X^\varphi(x)\right) = X(x) = \left(\psi_{|\mathfrak{K}_x}\right)^{-1}\left(X^\psi(x)\right),$$

thus

$$\delta^\psi\left(X^\psi\right) = g(x)\left(\delta^\varphi\left(X^\varphi\right)\right) \text{ and } X^\varphi(x) = g(x)^{-1}\left(X^\psi\right)$$

for a transition map $g : U \cap V \to \mathrm{Aut}(\mathfrak{k})$ of the principal bundle to which $\pi : \mathfrak{K} \to M$ is associated. By considering the commutative diagram

$$
\begin{array}{ccccc}
\mathfrak{k} = T_{\varphi(x)}\mathfrak{k} & \xleftarrow{\ T_x X^\varphi\ } & T_x M & \xrightarrow{\ T_x \xi\ } & \mathbb{R}^m \\
{\scriptstyle g(x)}\big\downarrow & & & {\scriptstyle T_x \eta}\nearrow & \\
\mathfrak{k} = T_{\psi(x)}\mathfrak{k} & \xleftarrow[\ T_x X^\psi\ ]{} & T_x M & &
\end{array}
\ ,
$$

we may calculate

$$\partial_{\xi_i} X^\varphi(x) = T_x X^\varphi \left(T_0 \xi^{-1}(e_i)\right) = \left(g(x)^{-1} \circ T_x X^\psi \circ T_0 \eta^{-1} \circ T_x \xi\right) \left(T_0 \xi^{-1}(e_i)\right)$$
$$= \left(g(x)^{-1} \circ T_x X^\psi\right) \left(T_0 \eta^{-1}(e_i)\right) = g(x)^{-1} \left(\partial_{\eta_i} X^\psi(x)\right)$$

and obtain, by the definition of $D, \widetilde{D}, S^1, \ldots, S^m, \widetilde{S}^1, \ldots, \widetilde{S}^m$:

$$\widetilde{D}\left(X^\psi(x)\right) + \sum_{i=1}^{m} \widetilde{S}^i \left(\partial_{\eta_i} X^\psi(x)\right) = \delta^\varphi\left(X^\varphi\right) = g(x)\left(\delta^\varphi\left(X^\varphi\right)\right)$$

$$= g(x) \left(D\left(X^\varphi(x)\right) + \sum_{i=1}^{m} S^i\left(\partial_{\xi_i} X^\varphi(x)\right)\right)$$

$$= \left(g(x) \circ D \circ g(x)^{-1}\right)\left(X^\psi(x)\right) + \sum_{i=1}^{m} \left(g(x) \circ S^i\right)\left(\partial_{\xi_i} X^\varphi(x)\right)$$

$$= \left(g(x) \circ D \circ g(x)^{-1}\right)\left(X^\psi(x)\right)$$

$$+ \sum_{i=1}^{m} \left(g(x) \circ S^i \circ g(x)^{-1}\right)\left(\partial_{\eta_i} X^\psi(x)\right).$$

$$(2.15)$$

For constant sections X this equation implies

$$\widetilde{D}\left(X^\psi(x)\right) = \left(g(x) \circ D \circ g(x)^{-1}\right)\left(X^\psi(x)\right),$$

thus $\widetilde{D} = g(x) \circ D \circ g(x)^{-1}$. Knowing this, Equation (2.15) also implies

$$\sum_{i=1}^{m} \widetilde{S}^i \left(\partial_{\eta_i} X^\psi(x)\right) = \sum_{i=1}^{m} \left(g(x) \circ S^i \circ g(x)^{-1}\right)\left(\partial_{\eta_i} X^\psi(x)\right)$$

and, by inserting appropriate sections for X, we may conclude that $\widetilde{S}^i = g(x) \circ S^i \circ g(x)^{-1}$ for all $i \in \{1, \ldots, m\}$. So the transitions $D \mapsto \widetilde{D}$ and $S^i \to \widetilde{S}^i$ are linear isomorphisms and we have, by Proposition 2.2.17, a vector bundle $\overline{p} : \mathcal{D}(\mathfrak{K}) \to M$ with fiber $\mathrm{Der}(\mathfrak{k}) \times \mathrm{Cent}(\mathfrak{k})^m$. There is an isomorphism of Lie algebras $\Phi : \mathrm{Der}(\Gamma(\mathfrak{K})) \to \Gamma(\mathcal{D}(\mathfrak{K}))$, defined as follows: If $D \in \mathrm{Der}(\Gamma(\mathfrak{K}))$, then $\Phi(D)$ is the map $M \to \mathcal{D}(\mathfrak{K})$, $x \mapsto \mathrm{ev}_x \circ D$. If $\mathfrak{X} \in \Gamma(\mathcal{D}(\mathfrak{K}))$, $X \in \Gamma(\mathfrak{K})$ and $x \in M$, then $\left(\Phi^{-1}(\mathfrak{X})X\right)_x = (\mathfrak{X}_x) X$.

We define $\sigma : \mathcal{D}(\mathfrak{K}) \to TM \otimes \mathrm{Cent}(\mathfrak{K})$ as follows. If $\delta \in \mathcal{D}_x(\mathfrak{K})$, then $\sigma(\delta) \in T_x M \otimes \mathrm{End}(\mathfrak{K}_x)$ is given by the following construction: We identify $T_x M \otimes \mathrm{End}(\mathfrak{K}_x) = \mathrm{Hom}(T_x M^*, \mathrm{End}(\mathfrak{K}_x))$ in the natural way, i.e., the element $\sum_i v_i \otimes A_i$ corresponds to the endomorphism $\alpha \mapsto \sum_i \alpha(v_i) \cdot A_i$. For any $\alpha \in T_x M^*$ and $w \in \mathfrak{K}_x$ let $a \in C^\infty(M, \mathbb{R})$ and $A \in \Gamma(\mathfrak{K})$ such that $T_x a = \alpha$ and $A(x) = w$. Then $\sigma(\delta)$ is pointwise defined by

$$(\sigma(\delta)(\alpha))(w) := \delta(aA) - a(x)\delta(A).$$

Now let us check that σ is well-defined and valued in $TM \otimes \mathrm{Cent}(\mathfrak{K})$ by using (2.10) and (2.14):

$$((\sigma(\delta)(\alpha))(w))^\varphi = (\delta(aA) - a(x)\delta(A))^\varphi$$
$$= D(a(x)A^\varphi(x)) + \sum_{i=1}^m S^i(\partial_{\xi_i}(aA^\varphi)(x))$$
$$- a(x)\left(D(A^\varphi(x)) + \sum_{i=1}^m S^i(\partial_{\xi_i}A^\varphi(x))\right)$$
$$= \sum_{i=1}^m S^i\left(\partial_{\xi_i}a(x) \cdot A^\varphi(x) + a(x)\partial_{\xi_i}A^\varphi(x) - a(x)\partial_{\xi_i}A^\varphi(x)\right)$$
$$= \sum_{i=1}^m S^i\left(\partial_{\xi_i}a(x) \cdot \varphi(w)\right) = \sum_{i=1}^m \partial_{\xi_i}a(x) \cdot S^i\left(\varphi(w)\right).$$

By leaving the local coordinates we see that $(\sigma(\delta)(\alpha))$ only depends on α and w and we also obtain $\sigma(\delta) \in T_xM \otimes \mathrm{Cent}(\mathfrak{K}_x)$. The above calculation also shows that σ maps exactly the elements without local part in $\mathrm{Cent}(\mathfrak{k})^m$ to zero, i.e., $\mathrm{im}\, i = \ker \sigma$.

We finally show the surjectivity of $\sigma : \mathcal{D}(\mathfrak{K}) \to TM \otimes \mathrm{Cent}(\mathfrak{K})$: Let $x \in M$ and (U, φ) a bundle chart in x with corresponding chart (U, ξ) of M and $\rho : M \to [0,1]$ a smooth map with $\mathrm{supp}\,\rho \subseteq U$ and $\rho_{|W} = 1$ for an x-neighborhood $W \subseteq U$ and $(S^1, \ldots, S^m) \in \mathrm{Cent}(\mathfrak{k})^m$. We set $f_i := \left(\varphi_{|\mathfrak{K}_x}\right)^{-1} \circ S^i \circ \varphi \in \mathrm{Cent}(\mathfrak{K}_x)$ and $\delta \in \mathcal{D}_x(\mathfrak{K})$ by $\delta(X) := \rho \cdot \left(\varphi_{|\mathfrak{K}_x}\right)^{-1}\left(\sum_{i=1}^m S^i\left(\partial_{\xi_i}X^\varphi(x)\right)\right)$ for $X \in \Gamma(\mathfrak{K})$. Then the local part of $\sigma(\delta)$ is $(S^1, \ldots, S^m)$. $\qquad\square$

Corollary 2.3.30. *The induced short sequence*

$$0 \longrightarrow \Gamma(\mathrm{Der}\,(\mathfrak{K})) \overset{i}{\longrightarrow} \Gamma(\mathcal{D}(\mathfrak{K})) \overset{\sigma}{\longrightarrow} \Gamma(TM \otimes \mathrm{Cent}(\mathfrak{K})) \longrightarrow 0,$$

where $(i(D))(x) := i(D(x))$ and $(\sigma(Z))(x) := \sigma(Z(x))$, is exact.

Definition 2.3.31. The map $\sigma : \mathcal{D}(\mathfrak{K}) \to TM \otimes \mathrm{Cent}(\mathfrak{K})$ from Theorem 2.3.29 is called the *symbol map*.

For a global description of $\mathrm{Der}(\Gamma(\mathfrak{K}))$ we first extend the notion of Lie connections. Fix a Lie connection $\nabla : \Gamma(TM) \to \mathrm{Der}(\Gamma(\mathfrak{K}))$.

Remark 2.3.32. The symbol of a derivation ∇_X of $\Gamma(\mathfrak{K})$ for a vector field $X \in \Gamma(TM)$ is $X \otimes \mathbf{1}$: Let $x \in M$ be a point, $a \in C^\infty(M, \mathbb{R})$ a smooth map and $A \in \Gamma(\mathfrak{K})$ a section of $\mathfrak{K}$. Then we identify $\nabla_X \in \mathrm{Der}(\Gamma(\mathfrak{K}))$ with $(y \mapsto \mathrm{ev}_y \circ \nabla_X) \in \Gamma(\mathcal{D}(\mathfrak{K}))$ and calculate:

$$[\sigma\left(\nabla_X(x)\right)(T_xa)]\left(A_x\right) = (\nabla_X(x))(aA) - a(x)(\nabla_X(x))(A) = (\nabla_X(aA))(x) - a(x)(\nabla_X A)(x)$$
$$= ((X.a)A + a\nabla_X(A))(x) - (a\nabla_X A)(x) = ((X.a)A)(x)$$
$$= (T_xa)(X_x) \cdot A_x$$

and therefore

$$\sigma\left(\nabla_X(x)\right) = X_x \otimes \mathbf{1}_x.$$

Proposition 2.3.33. *If $\mathfrak{k}$ is perfect or centerfree and ∇, ∇' are Lie connections on $\mathfrak{K}$, then there is a section $D \in \Gamma(\mathrm{Der}(\mathfrak{K}))$ such that $\nabla_X \xi = \nabla'_X \xi + i(D)(\xi)$ for all $X \in \Gamma(TM)$, $\xi \in \Gamma(\mathfrak{K})$. If $\mathfrak{k}$ is semisimple, then there is a section $\zeta \in \Gamma(\mathfrak{K})$ such that $\nabla_X \xi = \nabla'_X \xi + [\zeta, \xi]$ for all $X \in \Gamma(TM)$, $\xi \in \Gamma(\mathfrak{K})$.*

Proof. By Remark 2.3.32, $\sigma\left((\nabla - \nabla')_X\right) = 0$ for all $X \in \Gamma(TM)$. So Corollary 2.3.30 yields $i(D) = \nabla - \nabla'$ for some $D \in \Gamma(\mathrm{Der}(\mathfrak{K}))$. If $\mathfrak{k}$ is semisimple, then $\mathrm{ad}(\mathfrak{K}) = \mathrm{Der}(\mathfrak{K})$ by Remark 2.3.9, so there is a section $\zeta \in \Gamma(\mathfrak{K})$ with $i(\mathrm{ad}(\zeta)) = \nabla - \nabla'$, so $\nabla_X \xi = \nabla'_X \xi + [\zeta, \xi]$ for all $X \in \Gamma(TM)$, $\xi \in \Gamma(\mathfrak{K})$. $\qquad\square$

Remark 2.3.32 and the short exact sequence of Corollary 2.3.30 show us that some derivations of $\Gamma(\mathfrak{K})$ cannot be described by the Lie connection ∇. It is thus necessary to extend it in an appropriate way.

Definition 2.3.34. The *extension* of a Lie connection ∇ to $\Gamma(TM \otimes \mathrm{Cent}(\mathfrak{K}))$ is the linear map

$$\nabla : \Gamma(TM \otimes \mathrm{Cent}(\mathfrak{K})) \longrightarrow \mathrm{Der}(\Gamma(\mathfrak{K})) = \Gamma(\mathcal{D}(\mathfrak{K}))$$

defined by mapping $(X \otimes S)_x = X_x \otimes S_x$ to $S_x \circ (\nabla_X)_x$ for $x \in M$ and $X \otimes S \in \Gamma(TM \otimes \mathrm{Cent}(\mathfrak{K}))$.

Theorem 2.3.35. *Fix a Lie connection ∇ on $\mathfrak{K}$. If $\mathfrak{k}$ is perfect or centerfree, then a linear map $D : \Gamma(\mathfrak{K}) \to \Gamma(\mathfrak{K})$ is a derivation of $\Gamma(\mathfrak{K})$ if and only if there are sections $\mathcal{Y} \in \Gamma(TM \otimes \mathrm{Cent}(\mathfrak{K}))$ and $D_0 \in \Gamma(\mathrm{Der}(\mathfrak{K}))$ such that $D = \nabla_{\mathcal{Y}} + i(D_0)$. In this case the decomposition is unique.*

Proof. Obviously, for any $\mathcal{Y} \in \Gamma(TM \otimes \mathrm{Cent}(\mathfrak{K}))$ and $D_0 \in \Gamma(\mathrm{Der}(\mathfrak{K}))$ the linear map $\nabla_{\mathcal{Y}} + i(D_0)$ is a derivation of $\Gamma(\mathfrak{K})$.

On the other hand, fix $D \in \mathrm{Der}(\Gamma(\mathfrak{K})) = \Gamma(\mathcal{D}(\mathfrak{K}))$ and set $\mathcal{Y} := \sigma(D) \in \Gamma(TM \otimes \mathrm{Cent}(\mathfrak{K}))$. Since $\sigma(\nabla_{\mathcal{Y}}) = \mathcal{Y}$, the linear map $D - \nabla_{\mathcal{Y}} : \Gamma(\mathfrak{K}) \to \Gamma(\mathfrak{K})$ has symbol 0 and thus identifies with a differential operator of order 0 which is a section of $\mathrm{End}(\mathfrak{K})$. Since D and $\nabla_{\mathcal{Y}}$ are derivations of $\Gamma(\mathfrak{K})$, this section takes its values in $\mathrm{Der}(\Gamma(\mathfrak{K}))$. Therefore $D - \nabla_{\mathcal{Y}}$ can be written as $i(D_0)$ for a $D_0 \in \Gamma(\mathrm{Der}(\mathfrak{K}))$.

The uniqueness of the decomposition follows from the fact that $\mathcal{Y} = \sigma(D)$ is the only element in $\Gamma(TM \otimes \mathrm{Cent}(\mathfrak{K}))$ such that $\sigma(\nabla_{\mathcal{Y}}) = \mathcal{Y}$. $\qquad\square$

2.3.3. Centroid

In this subsection we will calculate the centroid of $\Gamma^k(\mathfrak{K})$ for $\mathfrak{k} \in \mathbb{N}_0 \cup \{\infty\}$. We begin with a technical lemma.

Lemma 2.3.36. *Let $X \in \Gamma(\mathfrak{K})$ be a section which is zero on an open set $U \subseteq M$ and $x \in U$. Then there exists a compact x-neigborhood Q and there are $D_t \in \mathrm{Der}(\Gamma(\mathfrak{K}))$, $X_t \in \Gamma(\mathfrak{K})$ with $X_{t|Q} = \mathbf{0}$ for $t \in \{1, \ldots, r\}$ such that*

$$X = \sum_{t=1}^{r} D_t(X_t).$$

Proof. Let Q be a compact x-neigborhood contained in U. There is a finite bundle atlas $(V_i, \psi_i)_{i \in F}$ of $\mathfrak{K}$ and a finite atlas $(V_i, \xi_i)_{i \in F}$ of M, which refines the open cover $\{U, M \backslash Q\}$ (cf. Remark 2.2.3) and we may assume that the $\xi(V_i)$'s are disjoint unions of convex open balls in $\mathbb{R}^{\dim(M)}$. Let $(\rho_i : M \to [0,1])_{i \in F}$ be a partition of unity such that $\operatorname{supp}(\rho_i)$ is a compact subset of V_i for all $i \in F$.

For any $i \in F$ and $Y \in \Gamma\left(\pi^{-1}(V_i)\right)$ recall that $Y^{\psi_i} : \mathbb{R}^m \supseteq \xi_i(V_i) \to \mathfrak{k}$ denotes the mapping such that the following diagram commutes:

$$
\begin{array}{ccc}
\mathfrak{K} \supseteq \pi^{-1}(V_i) & \xrightarrow{(\xi_i \circ \pi, \psi_i)} & \xi_i(V_i) \times \mathfrak{k} \\
{\scriptstyle Y}\Big\uparrow & & \Big\uparrow{\scriptstyle (\mathrm{id}, Y^{\psi_i})} \\
M \supseteq V_i & \xrightarrow{\quad \xi_i \quad} & \xi_i(V_i).
\end{array}
$$

The Lie bracket on $\Gamma\left(\pi^{-1}(V_i)\right)$ corresponds to the natural pointwise Lie bracket on $\mathrm{C}^\infty(\xi_i(V_i), \mathfrak{k})$, i.e., $[Y, Z]^{\psi_i} = [Y^{\psi_i}, Z^{\psi_i}]$ for $Y, Z \in \Gamma\left(\pi^{-1}(V_i)\right)$, for all $i \in F$. We set $Y := \rho_i X$ and choose a section $X^i \in \Gamma\left(\pi^{-1}(V_i)\right)$ by requiring $\partial_{(\xi_i)_1}(X^i)^{\psi_i} = Y^{\psi_i}$. With $D^i : \Gamma\left(\pi^{-1}(V_i)\right) \to \Gamma\left(\pi^{-1}(V_i)\right)$ defined by $(D^i)^{\psi_i} := \partial_{(\xi_i)_1}$ one obtains

$$\rho_i X = Y = D^i(X^i).$$

Note that in the case of $\operatorname{supp}(\rho_i) \subseteq U$ we can choose $X^i = \mathbf{0}$ because $X_{|U} = \mathbf{0}$ and each V_i is a subset of either U or $M \backslash Q$. We obtain:

$$\sum_{i \in F} D^i(X^i) = \sum_{i \in F} \rho_i X = X. \qquad \square$$

Theorem 2.3.37. *Any endomorphism $T : \Gamma(\mathfrak{K}) \to \Gamma(\mathfrak{K})$ with $T \circ \operatorname{Der}(\Gamma(\mathfrak{K})) \subseteq \operatorname{Der}(\Gamma(\mathfrak{K}))$ is a differential operator of order 0 and can be identified with a smooth section of $\operatorname{End}(\mathfrak{K})$.*

Proof. We want to show that T is local. Let $X \in \Gamma(\mathfrak{K})$ be zero on an open x-neighborhood $U \subseteq M$. Lemma 2.3.36 yields that there exist $D_1, \ldots, D_r \in \operatorname{Der}(\Gamma(\mathfrak{K}))$ and $X_1, \ldots, X_r \in \Gamma(\mathfrak{K})$, each X_t zero on an x-neighborhood Q, such that $X = \sum_{t=1}^r D_t(X_t)$. We obtain

$$TX_{|Q} = \sum_{t=1}^r \left((T \circ D_t)(X_t)\right)_{|Q} = \sum_{t=1}^r \mathbf{0} = \mathbf{0} \tag{2.16}$$

because, by Theorem 2.3.29, any $D \in \operatorname{Der}(\Gamma(\mathfrak{K}))$ can be identified with a section of $\mathcal{D}(\mathfrak{K})$ by $D \mapsto X_D$, where $X_D(x) := \mathrm{ev}_x \circ D$, and hence is local. By (2.16) we conclude that T is local.

Theorem 2.3.21 yields that T is a differential operator. By Theorem 2.3.29, any derivation $D \in \operatorname{Der}(\Gamma(\mathfrak{K}))$ is a differential operator of order at most one and so is any $T \circ D$. If T were of order $n > 0$, then there would be a bundle chart (U, φ) with corresponding chart (U, ξ) of M and functions $f_\alpha \in \mathrm{C}^\infty(U, \operatorname{End}(\mathfrak{k}))$, $\alpha \in \mathbb{N}_0^m$, $|\alpha| \leq n$, such that for all $X \in \Gamma\left(\pi^{-1}(U)\right)$ we would have

$$(TX)^\varphi = \sum_{|\alpha| \leq n} f_\alpha \cdot \left(\partial_\xi^\alpha X^\varphi\right)$$

and there would be $x' \in U$, $u \in \mathfrak{k}$ and $\alpha' = (\alpha'_1, \ldots, \alpha'_m) \in \mathbb{N}^m$ with $|\alpha'| = n$ such that $f_{\alpha'}(x')(u) \neq 0$. We may assume that $\xi(x') = 0$. Then $D \in \mathrm{Der}\left(\Gamma\left(\pi^{-1}(U)\right)\right)$ is given by $D^{\varphi} := \partial_{\xi_1}$ and $X \in \Gamma\left(\pi^{-1}(U)\right)$ by $X^{\varphi}(x) := \xi_1(x) \cdot \prod_{i=1}^m \xi_i(x)^{\alpha'_i} \cdot u$ we would obtain

$$(T(DX))^{\varphi} = \sum_{|\alpha| \leq n} f_\alpha \cdot \left(\partial_\xi^\alpha (DX)^\varphi\right) = \sum_{|\alpha| \leq n} f_\alpha \cdot \underbrace{\left(\partial_\xi^\alpha \partial_{\xi_1} X^\varphi\right)}_{=0 \text{ in } x', \text{ if } \alpha \neq \alpha'}$$
$$\implies (T(DX))^\varphi(x') = f_{\alpha'}(x')\left(\alpha'! \cdot u\right) \neq 0,$$

contradicting to the facts that $T \circ D$ is a differential operator of order at most one and $j^1_{x'}(X) = 0$. So T is of order 0. $\qquad\square$

Theorem 2.3.38. *If $\mathfrak{k}$ is perfect or centerfree, then we have* $\mathrm{Cent}\left(\Gamma^k(\mathfrak{K})\right) \cong \Gamma^k\left(\mathrm{Cent}(\mathfrak{K})\right)$ *as associative algebras for $k \in \mathbb{N}_0 \cup \{\infty\}$.*

Proof. In the case of $k = \infty$ we use Lemma 2.1.17.2 and Theorem 2.3.37 to see that every $T \in \mathrm{Cent}(\Gamma(\mathfrak{K}))$ identifies with a smooth section of $\mathrm{End}(\mathfrak{K})$.

For $k \in \mathbb{N}_0$ we start with showing that any $A \in \mathrm{Cent}(\Gamma^k(\mathfrak{K}))$ is local: Let $X \in \Gamma^k(\mathfrak{K})$ be zero on an open set $U \subseteq M$.

- Suppose $\mathfrak{z}(\mathfrak{k}) = 0$. Then for all $x \in U$ and $Y \in \Gamma^k(\mathfrak{K})$ we have:

$$[(AX)_x, Y_x] = [X_x, (AY)_x] = 0.$$

  This implies $(AX)_x \in \mathfrak{z}(\mathfrak{K}_x) = 0$ for all $x \in U$.

- Suppose $[\mathfrak{k}, \mathfrak{k}] = \mathfrak{k}$. Then, by Proposition 2.3.17, the Lie algebra $\Gamma^k(\mathfrak{K})$ is also perfect. So there exist sections $Y_1, Z_1, \ldots, Y_r, Z_r \in \Gamma^k(\mathfrak{K})$ such that $X = \sum_{t=1}^r [Y_t, Z_t]$. Now, if $x \in U$ then there is an x-neighborhood $W \subseteq U$ and a smooth map $\rho : M \to [0,1]$ such that $\rho_{|W} = 1$ and $\rho_{M \setminus U} = 0$. Then $X = (1 - \rho) \cdot X$ because $X_{|U} = 0$. We obtain $X = (1 - \rho) \cdot X = (1 - \rho) \cdot \sum_{t=1}^r [Y_t, Z_t] = \sum_{t=1}^r [Y_t, (1 - \rho) \cdot Z_t]$, yielding

$$(AX)_x = \sum_{t=1}^r [(AY_t)(x), (1 - \rho)(x) \cdot Z_t(x)] = 0.$$

In both cases we have $(AX)_{|U} = 0$, thus A is local. So we can use Theorem 2.3.23 to see that all $T \in \mathrm{Cent}(\Gamma^k(\mathfrak{K}))$ identify with C^k-sections of $\mathrm{End}(\mathfrak{K})$.

We know that, for $k \in \mathbb{N}_0 \cup \{\infty\}$, the linear map $\Psi : \mathrm{Cent}\left(\Gamma^k(\mathfrak{K})\right) \to \Gamma^k\left(\mathrm{End}(\mathfrak{K})\right)$ is well-defined by $\Psi(A)(X_x) := (AX)_x$ for $X \in \Gamma^k(\mathfrak{K})$, $x \in M$. Since we have

$$\Psi(A)[X_x, Y_x] = (A[X, Y])_x = ([AX, Y])_x = [\Psi(A)(X_x), Y_x]$$

for $X, Y \in \Gamma^k(\mathfrak{K})$ and $x \in M$, the image of Ψ is contained in $\Gamma^k(\mathrm{Cent}(\mathfrak{K}))$. Furthermore, the map Ψ is compatible with the composition of functions on $\mathrm{Cent}\left(\Gamma^k(\mathfrak{K})\right)$ and $\Gamma^k(\mathrm{Cent}(\mathfrak{K}))$:

$$\Psi(A \circ A')(X_x) = \left((A \circ A')X\right)_x = \left(A(A'X)\right)_x = (\Psi(A) \circ \Psi(A'))(X_x).$$

Finally, the function $\Phi : \Gamma^k(\mathrm{Cent}(\mathfrak{K})) \to \mathrm{Cent}\left(\Gamma^k(\mathfrak{K})\right)$ defined by $\left(\Phi\left(\mathfrak{A}\right)(X)\right)_x := \mathfrak{A}_x(X_x)$ for a section $\mathfrak{A} \in \Gamma^k(\mathrm{Cent}(\mathfrak{K}))$, $X \in \Gamma^k(\mathfrak{K})$, $x \in M$, is the inverse of Ψ. We conclude that Ψ is an isomorphism of associative algebras. $\qquad\square$

Theorem 2.3.39. *Let $\pi : \mathfrak{K} \to M$ be a Lie algebra bundle with perfect or centerfree fiber $\mathfrak{k}$ and $k \in \mathbb{N}_0 \cup \{\infty\}$.*

1. *The decomposition $\mathrm{Cent}(\mathfrak{k}) = \mathrm{N}(\mathfrak{k}) \oplus \mathrm{S}(\mathfrak{k})$ into nilpotent and semisimple endomorphisms from Lemma 2.1.17.4 is $\mathrm{Aut}(\mathfrak{k})$-stable with respect to the conjugation action of $\mathrm{Aut}(\mathfrak{k})$ on $\mathrm{Cent}(\mathfrak{k})$.*

2. *There is a decomposition $\Gamma^k(\mathrm{Cent}(\mathfrak{K})) = \Gamma^k(\mathrm{N}(\mathfrak{K})) \oplus \Gamma^k(\mathrm{S}(\mathfrak{K}))$ into associative subalgebras, where $\mathrm{N}(\mathfrak{K})$, $\mathrm{S}(\mathfrak{K})$ denote the corresponding subbundles $\mathfrak{K}(\mathrm{N}(\mathfrak{k}))$, $\mathfrak{K}(\mathrm{S}(\mathfrak{k}))$ of $\mathrm{Cent}(\mathfrak{K})$, respectively (cf. Definition 2.3.8.2).*

3. *For $\mathrm{S}(\mathfrak{k}) = \mathbb{K} \cdot \mathbf{1}$ we have $\Gamma^k(\mathrm{S}(\mathfrak{K})) = \mathrm{C}^k(M, \mathbb{K}) \cdot \mathbf{1}$. In particular, if $\mathfrak{k}$ is central simple, then $\mathrm{Cent}\left(\Gamma^k(\mathfrak{K})\right) \cong \Gamma^k(\mathrm{Cent}(\mathfrak{K})) = \Gamma^k(\mathrm{S}(\mathfrak{K})) = \mathrm{C}^k(M, \mathbb{K}) \cdot \mathbf{1}$.*

Proof. Let $\rho : \mathrm{P}(\mathfrak{K}) \to M$ be an $\mathrm{Aut}(\mathfrak{k})$-principal bundle to which $\mathfrak{K}$ is associated.

1. This immediately follows from the definition of nilpotent and semisimple endomorphisms.

2. The vector bundle $\mathrm{Cent}(\mathfrak{K})$ is associated to $\rho : \mathrm{P}(\mathfrak{K}) \to M$ with respect to the conjugation $\mathrm{Aut}(\mathfrak{k}) \times \mathrm{Cent}(\mathfrak{k}) \to \mathrm{Cent}(\mathfrak{k})$. Since the restricted actions $\mathrm{Aut}(\mathfrak{k}) \times \mathrm{N}(\mathfrak{k}) \to \mathrm{N}(\mathfrak{k})$ and $\mathrm{Aut}(\mathfrak{k}) \times \mathrm{S}(\mathfrak{k}) \to \mathrm{S}(\mathfrak{k})$ are well-defined, we obtain the corresponding vector bundles $\mathrm{N}(\mathfrak{K})$ and $\mathrm{S}(\mathfrak{K})$. The description (2.4) of sections gives us a decomposition

$$\Gamma^k(\mathrm{Cent}(\mathfrak{K})) \cong \Gamma^k(\mathrm{P}(\mathfrak{K}) \times_{\mathrm{Aut}(\mathfrak{k})} \mathrm{Cent}(\mathfrak{k})) \cong \Gamma^k(\mathrm{P}(\mathfrak{K}) \times_{\mathrm{Aut}(\mathfrak{k})} \mathrm{N}(\mathfrak{k})) \oplus \Gamma^k(\mathrm{P}(\mathfrak{K}) \times_{\mathrm{Aut}(\mathfrak{k})} \mathrm{S}(\mathfrak{k}))$$
$$\cong \Gamma^k(\mathrm{N}(\mathfrak{K})) \oplus \Gamma^k(\mathrm{S}(\mathfrak{K})).$$

The spaces $\Gamma^k(\mathrm{N}(\mathfrak{K}))$ and $\Gamma^k(\mathrm{S}(\mathfrak{K}))$ inherit the pointwise algebra structure of $\Gamma^k(\mathrm{Cent}(\mathfrak{K}))$ and are subalgebras, since, for all $m \in M$, the fibers $\mathrm{N}(\mathfrak{K})_m = \mathrm{N}(\mathfrak{K}_m)$ and $\mathrm{S}(\mathfrak{K})_m = \mathrm{S}(\mathfrak{K}_m)$, respectively, are subalgebras of $\mathrm{Cent}(\mathfrak{K})_m = \mathrm{Cent}(\mathfrak{K}_m)$ by Lemma 2.1.17.4.

3. For all $m \in M$, we have the isomorphism $\mathrm{S}(\mathfrak{K})_m \cong \mathrm{S}(\mathfrak{K}_m) \cong \mathbb{K} \cdot \mathbf{1}$. So $\mathrm{S}(\mathfrak{K})$ is a trivial bundle via the following equivalence:

$$M \times \mathbb{K} \supseteq \{m\} \times \mathbb{K} \longrightarrow \mathrm{S}(\mathfrak{K}_m) \cong \mathrm{S}(\mathfrak{K})_m \subseteq \mathrm{S}(\mathfrak{K})$$
$$(m, r) \longmapsto (v \mapsto rv). \qquad \square$$

Theorem 2.3.40. *Let $k \in \mathbb{N}_0 \cup \{\infty\}$ and $\mathfrak{k}$ be a perfect or centerfree Lie algebra. If $\mathfrak{k}$ is indecomposable, then $\Gamma^k(\mathfrak{K})$ is indecomposable.*

Proof. Let $\Gamma^k(\mathfrak{K}) = I_1 \oplus I_2$ be a decomposition into a direct sum of ideals and P^1 be the projection onto I_1 parallel to I_2. Note that $P^1 \in \mathrm{Cent}(\Gamma^k(\mathfrak{K})) = \Gamma^k(\mathrm{Cent}(\mathfrak{K}))$ because for any $X, Y \in \Gamma^k(\mathfrak{K})$ decomposing into $X_1, Y_1 \in I_1$ and $X_2, Y_2 \in I_2$ we have:

$$P^1[X, Y] = P^1([X_1, Y_1] + [X_2, Y_2]) = [X_1, Y_1] = [X_1 + X_2, Y_1] = [X, P^1 Y]. \qquad (2.17)$$

Since $P^1 \circ P^1 = P^1$ on $\Gamma^k(\mathfrak{K})$, each $P_x^1 \in \mathrm{Cent}(\mathfrak{K}_x)$, where $x \in M$, is a projection in $\mathrm{Cent}(\mathfrak{K}_x)$ and, by indecomposability of $\mathfrak{K}_x \cong \mathfrak{k}$, this implies $P_x^1 = \mathbf{0}$ or $P_x^1 = \mathbf{1}$. The map

$$M \longrightarrow \mathrm{Cent}(\mathfrak{K}_x) \subseteq \mathrm{Cent}(\mathfrak{K})$$
$$x \longmapsto P_x^1$$

is continuous and for each bundle atlas $(U_i, \varphi_i)_{i \in I}$, where each U_i is connected, the mapping $x \longmapsto \varphi_i \circ P_x^1$ has discrete image. But all this is only possible if $P^1 = \mathbf{0}$ or $P^1 = \mathbf{1}$, i.e., $I_1 = \mathbf{0}$ or $I_2 = \mathbf{0}$. So $\Gamma^k(\mathfrak{K})$ is indecomposable. $\qquad \square$

2.3.4. Decompositions

Now we will prove some facts on decompositions. Throughout this subsection, let $k \in \mathbb{N}_0 \cup \{\infty\}$ and $\pi : \mathfrak{K} \to M$ be a Lie algebra bundle with finite-dimensional fiber $\mathfrak{k}$ such that there is the following decomposition into indecomposable non-zero ideals:

$$\mathfrak{k} = \bigoplus_{i=1}^{n} \mathfrak{k}_i = \bigoplus_{j=1}^{\ell} \bigoplus_{k_j=1}^{m_j} \mathfrak{k}^{(j)} = \bigoplus_{j=1}^{\ell} m_j \mathfrak{k}^{(j)}, \qquad (2.18)$$

where $m_j \in \mathbb{N}_0$ and $\mathfrak{k}^{(j)} \not\cong \mathfrak{k}^{(j')}$ as Lie algebras for indices $j \neq j'$, and this decomposition is unique except for the order. Proposition 2.1.26 gives us a criterion when this is the case.

If $\mathfrak{k}$ is perfect or centerfree, then Remark 2.1.21 and Propositions 2.3.15 and 2.3.17 imply that any decomposition

$$\Gamma^k(\mathfrak{K}) = \bigoplus_{i=1}^{n} I_i \qquad (2.19)$$

into indecomposable ideals is unique except for the order. We are interested in decompositions

$$\Gamma^k(\mathfrak{K}) = \bigoplus_{i=1}^{n} \Gamma^k(\mathfrak{K}^i) \qquad (2.20)$$

into ideals, for subbundles $\pi_i = \pi_{|\mathfrak{K}^i} : \mathfrak{K}^i \to M$ with respective fibers $\mathfrak{k}_i$ such that $\pi = \bigoplus_{i=1}^{n} \pi_{|\mathfrak{K}^i}$. If $\mathfrak{k}$ is perfect or centerfree, then the $\Gamma^k(\mathfrak{K}^i)$'s are indecomposable, too (cf. Theorem 2.3.40). First we note that each decomposition (2.19) is in fact a decomposition (2.20), if $\mathfrak{k}$ is perfect.

Proposition 2.3.41. *If $\mathfrak{k}$ is perfect, then there is a bijection between the decompositions of the Lie algebra bundle $\pi : \mathfrak{K} \to M$ into subbundles and the decompositions of the Lie algebra $\Gamma^k(\mathfrak{K})$ into ideals. In particular, $\mathfrak{K}$ is indecomposable if and only if $\Gamma(\mathfrak{K})$ is so.*

Proof. The Theorem of Serre-Swan (cf., e.g., Theorem 3.2.14 of [Wag11]) gives an equivalence

$$\mathcal{G} : \mathbb{V} \longmapsto \Gamma^k(\mathbb{V})$$

between the category of vector bundles over M and the finitely generated projective $C^k(M, \mathbb{K})$-modules.

Let $\mathfrak{K} = \mathfrak{K}^1 \oplus \mathfrak{K}^2$, where $\pi_i = \pi_{|\mathfrak{K}^i} : \mathfrak{K}^i \to M$ is a subbundle with fiber $\mathfrak{k}_i$ for $i = 1, 2$. Then $\Gamma^k(\mathfrak{K}) \cong \Gamma^k(\mathfrak{K}^1) \oplus \Gamma^k(\mathfrak{K}^2)$ as Lie algebras by Proposition 2.3.7.

Let $\mathcal{G}(\mathfrak{K}) = \Gamma^k(\mathfrak{K}) = I_1 \oplus I_2$ be a decomposition into ideals. This is also a decomposition into $C^k(M, \mathbb{K})$-modules, since $\mathrm{Cent}(\Gamma^k(\mathfrak{K})) \supseteq C^k(M, \mathbb{K}) \cdot \mathbf{1}$, by Remark 2.1.20 and the perfectness of $\Gamma^k(\mathfrak{K})$. Since I_1 and I_2 are complemented submodules of the finitely generated projective module $\mathcal{G}(\mathfrak{K})$, they are finitely generated and projective, too. So there are vector subbundles $\mathbb{V}_1, \mathbb{V}_2$ of $\mathfrak{K}$ such that $\mathfrak{K} = \mathbb{V}_1 \oplus \mathbb{V}_2$ and $\mathcal{G}(\mathbb{V}_i) = \Gamma^k(\mathbb{V}_i) = I_i$ for $i = 1, 2$. $\square$

Remark 2.3.42. A decomposition (2.20) exists for trivial bundles $\mathfrak{K} = M \times \mathfrak{k} \to M$:
We have $\Gamma^k(\mathfrak{K}) = C^k(M, \mathfrak{k})$ and, for $\mathfrak{K}^i := M \times \mathfrak{k}_i$, we obtain the decomposition $C^k(M, \mathfrak{k}) = \bigoplus_{i=1}^{n} \Gamma^k(\mathfrak{K}^i) = \bigoplus_{i=1}^{n} C^k(M, \mathfrak{k}_i)$.

Now we want to approach this problem for general finite-dimensional Lie algebra bundles with fiber $\mathfrak{k}$ which has a unique decomposition (2.18).

Definition 2.3.43. Let $\mathcal{M}(\mathfrak{k}) := \{\mathfrak{k}_1, \ldots, \mathfrak{k}_n\}$ be the set of indecomposable ideals of $\mathfrak{k}$ and $\mathcal{S}(\mathfrak{k}) := \prod_{j=1}^{\ell} \mathcal{S}_{m_j}$ the product of symmetric groups corresponding to the m_j's in (2.18).

Definition 2.3.44. Define the subgroup

$$\mathrm{H}(\mathfrak{k}) := \mathrm{Aut}(\mathfrak{k}; \mathfrak{k}_1, \ldots, \mathfrak{k}_n) := \bigcap_{i=1}^{n} \{f \in \mathrm{Aut}(\mathfrak{k}) \mid f(\mathfrak{k}_i) = \mathfrak{k}_i\},$$

which does not depend on the decomposition (2.18) by Proposition 2.1.26. It is even a normal subgroup since, for all $f \in \mathrm{H}(\mathfrak{k})$ and $g \in \mathrm{Aut}(\mathfrak{k})$, there is a permutation $\sigma \in \mathcal{S}_n$ with $g(\mathfrak{k}_i) = \mathfrak{k}_{\sigma^{-1}(i)}$ for all $i \in \{1, \ldots, n\}$ by Remark 2.1.28. So $gfg^{-1}(\mathfrak{k}_i) = gf(\mathfrak{k}_{\sigma(i)}) = g(\mathfrak{k}_{\sigma(i)}) = \mathfrak{k}_i$.

Lemma 2.3.45. *The subgroup* $\mathrm{H}(\mathfrak{k}) \leq \mathrm{Aut}(\mathfrak{k}) \subseteq \mathrm{GL}(\mathfrak{k})$ *is open. In particular,* $\mathrm{H}(\mathfrak{k}) \supseteq \mathrm{Aut}(\mathfrak{k})_0$.

Proof. The complement of $\mathrm{H}(\mathfrak{k})$ in $\mathrm{Aut}(\mathfrak{k})$ is the closed set

$$\bigcup_{\sigma \in \mathcal{S}_n \backslash \{\mathrm{id}_n\}} \{f \in \mathrm{Aut}(\mathfrak{k}) \mid f(\mathfrak{k}_i) = \mathfrak{k}_{\sigma(i)} \text{ for all } i = 1, \ldots, n\}. \qquad \square$$

Remark 2.3.46. We choose a permutation $\sigma = (\sigma_j)_{j=1}^{\ell} \in \mathcal{S}(\mathfrak{k})$. Then there is an automorphism $f_\sigma : \mathfrak{k} \to \mathfrak{k}$ such that $f_\sigma\left(\mathfrak{k}^{(k_j,j)}\right) = \mathfrak{k}^{(\sigma_j(k_j),j)}$ for the components $\mathfrak{k}^{(1,j)} \oplus \ldots \oplus \mathfrak{k}^{(m_j,j)} = m_j \mathfrak{k}^{(j)}$ for all $j \in \{1, \ldots \ell\}$.

By the preceding, the group $\mathcal{S}(\mathfrak{k})$ can be embedded into $\mathrm{Aut}(\mathfrak{k})$. Clearly, the subgroups $\mathcal{S}(\mathfrak{k})$ and $\mathrm{H}(\mathfrak{k})$ of $\mathrm{Aut}(\mathfrak{k})$ have a trivial intersection, $\mathrm{H}(\mathfrak{k})$ is invariant under conjugation with elements in $\mathcal{S}(\mathfrak{k})$, and $\mathrm{Aut}(\mathfrak{k}) = \mathrm{H}(\mathfrak{k}) \cdot \mathcal{S}(\mathfrak{k})$. Thus $\mathrm{Aut}(\mathfrak{k}) = \mathrm{H}(\mathfrak{k}) \rtimes \mathcal{S}(\mathfrak{k})$.

In order to state a theorem on decompositions of Lie algebras of sections, we use notions of non-abelian Čech cohomology (cf. Section B.3).

Remark 2.3.47. The exact sequence of Lie groups $1 \to \mathrm{H}(\mathfrak{k}) \xrightarrow{\iota} \mathrm{Aut}(\mathfrak{k}) \xrightarrow{q} \mathcal{S}(\mathfrak{k}) \to 1$ induces, by Theorem B.3.5, the following exact sequence of pointed spaces:

$$\ldots \to \check{\mathrm{H}}^1(M, \underline{\mathrm{H}(\mathfrak{k})}) \xrightarrow{[\iota]} \check{\mathrm{H}}^1(M, \underline{\mathrm{Aut}(\mathfrak{k})}) \xrightarrow{[q]} \check{\mathrm{H}}^1(M, \underline{\mathcal{S}(\mathfrak{k})}). \tag{2.21}$$

Theorem 2.3.48. *If* $\mathfrak{k}$ *is a finite-dimensional Lie algebra which has a unique decomposition* (2.18), *and the Lie algebra bundle* $\pi : \mathfrak{K} \to M$ *with fiber* $\mathfrak{k}$ *is associated to the* $\mathrm{Aut}(\mathfrak{k})$*-principal bundle* $\rho : \mathrm{P}(\mathfrak{K}) \to M$, *then the following statements are equivalent:*

1. *There is a decomposition* (2.20) *of* $\Gamma^k(\mathfrak{K})$.

2. $[\mathrm{P}(\mathfrak{K})] \in \check{\mathrm{H}}^1(M, \underline{\mathrm{Aut}(\mathfrak{k})})$ *is in the image of* $[\iota] : \check{\mathrm{H}}^1(M, \underline{\mathrm{H}(\mathfrak{k})}) \to \check{\mathrm{H}}^1(M, \underline{\mathrm{Aut}(\mathfrak{k})})$.

3. $[\mathrm{P}(\mathfrak{K})] \in \check{\mathrm{H}}^1(M, \underline{\mathrm{Aut}(\mathfrak{k})})$ *is in the kernel of* $[q] : \check{\mathrm{H}}^1(M, \underline{\mathrm{Aut}(\mathfrak{k})}) \to \check{\mathrm{H}}^1(M, \underline{\mathcal{S}(\mathfrak{k})})$. *Here, the map* $[q]$ *is a morphism in the category of pointed sets, so the kernel is* $[q]^{-1}([M \times \mathcal{S}(\mathfrak{k})])$.

Proof.

- Let $\Gamma^k(\mathfrak{K}) = \bigoplus_{i=1}^n \Gamma^k(\mathfrak{K}^i)$ for subbundles $\pi_i : \mathfrak{K}^i \to M$ with fiber $\mathfrak{k}_i$ such that $\pi = \bigoplus_{i=1}^n \pi_i$. We will show the existence of an $H(\mathfrak{k})$-valued cocycle for $\pi : \mathfrak{K} \to M$:

  For $i \in \{1, \ldots, n\}$ the bundle $\pi_i : \mathfrak{K}^i \to M$ possesses an $\mathrm{Aut}(\mathfrak{k}_i)$-valued 1-cocycle, the sum of these bundles is $\pi : \mathfrak{K} \to M$ and, as seen in Proposition 2.3.6, the automorphisms in the 1-cocycle of the sum of Lie algebra bundles leave the summands invariant, i.e., the 1-cocycle has values in $H(\mathfrak{k}) = \mathrm{Aut}(\mathfrak{k}; \mathfrak{k}_1, \ldots, \mathfrak{k}_n)$ only.

- The equivalence of 2. and 3. is given by the exact sequence (2.21).

- Let $[\mathrm{P}(\mathfrak{K})]$ be in the image of $[\iota] : \check{\mathrm{H}}^1(M, \underline{\mathrm{H}(\mathfrak{k})}) \to \check{\mathrm{H}}^1(M, \underline{\mathrm{Aut}(\mathfrak{k})})$, i.e., there is a bundle atlas $(U_i, \varphi_i)_{i \in I}$ of $\rho : \mathrm{P}(\mathfrak{K}) \to M$ such that the corresponding 1-cocycle (g_{ji}) has values in $H(\mathfrak{k})$ only. Since $\mathfrak{K}$ is associated to this principal bundle with respect to the natural action $\lambda : \mathrm{Aut}(\mathfrak{k}) \times \mathfrak{k} \to \mathfrak{k}$, we conclude that it is also associated with respect to the restricted action $\lambda' : \mathrm{H}(\mathfrak{k}) \times \mathfrak{k} \to \mathfrak{k}$ which respects the decomposition (2.18) of $\mathfrak{k}$. For $i \in \{1, \ldots, n\}$, let $\lambda_i : \mathrm{H}(\mathfrak{k}) \times \mathfrak{k}_i \to \mathfrak{k}_i$ be the induced action on $\mathfrak{k}_i$ and $\pi_i : \mathfrak{K}^i \to M$ the corresponding Lie algebra bundle. The description (2.4) of sections gives us a decomposition

$$\bigoplus_{i=1}^n \Gamma^k(\mathfrak{K}^i) \cong \bigoplus_{i=1}^n \Gamma^k(\mathrm{P}(\mathfrak{K}) \times_{\mathrm{H}(\mathfrak{k})} \mathfrak{k}_i) \cong \Gamma^k(\mathrm{P}(\mathfrak{K}) \times_{\mathrm{H}(\mathfrak{k})} \mathfrak{k}) \cong \Gamma^k(\mathfrak{K}). \qquad \square$$

In general, we cannot expect that a decomposition (2.20) exists, but we will use the preceding theorem and Proposition 2.2.26 to show that there exists a finite covering $\widehat{q} : \widehat{M} \to M$ such that the section algebra of the pullback bundle $\widehat{q}^*\mathfrak{K}$ decomposes.

Theorem 2.3.49. *Let $\mathfrak{k}$ be a finite-dimensional Lie algebra which has a unique decomposition (2.18) and $\pi : \mathfrak{K} \to M$ be a Lie algebra bundle with fiber $\mathfrak{k}$ associated to the $\mathrm{Aut}(\mathfrak{k})$-principal bundle $\rho : \mathrm{P}(\mathfrak{K}) \to M$. Then there is a finite covering $\widehat{q} : \widehat{M} \to M$ such that, for the pullback Lie algebra bundle $\widehat{q}^*\mathfrak{K} =: \mathfrak{G}$, there is a decomposition*

$$\Gamma^k(\mathfrak{G}) = \bigoplus_{i=1}^n \Gamma^k(\mathfrak{G}^i).$$

into indecomposable ideals, where each $\mathfrak{G}^i \to \widehat{M}$ is a subbundle of $\mathfrak{G}$ with fiber $\mathfrak{k}_i$.

Proof. Let $\widehat{q} : \widehat{M} \to M$ be the finite covering of M from Proposition 2.2.26 such that $Q/\mathrm{H}(\mathfrak{k})$ is trivial for $Q := \widehat{q}^*\mathrm{P}(\mathfrak{K})$. So $[Q] = \left[\widehat{M} \times \mathcal{S}(\mathfrak{k})\right] \in \check{\mathrm{H}}^1(\widehat{M}, \underline{\mathcal{S}(\mathfrak{k})})$ is the distinguished element. Since the Lie algebra bundle $\mathfrak{G} := \widehat{q}^*\mathfrak{K}$ is associated to Q, Theorem 2.3.48 finishs the proof. $\square$

Example 2.3.50. Let $\mathfrak{s}$ be a simple Lie algebra, $\mathfrak{k} := \mathfrak{s} \oplus \mathfrak{s}$ and $M := \mathbb{S}^1$ with fundamental group $\pi_1(\mathbb{S}^1) = \mathbb{Z}$ and universal covering bundle $\mathbb{R} \to \mathbb{S}^1 \cong \mathbb{R}/\mathbb{Z}$. For any group morphism $\alpha : \mathbb{Z} \to \mathrm{Aut}(\mathfrak{k})$, let $\mathfrak{K}_\alpha := \mathbb{R} \times_\alpha \mathfrak{k}$ be the associated Lie algebra bundle. If $\alpha(1)$ is the flip $\mathfrak{s} \oplus \mathfrak{s} \to \mathfrak{s} \oplus \mathfrak{s}$, $(x, y) \mapsto (y, x)$, then $\mathfrak{K}$ and thus $\Gamma^k(\mathfrak{K})$ is indecomposable. For $\widehat{q} : \mathbb{R}/2\mathbb{Z} \to \mathbb{R}/\mathbb{Z}$, $[r] \mapsto [r]$ the bundle $\mathfrak{G} := \widehat{q}^*\mathfrak{K}_\alpha$ decomposes into subbundles $\mathfrak{G}_1$, $\mathfrak{G}_2$, each with fiber $\mathfrak{s}$.

2.3.5. Complex Structures

A statement on complex structures of real Lie algebras $\Gamma^k(\mathfrak{K})$, is given by the following proposition. By $\mathcal{C}_2$ we mean the cyclic group of two elements.

Proposition 2.3.51. *Let $\pi : \mathfrak{K} \to M$ be a finite-dimensional Lie algebra bundle with indecomposable fiber $\mathfrak{k}$, which is perfect or centerfree. Then we have the following:*

1. *M has a 2-fold connected covering $\widehat{q} : \widehat{M} \to M$ if and only if $\check{\mathrm{H}}^1\left(M, \underline{\mathcal{C}_2}\right) \cong \mathrm{Hom}(\pi_1(M), \mathcal{C}_2)$ (cf. Remark B.3.3.2) does not vanish.*

2. *$\Gamma^k(\mathfrak{K})$ admits at most two complex structures.*

3. *If $\check{\mathrm{H}}^1\left(M, \underline{\mathcal{C}_2}\right) \cong \mathrm{Hom}(\pi_1(M), \mathcal{C}_2)$ is trivial, then $\Gamma^k(\mathfrak{K})$ admits a complex structure if and only if $\mathfrak{k}$ does.*

4. *If $\Gamma^k(\mathfrak{K})$ does not admit a complex structure but $\mathfrak{k}$ does, then for some 2-fold covering $\widehat{q} : \widehat{M} \to M$, the Lie algebra $\Gamma^k(\widehat{q}^*\mathfrak{K})$ admits a complex structure.*

Proof.

1. The group $\mathrm{Hom}(\pi_1(M), \mathcal{C}_2)$ is trivial if and only if every $\mathcal{C}_2$-principal bundle over M is trivial, i.e., the total space of any $\mathcal{C}_2$-principal bundle is disconnected.

2. A complex structure $J \in \mathrm{Cent}\left(\Gamma^k(\mathfrak{K})\right) = \Gamma^k(\mathrm{Cent}(\mathfrak{K}))$ induces a complex structure J_x on each fiber $\mathrm{Cent}(\mathfrak{K}_x)$, thus on $\mathfrak{k}$. Lemma 2.1.29.2 implies that $\mathfrak{k}$ possesses at most two complex structures, thus, by the connectedness of M, there can be at most two different C^k-complex structures, namely J and $-J$, on $\Gamma^k(\mathfrak{K})$.

3. Let J_0 be a complex structure on $\mathfrak{k}$. We turn $\mathfrak{k}$ into the complex Lie algebra $\mathfrak{k}^{\mathbb{C}}$ by defining the multiplication by a complex scalar via $(a + bi) \cdot x := ax + bJ_0(x)$ for $a, b \in \mathbb{R}$ and $x \in \mathfrak{k}$. Let $\mathrm{Aut}(\mathfrak{k})$ be the group of real automorphisms of the real Lie algebra $\mathfrak{k}$ and $\mathrm{Aut}\left(\mathfrak{k}^{\mathbb{C}}\right)$ the group of complex automorphisms of the complex Lie algebra $\mathfrak{k}^{\mathbb{C}}$. If $\sigma \in \mathrm{Aut}\left(\mathfrak{k}\right)$, $a, b \in \mathbb{R}$ and $x \in \mathfrak{k}$, then

$$\sigma((a + bi) \cdot x) = \sigma(ax + bJ_0(x)) = a\sigma(x) + b\sigma(J_0(x)).$$

Thus $\sigma \in \mathrm{Aut}\left(\mathfrak{k}\right)$ is in $\mathrm{Aut}(\mathfrak{k}^{\mathbb{C}})$ if and only if $\sigma \circ J_0 = J_0 \circ \sigma$. Otherwise we have the identity $\sigma \circ J_0 \circ \sigma^{-1} = -J_0$ because $\left(\sigma \circ J_0 \circ \sigma^{-1}\right)^2 = -\mathbf{1}$ and J_0, $-J_0$ are the only complex structures on $\mathfrak{k}$ by Lemma 2.1.29.2. So $\mathrm{Aut}(\mathfrak{k}^{\mathbb{C}})$ is a subgroup of index 2 of $\mathrm{Aut}(\mathfrak{k})$, thus normal and $\mathrm{Aut}\left(\mathfrak{k}\right)/\mathrm{Aut}(\mathfrak{k}^{\mathbb{C}}) \cong \mathcal{C}_2$. Since $\check{\mathrm{H}}^1(M, \mathcal{C}_2)$ is trivial, we have the isomorphism $\check{\mathrm{H}}^1(M, \underline{\mathrm{Aut}(\mathfrak{k})}) \cong \check{\mathrm{H}}^1(M, \underline{\mathrm{Aut}(\mathfrak{k}^{\mathbb{C}})})$ by Theorem B.3.5, yielding the existence of an $\mathrm{Aut}(\mathfrak{k}^{\mathbb{C}})$-valued cocycle of $\mathfrak{K}$ and this turns $\pi : \mathfrak{K} \to M$ into a Lie algebra bundle with complex fiber $\mathfrak{k}$ and $\Gamma^k(\mathfrak{K})$ into a complex Lie algebra.

   Conversely, if $\Gamma^k(\mathfrak{K})$ admits a complex structure, then, by Theorem 2.3.38, this induces a complex structure on each fiber of $\mathfrak{K}$, thus on $\mathfrak{k}$.

4. Let J_0 be a complex structure on $\mathfrak{k}$ and $\rho : P \to M$ be the $\operatorname{Aut}(\mathfrak{k})$-principal bundle to which the Lie algebra bundle $\pi : \mathfrak{K} \to M$ is associated. We apply Proposition 2.2.26 to the normal subgroup $N := \operatorname{Aut}(\mathfrak{k}^\mathbb{C})$ of $G := \operatorname{Aut}(\mathfrak{k})$, obtaining a 2-fold covering $\widehat{q} : \widehat{M} \to M$ such that the $\mathcal{C}_2$-principal bundle $\widehat{q}^*(P/N) \to \widehat{M}$ is trivial. By Theorem B.3.5, there is an $\operatorname{Aut}(\mathfrak{k}^\mathbb{C})$-valued cocycle of $\widehat{q}^*\mathfrak{K}$ and this turns $\widehat{q}^*\mathfrak{K} \to \widehat{M}$ into a Lie algebra bundle with complex fiber $\mathfrak{k}$ and $\Gamma^k(\widehat{q}^*\mathfrak{K})$ into a complex Lie algebra. $\qquad\square$

Example 2.3.52. For the manifold $M := \mathbb{S}^1$, the real simple Lie algebra $\mathfrak{k} := \mathfrak{sl}(2,\mathbb{C})^\mathbb{R}$ and any group morphism $\alpha : \pi_1(\mathbb{S}^1) \cong \mathbb{Z} \to \operatorname{Aut}(\mathfrak{k})$, we define the Lie algebra bundle $\mathfrak{K}_\alpha := \mathbb{R} \times_\alpha \mathfrak{k}$ for the universal covering bundle $\mathbb{R} \to \mathbb{S}^1 \cong \mathbb{R}/\mathbb{Z}$. If $\alpha(1) \notin \operatorname{Aut}(\mathfrak{k}^\mathbb{C}) = \operatorname{Aut}(\mathfrak{sl}(2,\mathbb{C}))$, then $\Gamma^k(\mathfrak{K}_\alpha)$ has no complex structure, but for $\widehat{q} : \mathbb{R}/2\mathbb{Z} \to \mathbb{R}/\mathbb{Z}$, $[r] \mapsto [r]$, the Lie algebra $\Gamma^k(\widehat{q}^*\mathfrak{K}_\alpha)$ admits a complex structure.

2.3.6. Isomorphisms

In the light of Theorem 2.3.48, the discussion of the isomorphisms of Lie algebras of C^k-sections reduces to the case of indecomposable Lie algebras. The main key to the analysis of the isomorphisms is Proposition 2.3.55 below, which is in the spirit and uses techniques of the famous Theorem of Pursell and Shanks (cf. [SP54]) stating that isomorphic Lie algebras of smooth vector fields come from diffeomorphic manifolds. In the following, M, N are smooth, hausdorff, paracompact and connected manifolds of positive dimensions m, n, respectively.

Lemma 2.3.53. For $k \in \mathbb{N}_0 \cup \{\infty\}$ the topology on M is the same as the initial topology of the maps in $\mathrm{C}^k(M, \mathbb{K})$.

Proof. Let $\mathcal{O}_M$ denote the topology on M. The initial topology of the maps in $\mathrm{C}^k(M, \mathbb{K})$, denoted by $\mathcal{O}_i$, is the coarsest topology on M such that each $f \in \mathrm{C}^k(M, \mathbb{K})$ is continuous, so that $\mathcal{O}_i \subseteq \mathcal{O}_M$. Let V be a neighbourhood of $x \in M$ with respect to $\mathcal{O}_M$. Then, by paracompactness of M, there is a smooth function $\rho : M \to [0,1] \subseteq \mathbb{K}$ such that $\rho_{|M \setminus V} = \mathbf{0}$ and $\rho_{|W} = \mathbf{1}$ for a smaller x-neighbourhood $W \subseteq V$. But this yields $W \subseteq \rho^{-1}(B(0.5; 1)) \subseteq V$, where $B(r; z)$ denotes the open ball with radius r around z in $\mathbb{K}$. Thus V is also a neighbourhood of $x \in M$ with respect to $\mathcal{O}_i$ and we conclude $\mathcal{O}_M \subseteq \mathcal{O}_i$, so $\mathcal{O}_i = \mathcal{O}_M$. $\qquad\square$

Lemma 2.3.54. There exists a non-constant proper map $\rho \in \mathrm{C}^\infty(M, \mathbb{K})$, i.e., ρ-preimages of compact subspaces are compact.

Proof. By the Whitney Embedding Theorem (cf. Theorem II.10.8 of [Bre93]), there is an $\ell \in \mathbb{N}$ and a smooth embedding $\phi : M \to \mathbb{R}^\ell$ such that $\phi(M) \subseteq \mathbb{R}^\ell$ is closed. For any $x_0 \in \mathbb{R}^\ell$ the map $\rho : M \to \mathbb{R}^\ell$, $m \mapsto \|\phi(m) - x_0\|_2^2$ is proper. $\qquad\square$

Proposition 2.3.55. Let $v : \mathrm{C}^k(M, \mathbb{K}) \to \mathrm{C}^\ell(N, \mathbb{K})$ be an isomorphism of associative algebras for some $k, \ell \in \mathbb{N} \cup \{\infty\}$. Then $k = \ell$ and there is a C^k-diffeomorphism $\lambda : M \to N$ such that $v(f) = f \circ \lambda^{-1}$ for all $f \in \mathrm{C}^k(M, \mathbb{K})$.

Proof. We set $\mathcal{A} := \mathcal{A}^M := \mathrm{C}^k(M, \mathbb{K})$ and $\mathcal{A}_U := \mathcal{A}_U^M := \{f \in \mathcal{A}|\operatorname{supp} f \subseteq U\}$ for open $U \subseteq M$. Furthermore, for $x_0 \in M$ we write $\mathcal{N}_{x_0} := \mathcal{N}_{x_0}^M := \{f \in \mathcal{A}|f(x_0) = 0\}$. The symbols $\mathcal{A}^N$, $\mathcal{A}_V^N$ for open $V \subseteq N$ and $\mathcal{N}_{y_0}^N$ for $y_0 \in N$ are understood in the obvious analogous manner.

Since any $\mathcal{N}_{x_0}$ obviously is a vector subspace of $\mathcal{A}$ and $\mathcal{A} \cdot \mathcal{N}_{x_0} \subseteq \mathcal{N}_{x_0}$, it is even an ideal of $\mathcal{A}$. Its codimension[5] is 1 because, evidently, we have $\mathcal{N}_{x_0} \oplus \mathbb{K} \cdot \mathbf{1} = \mathcal{A}$. We will show that in fact every ideal $I \trianglelefteq \mathcal{A}$ of codimension 1 takes the form $\mathcal{N}_x$ for a some $x \in M$ and the proof will also show the analogous statement for the ideals $J \trianglelefteq \mathcal{A}^N$ with codimension 1. For this, it suffices to show $I \subseteq \mathcal{N}_x$ for some $x \in M$ because all $\mathcal{N}_x$ are ideals of codimension 1. Since, for $x_0 \neq x_1$, we have $(\mathcal{N}_{x_0} \cap \mathcal{N}_{x_1}) \oplus \mathbb{K} \cdot \mathbf{1} \neq \mathcal{A}$, the point $x \in M$ corresponding to the ideal I is uniquely determined.

Assume there exists an ideal $I \trianglelefteq \mathcal{A}$ of codimension 1 such that for all $x \in M$ we have $I \not\subseteq \mathcal{N}_x$, i.e., there exists $f_0 \in I$ with $f_0(x) \neq 0$. By continuity, there is even an admissible[6] open x-neighbourhood $U \subseteq M$ with $f_0(x') \neq 0$ for all $x' \in U$. Note that M can, in our case, be covered by admissible open sets, say $M = \bigcup_{\gamma \in \Gamma} U_\gamma$. If $U \subseteq M$ is open and $f \in \mathcal{A}_U$ then $\frac{f}{f_0}$ is a well-defined function in $\mathcal{A} = \mathrm{C}^k(M, \mathbb{K})$ and $f = f_0 \cdot \frac{f}{f_0} \in I$ because $f_0 \in I$ and I is an ideal of $\mathcal{A}$. Thus $\mathcal{A}_U \subseteq I$.

Now let $\varphi \in \mathcal{A}$ be a non-locally constant proper map, which exists by Lemma 2.3.54. Since $\mathrm{codim}_\mathcal{A} I = 1$ and $\dim(\mathbb{K} \cdot \varphi \oplus \mathbb{K} \cdot \varphi^2) = 2$, there are $a, b \in \mathbb{K}$ such that $a\varphi + b\varphi^2 =: f_1 \in I$ is not zero and $f_1^{-1}(\{0\})$ is compact because $f_1^{-1}(\{0\}) = (b\varphi)^{-1}(\{0\}) \cup \varphi^{-1}(\{-\frac{a}{b}\})$ and $b\varphi$ is proper in the case of $b \neq 0$ and $f_1^{-1}(\{0\}) = (a\varphi)^{-1}(\{0\})$ and $a\varphi$ is proper if $b = 0$. We define an open set $U_1 := f_1^{-1}(\mathbb{K} \setminus \{0\})$. The compact subset $f_1^{-1}(\{0\}) \subseteq M$ is covered by finitely many admissible open sets $U_2, U_3, \ldots, U_r \in (U_\gamma)_{\gamma \in \Gamma}$. Let $(V_\beta)_{\beta \in B}$ be a locally finite refinement of $(U_1, \ldots, U_r)$ such that there is a smooth partition of unity $(\rho_\beta : M \to [0,1])_{\beta \in B}$ such that $\mathrm{supp}(\rho_\beta) \subseteq V_\beta$ for all $\beta \in B$. Let $B_1, \ldots, B_r \subseteq B$ be subsets such that $B = \bigcup_{i=1}^r B_i$ and let $\beta \in B_i$ for some $i \in \{1, \ldots, r\}$ imply $V_\beta \subseteq U_i$. Then, for any $i \in \{1, \ldots r\}$ and $\beta \in B_i$, we have $\rho \cdot \mathcal{A} \subseteq \mathcal{A}_{U_i}$. We obtain

$$\mathcal{A} = \mathbf{1} \cdot \mathcal{A} = \sum_{\beta \in B} (\rho_\beta \cdot \mathcal{A}) \subseteq \sum_{i=1}^r \left(\sum_{\beta \in B_i} \rho_\beta \cdot \mathcal{A} \right) \subseteq \sum_{i=1}^r \mathcal{A}_{U_i} \subseteq I,$$

but this contradicts the fact that I has codimension 1. This shows that for any ideal $I \trianglelefteq \mathcal{A}$ with codimension 1 there exists $x \in M$ such that $I = \mathcal{N}_x$.

Since v is an isomorphism of associative algebras, $v(\mathcal{N}_x)$ is, for any $x \in M$, an ideal of $\mathcal{A}^N$ with codimension 1, thus equal to $\mathcal{N}_y^N$ for some unique $y \in N$. We write $y =: \lambda(x)$ and obtain a mapping $\lambda : M \to N$. A dual argumentation with v^{-1} instead of v leads to a mapping $\widetilde{\lambda} : N \to M$ such that $\lambda \circ \widetilde{\lambda} = \mathrm{id}_N$ and $\widetilde{\lambda} \circ \lambda = \mathrm{id}_M$, so λ is bijective. Note that we can perform the following calculation for $f \in \mathcal{A}$ and $y \in N$:

$$f\left(\lambda^{-1}(y)\right) = 0 \quad \Longrightarrow \quad f \in \mathcal{N}_{\lambda^{-1}(y)} \quad \Longrightarrow \quad v(f) \in \mathcal{N}_y^N \quad \Longrightarrow \quad v(f)(y) = 0.$$

But this already implies that $f \circ \lambda^{-1} = v(f)$ for any $f \in \mathcal{A}$ because we may replace the mapping f by $f - r \cdot \mathbf{1}$ for any $r \in \mathbb{K}$ in the above calculation. We will now show that λ is a C^k-diffeomorphism.

By Lemma 2.3.53, the topologies on M and N can be described as the initial topologies of the maps in $\mathrm{C}^k(M, \mathbb{K})$ and $\mathrm{C}^\ell(N, \mathbb{K})$, respectively. Since $v^{-1}(g) = g \circ \lambda \in \mathrm{C}^k(M, \mathbb{K})$ for

---

[5]The *codimension* $\mathrm{codim}_\mathcal{A} I$ of an ideal $I \trianglelefteq \mathcal{A} = \mathcal{A}^M$ or $J \trianglelefteq \mathcal{A}^N$ is meant to be the vector space dimension of the quotient algebra $\mathcal{A}/I$ or $\mathcal{A}^N/J$, respectively.

[6]We call an open set U *admissible* (for I), if there is a function $f \in I$ such that $f(y) \neq 0$ for all $y \in U$.

all $g \in C^{\ell}(N, \mathbb{K})$, the map $\lambda : M \to N$ is continuous. Let $(U_i', \varphi_i')_{i \in I}$ be a locally finite atlas of M and $(V_j', \psi_j')_{j \in J}$ a locally finite atlas of N. We modify the chart maps by multiplying them with the maps $\rho_i : M \to [0,1]$, $\mathrm{supp}(\rho_i) \subseteq U_i'$, and $\varpi_j : N \to [0,1]$, $\mathrm{supp}(\varpi_j) \subseteq V_j'$, respectively, of smooth partitions of unity. Then $\rho_{i|U_i} = 1$ and $\varpi_{j|V_j} = 1$ for open sets $U_i \subseteq U_i'$ and $V_j \subseteq V_j'$ still covering the whole manifold. We obtain new atlases, denoted by $(U_i, \varphi_i)_{i \in I}$ and $(V_j, \psi_j)_{j \in J}$, where each chart map is defined on the whole manifold. For $t \in \{1, \dots, n\}$, let e_t^* be the dual map $\mathbb{R}^n \to \mathbb{R}$ with respect to the canonical basis $(e_1, \dots, e_n)$ of $\mathbb{R}^n$, $i \in I$, $j \in J$ and define $A := \varphi_i \left(U_i \cap \lambda^{-1}(V_j) \right) \subseteq \mathbb{R}^m$ and $D := e_t^*(\psi_j(V_j \cap \lambda(U_i))) \subseteq \mathbb{R}$. The map $e_t^* \circ \psi_j \circ \lambda \circ (\varphi_i)^{-1} : A \to D$ is equal to

$$\underbrace{v^{-1} \left(e_t^* \circ \psi_j \right)}_{\in C^k(M, \mathbb{R})} \circ (\varphi_i)^{-1} \in C^k(A, D),$$

thus, since t, i and j were arbirarily chosen, $\lambda : M \to N$ is C^k.

By the symmetry of the arguments, λ^{-1} is a C^{ℓ}-map and, by the Inverse Mapping Theorem, even a $\max(k, \ell)$-times continuously differentiable diffeomorphism. Since the composition with λ turns $C^{\ell}(N, \mathbb{K})$-maps into $C^k(M, \mathbb{K})$-maps and λ^{-1} turns $C^k(M, \mathbb{K})$-maps into $C^{\ell}(N, \mathbb{K})$-maps, we have $C^k(M, \mathbb{K}) = C^{\ell}(M, \mathbb{K})$ and $C^k(N, \mathbb{K}) = C^{\ell}(N, \mathbb{K})$, thus $k = \ell$, completing the proof. $\square$

The following lemma will be used in the proof of Theorem 2.3.57.

Lemma 2.3.56.

1. *Let V be a vector space of finite dimension, (U, ξ) a chart of M and $T : U \to \mathrm{End}(V)$, $f : U \to V$ smooth functions. Then, for any multi-index $\alpha \in \mathbb{N}_0^m$, we have:*

$$\partial_\xi^\alpha (T \cdot f) = \sum_{\gamma \leq \alpha} \binom{\alpha}{\gamma} \left(\partial_\xi^\gamma T \cdot \partial_\xi^{\alpha - \gamma} f \right). \qquad (2.22)$$

2. *We have, for $\alpha \in \mathbb{N}_0^m$:*

$$\sum_{\gamma \leq \alpha} \frac{(-1)^{|\alpha - \gamma|}}{\gamma! (\alpha - \gamma)!} = \begin{cases} 1 & \text{if } \alpha = 0 \\ 0 & \text{otherwise} \end{cases} = \frac{\delta_{\alpha, 0}}{\alpha!}. \qquad (2.23)$$

Proof.

1. The proof is done by mathematical induction over $|\alpha|$. The claim is trivially true for $\alpha = 0$. For $|\alpha| = 1$ there exists a canonical basis vector e_i of $\mathbb{R}^m$ such that $\alpha = e_i$ and $\partial_\xi^\alpha = \partial_{\xi_i}$. Then

$$\partial_{\xi_i}(T \cdot f) = \partial_{\xi_i} T \cdot f + T \cdot \partial_{\xi_i} f$$

and the claim is also true. Now consider the claim shown for all multi-indices $\beta \in \mathbb{N}_0^m$ with $|\beta| = n > 0$ and fix $\alpha \in \mathbb{N}_0^m$ with $|\alpha| = n + 1$. There exists a canonical basis vector e_i of $\mathbb{R}^m$ such that $\beta := \alpha - e_i \in \mathbb{N}_0^m$ and $|\beta| = n$. By induction hypothesis, we have

$$\partial_\xi^\beta (T \cdot f) = \sum_{\gamma \leq \beta} \binom{\beta}{\gamma} \partial_\xi^\gamma T \cdot \partial_\xi^{\beta - \gamma} f,$$

thus

$$\partial_\xi^\alpha(T \cdot f) = \sum_{\gamma \leq \beta} \binom{\beta}{\gamma} \partial_{\xi_i} \left(\partial_\xi^\gamma T \cdot \partial_\xi^{\beta-\gamma} f \right) = \sum_{\gamma \leq \beta} \binom{\beta}{\gamma} \left(\partial_\xi^{\gamma+e_i} T \cdot \partial_\xi^{\beta-\gamma} f + \partial_\xi^\gamma T \cdot \partial_\xi^{\beta-\gamma+e_i} f \right)$$

$$= \sum_{\gamma \leq \beta} \binom{\beta}{\gamma} \partial_\xi^{\gamma+e_i} T \cdot \partial_\xi^{\beta-\gamma} f + \sum_{\gamma \leq \beta} \binom{\beta}{\gamma} \partial_\xi^\gamma T \cdot \partial_\xi^{\beta-\gamma+e_i} f$$

$$= \sum_{e_i \leq \delta \leq \alpha} \binom{\alpha-e_i}{\delta-e_i} \partial_\xi^\delta T \cdot \partial_\xi^{\alpha-\delta} f + \sum_{\gamma \leq \alpha-e_i} \binom{\alpha-e_i}{\gamma} \partial_\xi^\gamma T \cdot \partial_\xi^{\alpha-\gamma} f$$

$$= 1 \cdot \partial_\xi^0 T \cdot \partial_\xi^\alpha f + \sum_{e_i \leq \delta \leq \alpha-e_i} \left(\binom{\alpha-e_i}{\delta-e_i} + \binom{\alpha-e_i}{\delta} \right) \partial_\xi^\delta T \cdot \partial_\xi^{\alpha-\delta} f + 1 \cdot \partial_\xi^\alpha T \cdot \partial_\xi^0 f$$

$$= \sum_{\delta \leq \alpha} \binom{\alpha}{\delta} \left(\partial_\xi^\delta T \cdot \partial_\xi^{\alpha-\delta} f \right).$$

So the claim is true for all multi-indices $\alpha \in \mathbb{N}_0^m$.

2. For a given multi-index $\alpha \in \mathbb{N}_0^m$ we define a mapping $F_\alpha : \mathbb{R}^m \to \mathbb{R}$, $x \mapsto \prod_{i=1}^m x_i^{\alpha_i}$. By the Taylor Formula and the fact that $\partial^\gamma F = \mathbf{0}$ for each $\gamma \in \mathbb{N}_0^m$ with $\gamma_i > \alpha_i$ for an index $i \in \{1, \ldots, m\}$, we obtain:

$$F_\alpha(x-y) = \sum_{\gamma \leq \alpha} \frac{\prod_{i=1}^m (-y_i)^{\alpha_i}}{\gamma!} \cdot \partial^\gamma F_\alpha(x) + 0$$

$$= \sum_{\gamma \leq \alpha} \frac{1}{\gamma!} \cdot \left(\prod_{i=1}^m (-y_i)^{\alpha_i} \right) \cdot \left(\frac{\alpha!}{(\alpha-\gamma)!} \prod_{j=1}^m x_j^{\alpha_j - \gamma_j} \right).$$

This yields, by setting $x = y = (1, \ldots, 1) \in \mathbb{R}^m$:

$$\frac{\delta_{\alpha,0}}{\alpha!} = \frac{F_\alpha(0)}{\alpha!} = \sum_{\gamma \leq \alpha} \frac{1}{\gamma!} \cdot \left(\prod_{i=1}^m (-1)^{\alpha_i} \right) \cdot \left(\frac{1}{(\alpha-\gamma)!} \prod_{j=1}^m 1^{\alpha_j-\gamma_j} \right) = \sum_{\gamma \leq \alpha} \frac{(-1)^{|\alpha-\gamma|}}{\gamma!(\alpha-\gamma)!}.$$

$\square$

Theorem 2.3.57. *Let $\pi : \mathfrak{K} \to M$ and $\varpi : \mathfrak{G} \to N$ be two Lie algebra bundles with perfect or centerfree fibers $\mathfrak{k}$ and $\mathfrak{g}$, respectively, $\dim \mathfrak{k} = d$, $\dim \mathfrak{g} = e$ and $S(\mathfrak{k}) = \mathbb{K} \cdot \mathbf{1}$, $S(\mathfrak{g}) = \mathbb{K} \cdot \mathbf{1}$. Suppose there exists an isomorphism of Lie algebras $\mu : \Gamma^k(\mathfrak{K}) \to \Gamma^\ell(\mathfrak{G})$ for some $k, \ell \in \mathbb{N} \cup \{\infty\}$. Then $k = \ell$ and*

(a) if $k \in \mathbb{N}$, then μ is induced by a C^k-isomorphism of vector bundles $\kappa : \mathfrak{K} \to \mathfrak{G}$, i.e., if $\kappa' : M \to N$ is the map with $\kappa' \circ \pi = \varpi \circ \kappa$, then $\mu(X) = \kappa \circ X \circ (\kappa')^{-1}$ for all $X \in \Gamma^k(\mathfrak{K})$. In particular, the manifolds M, N are C^k-diffeomorphic and the Lie algebras $\mathfrak{k}, \mathfrak{g}$ are isomorphic.

(b) if $k = \infty$, then the manifolds M, N are smoothly diffeomorphic and the Lie algebras $\mathfrak{k}, \mathfrak{g}$ are isomorphic. After identifying the manifolds and the Lie algebras, μ turns into a linear

differential operator of order at most $d-1$ taking the following local form on a bundle chart (U, φ) of $\pi : \mathfrak{K} \to M$ with corresponding chart (U, ξ) of M:

$$A^\varphi \xmapsto{\mu^\varphi} \sum_{|\alpha| < d} \frac{1}{\alpha!} N^\alpha \cdot \left(\mu_0 \cdot \partial_\xi^\alpha A^\varphi \right), \tag{2.24}$$

where $\mu_0 : U \to \mathrm{Aut}(\mathfrak{g})$ and $N_1, \ldots, N_m \in \mathrm{C}^\infty(U, \mathrm{N}(\mathfrak{g}))$ are smooth functions. Again, we use the local description of differential operators (cf. Definition 2.3.12).

Proof. The isomorphism of Lie algebras $\mu : \Gamma^k(\mathfrak{K}) \to \Gamma^\ell(\mathfrak{G})$ induces an isomorphism of associative algebras $\widetilde{\mu} : \mathrm{Cent}\left(\Gamma^k(\mathfrak{K}) \right) \to \mathrm{Cent}\left(\Gamma^\ell(\mathfrak{G}) \right)$ by $\widetilde{\mu}(T) := \mu \circ T \circ \mu^{-1}$ for $T \in \mathrm{Cent}\left(\Gamma^k(\mathfrak{K}) \right)$. This identifies, by Theorem 2.3.38, with an isomorphism $\widetilde{\mu}_0 : \Gamma^k(\mathrm{Cent}(\mathfrak{K})) \to \Gamma^\ell(\mathrm{Cent}(\mathfrak{G}))$. We want to show that $\widetilde{\mu}_0$ induces an isomorphism of associative algebras $v : \mathrm{C}^k(M, \mathbb{K}) \to \mathrm{C}^\ell(N, \mathbb{K})$ by

$$\widetilde{\mu}_0(f \cdot \mathbf{1}) =: v(f) \cdot \mathbf{1} + N_f, \tag{2.25}$$

where $N_f \in \Gamma^\ell(\mathrm{N}(\mathfrak{G}))$ is nilpotent and this decomposition is unique by Theorem 2.3.39.2.

For constant functions $f = c \cdot \mathbf{1}$ we have $\widetilde{\mu}_0(f \cdot \mathbf{1}) = c \cdot \widetilde{\mu}_0(\mathbf{1}) = c \cdot \mathbf{1} = f \cdot \mathbf{1}$. By the uniqueness of the decomposition (2.25), we obtain $N_f = 0$ for constant functions f. The morphism property of v is shown by the following calculations:

$$\widetilde{\mu}_0(f \cdot \mathbf{1}) \cdot \widetilde{\mu}_0(g \cdot \mathbf{1}) = \widetilde{\mu}_0((f \cdot \mathbf{1}) \cdot (g \cdot \mathbf{1})) = \widetilde{\mu}_0(fg \cdot \mathbf{1}) = v(fg) \cdot \mathbf{1} + N_{fg}$$

and

$$\widetilde{\mu}_0(f \cdot \mathbf{1}) \cdot \widetilde{\mu}_0(g \cdot \mathbf{1}) = (v(f) \cdot \mathbf{1} + N_f) \cdot (v(g) \cdot \mathbf{1} + N_g)$$
$$= v(f)v(g) \cdot \mathbf{1} + \underbrace{v(f) \cdot N_g + v(g) \cdot N_f + N_f N_g}_{\text{nilpotent}}.$$

Furthermore, the Binomial Theorem yields, for $f \in \mathrm{C}^k(M, \mathbb{K})$ and $r \in \mathbb{N}$:

$$N_{f^r} = \sum_{t=1}^r \binom{r}{t} v(f)^{r-t} (N_f)^t.$$

Note that v is bijective because $\widetilde{\mu}_0$ on $\Gamma^k(\mathrm{S}(\mathfrak{K}))$ and $\widetilde{\mu}_0^{-1}$ on $\Gamma^\ell(\mathrm{S}(\mathfrak{G}))$ are injective. By applying Proposition 2.3.55 to $v : \mathrm{C}^k(M, \mathbb{K}) \to \mathrm{C}^\ell(N, \mathbb{K})$, we obtain $k = \ell$ and the existence of a C^k-diffeomorphism $\lambda : M \to N$ such that $v(f) = f \circ \lambda^{-1}$ for all $f \in \mathrm{C}^k(M, \mathbb{K})$.

We now identifiy the manifolds M, N via λ and the associative algebras $\mathrm{C}^k(M, \mathbb{K}), \mathrm{C}^k(N, \mathbb{K})$ via v, so that we have the isomorphisms

$$\mu : \Gamma^k(\mathfrak{K}) \longrightarrow \Gamma^k(\mathfrak{G}),$$

$$\widetilde{\mu} : \Gamma^k(\mathrm{Cent}(\mathfrak{K})) = \mathrm{Cent}\left(\Gamma^k(\mathfrak{K}) \right) \longrightarrow \Gamma^k(\mathrm{Cent}(\mathfrak{G})) = \mathrm{Cent}\left(\Gamma^k(\mathfrak{G}) \right).$$

For all sections $A \in \Gamma^k(\mathfrak{K})$, all points $x \in M$, all mappings $f \in \mathrm{C}^k(M, \mathbb{K})$ with $f(x) = 0$ and $\dim(\mathfrak{g}) = e$, we calculate:

$$(\mu(f^e A))_x = (\mu((f^e \cdot \mathbf{1})(A)))_x = ((\widetilde{\mu}(f^e \cdot \mathbf{1}))(\mu(A)))_x = ((f^e \cdot \mathbf{1} + N_{f^e})(\mu(A)))_x$$

$$= (f^e \mu(A))_x + (N_{f^e}(\mu(A)))_x = f(x)^e (\mu(A))_x + \sum_{t=1}^e \binom{e}{t} f(x)^{e-t} (N_f)_x^t (\mu(A))_x$$

$$= 0 + 0 = 0. \tag{2.26}$$

Suppose a section $A \in \Gamma^k(\mathfrak{K})$ is zero on an open set $U \subseteq M$, $x \in U$ and let $\rho : M \to [0,1]$ be a smooth function and $W \subseteq U$ a smaller x-neighbourhood such that $\rho_{|M \setminus U} = \mathbf{0}$ and $\rho_{|W} = \mathbf{1}$. Then $A = (1 - \rho)^e A$ and (2.26) shows $(\mu(A))_x = 0$. Since x was arbitrarily chosen, $\mu(A)$ is also zero on U. So $\mu : \Gamma^k(\mathfrak{K}) \to \Gamma^k(\mathfrak{G})$ is local and, by the Peetre Theorems 2.3.21 and 2.3.23, a differential operator.

If $k \in \mathbb{N}$, then Theorem 2.3.23 even implies that μ is a differential operator of order 0 inducing a bundle isomorphism $\kappa : \mathfrak{K} \to \mathfrak{G}$ (cf. Definition 2.2.7 and Remark 2.3.24). This proves (a).

Now assume $k = \infty$. Let $A \in \Gamma(\mathfrak{K})$ be a section with $j_x^{e-1}(A) = 0$. By Lemma 2.3.28, the section A locally takes the form

$$A = \sum_{i=1}^{r} f_i^e A_i$$

for functions $f_i \in C^\infty(M, \mathbb{K})$, $f_i(x) = 0$ and $A_i \in \Gamma(\mathfrak{K})$. By using relation (2.26), we see that

$$(\mu(A))_x = \sum_{i=1}^{r} [\mu(f_i^e A_i)]_x = \sum_{i=1}^{r} 0 = 0.$$

This proves that the order of the differential operator μ is at most $e - 1$ because $j_x^{e-1}(A) = 0$ already implies $(\mu(A))_x = 0$. Let (U, φ) be a bundle chart of $\pi : \mathfrak{K} \to M$ with corresponding chart (U, ξ) of M such that $\xi(U)$ is convex. We have the local forms $\mu^\varphi : C^k(U, \mathfrak{k}) \to C^k(U, \mathfrak{g})$, $\widetilde{\mu}^\varphi : C^k(U, \mathrm{Cent}(\mathfrak{k})) \to C^k(U, \mathrm{Cent}(\mathfrak{g}))$ and $N_f^\varphi \in C^k(U, \mathrm{N}(\mathfrak{g}))$ of μ, $\widetilde{\mu}$ and $N_f \in \Gamma^k(\mathrm{N}(\mathfrak{G}))$, respectively. Locally, the decomposition (2.25), turns into:

$$\widetilde{\mu}^\varphi(f \cdot \mathbf{1}) = f \cdot \mathbf{1} + N_f^\varphi.$$

The Taylor Formula yields, for each $x \in U$ and $A \in \Gamma(\mathfrak{K})$:

$$j_x^{e-1}\left(y \longmapsto A^\varphi(y) - \sum_{|\alpha| < e} \frac{\prod_{i=1}^{m}(\xi_i(y) - \xi_i(x))^{\alpha_i}}{\alpha!} \cdot \partial_\xi^\alpha A^\varphi(x) \right) = 0. \tag{2.27}$$

For $x, y \in U$, $i \in \{1, \ldots, m\}$ and $\alpha \in \mathbb{N}^m$ with $|\alpha| < e$ we write:

$$\xi_{i,x}(y) := \xi_i(y) - \xi_i(x),$$

$$\Xi_{\alpha,x}(y) := \prod_{i=1}^{m}(\xi_i(y) - \xi_i(x))^{\alpha_i},$$

$$\phi_{\alpha,x}(y) := \prod_{i=1}^{m}(\xi_i(y) - \xi_i(x))^{\alpha_i} \cdot \partial_\xi^\alpha A^\varphi(x).$$

Then we define smooth mappings $N_1, \ldots, N_m : U \to \mathrm{N}(\mathfrak{g})$ and $\mu_0 : U \to \mathrm{Hom}(\mathfrak{k}, \mathfrak{g})$ (in the sense of Lie algebra morphisms), by setting for $x \in U$, $u \in \mathfrak{k}$ and the constant map $c_u : U \to \mathfrak{g}$, $y \mapsto u$:

$$\mu_0(x)(u) := \mu^\varphi(c_u)(x)$$

and

$$N_i(x) := N^\varphi_{\xi_{i,x}}(x) = \widetilde{\mu}^\varphi(\xi_{i,x} \cdot \mathbf{1})(x) - (\xi_{i,x} \cdot \mathbf{1})(x)$$
$$= \widetilde{\mu}^\varphi(\xi_{i,x} \cdot \mathbf{1})(x)$$

We calculate:

$$\mu^\varphi(\phi_{\alpha,x}) = \mu^\varphi\left(y \mapsto \left(\prod_{i=1}^{m}(\xi_i(y) - \xi_i(x))^{\alpha_i} \cdot \mathbf{1}\right)(\partial_\xi^\alpha A^\varphi(x))\right)$$

$$= \widetilde{\mu}^\varphi(\Xi_{\alpha,x} \cdot \mathbf{1}) \cdot \mu^\varphi\left(y \mapsto \partial_\xi^\alpha A^\varphi(x)\right)$$

$$= \left(\prod_{i=1}^{m}\underbrace{\widetilde{\mu}^\varphi\left(\xi_{i,x}^{\alpha_i} \cdot \mathbf{1}\right)}_{=N_i^{\alpha_i}}\right) \cdot (\mu_0 \cdot \partial_\xi^\alpha A^\varphi)$$

$$= \left(\prod_{i=1}^{m} N_i^{\alpha_i}\right) \cdot (\mu_0 \cdot \partial_\xi^\alpha A^\varphi) = N^\alpha \cdot (\mu_0 \cdot \partial_\xi^\alpha A^\varphi).$$

Thus, by (2.27) and the fact that μ is of order at most $e - 1$, we have the local form

$$A^\varphi \overset{\mu^\varphi}{\longmapsto} \sum_{|\alpha|<e} \frac{1}{\alpha!} N^\alpha \cdot (\mu_0 \cdot \partial_\xi^\alpha A^\varphi). \tag{2.28}$$

It remains to show that each $\mu_0(x)$, where $x \in U$, is bijective. This will be done in two steps. First, we verify the following identitiy for $A \in \Gamma(\mathfrak{K})$ and $B \in \Gamma(\mathfrak{G})$ with $\mu(A) = B$ (implying $\mu^\varphi(A^\varphi) = B^\varphi$):

$$P(A) := \sum_{|\alpha|<e} \frac{(-1)^{|\alpha|}}{\alpha!} \partial_\xi^\alpha(N^\alpha \cdot (\mu_0 \cdot A^\varphi)) = \sum_{|\alpha|<e} \frac{(-1)^{|\alpha|}}{\alpha!} \partial_\xi^\alpha(N^\alpha \cdot B^\varphi) =: Q(B). \tag{2.29}$$

Note:

$$\text{If } S_1, \ldots, S_m \in \mathrm{N}(\mathfrak{g}) \text{ and } |\alpha| \geq e, \text{ then } S^\alpha = S_1^{\alpha_1} \circ \ldots \circ S_m^{\alpha_m} = 0 \tag{2.30}$$

because S^α can be written as a sum of $|\alpha|$-th powers of linear combinations of the S_i and the e-th power of a nilpotent morphism cotained in $\mathrm{N}(\mathfrak{g})$ is zero. Therefore we can perform the following calculations:

$$\sum_{|\alpha|<e} \frac{(-1)^{|\alpha|}}{\alpha!} \partial_\xi^\alpha(N^\alpha \cdot B^\varphi) \overset{(2.28)}{=} \sum_{|\alpha|<e} \frac{(-1)^{|\alpha|}}{\alpha!} \partial_\xi^\alpha\left(N^\alpha \cdot \sum_{|\beta|<e} \frac{1}{\beta!} N^\beta \cdot \left(\mu_0 \cdot \partial_\xi^\beta A^\varphi\right)\right)$$

$$= \sum_{\substack{|\alpha|<e \\ |\beta|<e}} \frac{(-1)^{|\alpha|}}{\alpha!\beta!} \partial_\xi^\alpha\left(N^{\alpha+\beta} \cdot \left(\mu_0 \cdot \partial_\xi^\beta A^\varphi\right)\right)$$

$$\overset{(2.30)}{=} \sum_{|\alpha+\beta|<e} \frac{(-1)^{|\alpha|}}{\alpha!\beta!} \partial_\xi^\alpha\left(\left(N^{\alpha+\beta} \circ \mu_0\right) \cdot \partial_\xi^\beta A^\varphi\right)$$

$$\overset{(2.22)}{=} \sum_{\substack{|\alpha+\beta|<e \\ \gamma \leq \alpha}} \frac{(-1)^{|\alpha|}}{\beta!\gamma!(\alpha-\gamma)!} \partial_\xi^\gamma\left(N^{\alpha+\beta} \circ \mu_0\right) \cdot \partial_\xi^{\alpha+\beta-\gamma} A^\varphi.$$

Now we perform the following substitutions: $\alpha' := \alpha + \beta - \gamma$, $\beta' := \gamma$ and $\gamma' := \beta$, thus $\alpha + \beta = \alpha' + \beta'$, $\alpha - \gamma = \alpha' - \gamma'$ and $\gamma \leq \alpha \iff \gamma' \leq \alpha'$. And so we can finally show (2.29):

$$\sum_{|\alpha|<e} \frac{(-1)^{|\alpha|}}{\alpha!} \partial_\xi^\alpha \left(N^\alpha \cdot B^\varphi\right) = \sum_{\substack{|\alpha+\beta|<e \\ \gamma \leq \alpha}} \frac{(-1)^{|\alpha|}}{\beta!\gamma!(\alpha-\gamma)!} \partial_\xi^\gamma \left(N^{\alpha+\beta} \circ \mu_0\right) \cdot \partial_\xi^{\alpha+\beta-\gamma} A^\varphi$$

$$= \sum_{\substack{|\alpha'+\beta'|<e \\ \gamma' \leq \alpha'}} \frac{(-1)^{|\alpha'-\gamma'|} \cdot (-1)^{\beta'}}{\gamma'!\beta'!(\alpha'-\gamma')!} \partial_\xi^{\beta'} \left(N^{\alpha'+\beta'} \circ \mu_0\right) \cdot \partial_\xi^{\alpha'} A^\varphi$$

$$= \sum_{|\alpha'+\beta'|<e} \left(\sum_{\gamma' \leq \alpha'} \frac{(-1)^{|\alpha'-\gamma'|}}{\gamma'!(\alpha'-\gamma')!}\right) \cdot \frac{(-1)^{\beta'}}{\beta'!} \partial_\xi^{\beta'} \left(N^{\alpha'+\beta'} \circ \mu_0\right) \cdot \partial_\xi^{\alpha'} A^\varphi$$

$$\overset{(2.23)}{=} \sum_{|\beta'|<e} \frac{(-1)^{|\beta'|}}{\beta'!} \partial_\xi^{\beta'} \left(N^{\beta'} \cdot (\mu_0 \cdot A^\varphi)\right).$$

By the definition of P and μ_0, if $A^\varphi(x) = u = A'^\varphi(x)$ for $A, A' \in \Gamma(\mathfrak{K})$ and $x \in U$, then:

$$P(A)(x) = \sum_{|\alpha|<e} \frac{(-1)^{|\alpha|}}{\alpha!} \partial_\xi^\alpha \left(N^\alpha(x) \left(\mu_0(x) \left(A^\varphi(x)\right)\right)\right)$$

$$= \sum_{|\alpha|<e} \frac{(-1)^{|\alpha|}}{\alpha!} \partial_\xi^\alpha \left(N^\alpha(x) \left(\mu^\varphi(c_u)(x)\right)\right) = P(A')(x).$$

So $P(A)(x) \in \mathfrak{g}$ depends on A^φ only via $A^\varphi(x) \in \mathfrak{k}$ and we can define a linear map $P_x : \mathfrak{k} \to \mathfrak{g}$ by setting $P_x(A^\varphi(x)) := P(A)(x)$ for $x \in U$. We may also define linear maps $Q_x : \mathfrak{g} \to \mathfrak{g}$ for $x \in U$ by $Q_x(B^\varphi(x)) := Q(B)(x)$ for $x \in U$ due to an analogous element as above with the N^α instead of μ_0. Since any Q_x is a sum of $\mathbf{1}$ and a nilpotent linear map (see the sum in the third term of (2.29) evaluated in x), it is bijective. We define a smooth mapping $\eta_0 : U \to \mathrm{Hom}(\mathfrak{g}, \mathfrak{k})$ (in the sense of Lie algebra morphisms) by setting for $x \in U$, $v \in \mathfrak{g}$ and the constant map $c_v : U \to \mathfrak{g}$, $x' \to v$:

$$\eta_0(x)(v) := \left(\mu^{-1}\right)^\varphi (c_v)(x).$$

We now fix $x \in U$ and $v \in \mathfrak{g}$. Since μ is surjective, there exists $A \in \Gamma(\mathfrak{K})$ such that $\mu^\varphi(A^\varphi) = c_v$, thus $A^\varphi(x) = \eta_0(x)(v)$. By (2.29), we have $P_x(\eta_0(x)(v)) = Q_x(v)$. The injectivity of Q_x implies the injectivity of $\eta_0(x) : \mathfrak{g} \to \mathfrak{k}$, yielding $\dim(\mathfrak{g}) = e \leq d = \dim(\mathfrak{k})$. By the symmetry of the arguments, $\mu_0(x) : \mathfrak{k} \to \mathfrak{g}$ is also injective, yielding $d \leq e$. So $\mu_0(x)$ is an isomorphism of Lie algebras. $\square$

The following corollary generalizes Theorem 2.10 of [HG09] by Groß and Heintze, stating that any isomorphism of loop algebras is "standard".

Corollary 2.3.58. *Consider $\mathfrak{k}$ and $\mathfrak{g}$ central simple, thus $\mathrm{Cent}(\mathfrak{k}) = \mathrm{S}(\mathfrak{k}) = \mathbb{K} \cdot \mathbf{1}$ and $\mathrm{S}(\mathfrak{g}) = \mathbb{K} \cdot \mathbf{1}$. Then for any $\mu \in \mathrm{Iso}(\Gamma(\mathfrak{K}), \Gamma(\mathfrak{G}))$ there is a C^∞-diffeomorphism $\lambda : M \to N$ such that μ can be identified with some $\mu_0 \in \Gamma(\mathrm{Iso}_\lambda(\mathfrak{K}, \mathfrak{G}))$, i.e., for all $x \in M$ the map $\mu_0(x) : \mathfrak{K}_x \to \mathfrak{G}_{\lambda(x)}$ is an isomorphism of Lie algebras with $\mu_0(x)(X_x) = \mu(X)_{\lambda(x)}$ for all $X \in \Gamma(\mathfrak{K})$.*

Corollary 2.3.59. *Consider $M = N$ and $\mathfrak{k} = \mathfrak{g}$ central simple, thus $\mathrm{Cent}(\mathfrak{g}) = \mathrm{S}(\mathfrak{g}) = \mathbb{K} \cdot \mathbf{1}$. Then for any $\mu \in \mathrm{Iso}\left(\Gamma^k(\mathfrak{K}), \Gamma^k(\mathfrak{G})\right)$ there is a C^k-diffeomorphism $\Lambda(\mu) := \lambda : M \to M$ such that μ can be identified with some $\mu_0 \in \Gamma^k\left(\mathrm{Iso}_\lambda(\mathfrak{K}, \mathfrak{G})\right)$, i.e., the map $\mu_0(x) : \mathfrak{K}_x \to \mathfrak{G}_{\lambda(x)}$ is an isomorphism of Lie algebras with $\mu_0(x)(X_x) = \mu(X)_{\lambda(x)}$ for all $x \in M$, $X \in \Gamma^k(\mathfrak{K})$. There is an isomorphism of Lie algebra bundles as follows:*

$$\mathfrak{K} \longrightarrow \mathfrak{G}$$
$$v \longmapsto \mu_0(\pi(v))(v).$$

In particular, $\Gamma^k(\mathfrak{K})$ and $\Gamma^k(\mathfrak{G})$ are isomorphic Lie algebras if and only if the bundles $\pi : \mathfrak{K} \to M$ and $\varpi : \mathfrak{G} \to M$ are isomorphic (cf. Proposition 2.3.3).

Corollary 2.3.60. *Consider $\mathfrak{k}$ central simple. Then for any $\mu \in \mathrm{Aut}\left(\Gamma^k(\mathfrak{K})\right)$ there is a C^k-diffeomorphism $\lambda : M \to M$ such that μ can be identified with some $\mu_0 \in \Gamma^k\left(\mathrm{Aut}_\lambda(\mathfrak{K})\right)$. The bundle $\mathrm{Aut}_\lambda(\mathfrak{K})$ is equivalent to $\mathrm{Aut}(\mathfrak{K})$ by $\left(f : \mathfrak{K}_x \to \mathfrak{K}_{\lambda(x)}\right) \longmapsto \left(\mu_0(x)^{-1} \circ f : \mathfrak{K}_x \to \mathfrak{K}_x\right)$.*

The map Λ from Corollary 2.3.59 provides us with an interesting short exact sequence.

Proposition 2.3.61. *If $\pi : \mathfrak{K} \to M$ is a Lie algebra bundle with central simple fiber $\mathfrak{k}$, then there is the following sequence of groups:*

$$1 \longrightarrow \Gamma^k\left(\mathrm{Aut}(\mathfrak{K})\right) \overset{\Psi}{\cong} \mathrm{Aut}_{C^k(M,\mathbb{K})}\left(\Gamma^k(\mathfrak{K})\right) \hookrightarrow \mathrm{Aut}\left(\Gamma^k(\mathfrak{K})\right) \overset{\Lambda}{\longrightarrow} \mathrm{Diff}^k(M)_{\mathfrak{K}} \longrightarrow 1. \tag{2.31}$$

Here, $\mathrm{Diff}^k(M)_{\mathfrak{K}}$ denotes the set of C^k-diffeomorphisms λ of M that may come from an automorphism of $\Gamma^k(\mathfrak{K})$. Note that $\mathrm{Aut}_{C^k(M,\mathbb{K})}\left(\Gamma^k(\mathfrak{K})\right) = \mathrm{Aut}_{\mathrm{Cent}\left(\Gamma^k(\mathfrak{K})\right)}\left(\Gamma^k(\mathfrak{K})\right)$ by Theorem 2.3.39.

Proof. The map $\Lambda : \mathrm{Aut}\left(\Gamma^k(\mathfrak{K})\right) \to \mathrm{Diff}^k(M)_{\mathfrak{K}}$ is defined in Corollary 2.3.59 and is, by definition, surjective. It is also a group morphism, since for $\mu, \nu \in \mathrm{Aut}\left(\Gamma^k(\mathfrak{K})\right)$, $x \in M$, $X \in \Gamma^k(\mathfrak{K})$ we have:

$$(\mu \circ \nu)(X)_{\Lambda(\mu \circ \nu)(x)} = (\mu \circ \nu)_0(x)(X_x) = \mu_0(x)(\nu_0(x)(X_x)) = \mu_0(x)(\nu(X)_{\Lambda(\nu)(x)})$$
$$= \mu(\nu(X))_{\Lambda(\mu)(\Lambda(\nu)(x))}.$$

The kernel of Λ is, by Corollary 2.3.59, exactly the subgroup $S \leq \mathrm{Aut}\left(\Gamma^k(\mathfrak{K})\right)$ which is the preimage of $\Gamma^k\left(\mathrm{Aut}(\mathfrak{K})\right)$ with respect to $\mu \mapsto \mu_0$. We will show that $S = \mathrm{Aut}_{C^k(M,\mathbb{K})}\left(\Gamma^k(\mathfrak{K})\right)$ and that there is a group isomorphism $\Psi : \Gamma^k\left(\mathrm{Aut}(\mathfrak{K})\right) \to \mathrm{Aut}_{C^k(M,\mathbb{K})}\left(\Gamma^k(\mathfrak{K})\right)$ with $\Psi(\mu_0) = \mu$ for all $\mu \in \mathrm{Aut}_{C^k(M,\mathbb{K})}\left(\Gamma^k(\mathfrak{K})\right)$.

If $\chi \in \Gamma^k(\mathrm{Aut}(\mathfrak{K}))$, then $\mu(X)_x := \chi(x)(X_x)$ defines an element $\mu \in \mathrm{Aut}_{C^k(M,\mathbb{K})}\left(\Gamma^k(\mathfrak{K})\right)$ because $\mu(fX)_x = \chi(x)(f(x)X_x) = f(x)\mu(X)_x$ for $x \in M$, $X \in \Gamma^k(\mathfrak{K})$, $f \in C^k(M, \mathbb{K})$, and, obviously, $\mu_0 = \chi$. Starting with $\mu \in \mathrm{Aut}_{C^k(M,\mathbb{K})}\left(\Gamma^k(\mathfrak{K})\right)$, we have a section $\mu_0 \in \Gamma^k(\mathrm{Aut}_\lambda(\mathfrak{K}))$ for some diffeomorphism $\Lambda(\mu) = \lambda : M \to M$. If λ was not the identity on M, there would be an $x \in M$ such that $x \neq \lambda(x) = y \in M$ and also a smooth map $f \in C^k(M, \mathbb{K})$ such that $f(x) = 0$ and $f(y) = 1$, thus

$$\mu_0(x)(X_x) = f(y)\mu_0(x)(X_x) = f(y)\mu(X)_y = \mu(fX)_y = \mu_0(x)((fX)_x) = f(x)\mu_0(x)(X_x) = 0$$

for all $X \in \Gamma^k(\mathfrak{K})$, which is absurd, so $\lambda = \mathrm{id}_M$ and $S = \mathrm{Aut}_{C^k(M,\mathbb{K})}\left(\Gamma^k(\mathfrak{K})\right)$.

Now we define $\Psi : \Gamma^k\left(\mathrm{Aut}(\mathfrak{K})\right) \to \mathrm{Aut}_{\mathrm{C}^k(M,\mathbb{K})}\left(\Gamma^k(\mathfrak{K})\right)$ by $\Psi(\chi)(X)_x := \chi(x)(X_x)$, which is a group morphism, as the following calculation shows:

$$\Psi(\chi \cdot \eta)(X)_x = (\chi \cdot \eta)(x)(X_x) = \chi(x)\left(\eta(x)(X_x)\right) = \chi(x)\left(\Psi(\eta)(X)_x\right) = \Psi(\chi)(\Psi(\eta)(X))_x.$$

Obviously, $\Psi(\mu_0) = \mu$ for all $\mu \in \mathrm{Aut}_{\mathrm{C}^k(M,\mathbb{K})}\left(\Gamma^k(\mathfrak{K})\right)$ and Ψ is surjective. If χ is in the kernel of Ψ, then, for all $x \in M$, $X \in \Gamma^k(\mathfrak{K})$, we have $\chi(x)(X_x) = \Psi(\chi)(X)_x = X_x$, thus $\chi(x) = \mathrm{id}_{\mathfrak{K}_x}$ and $\chi = \mathbf{1}$. So Ψ is a group isomorphism. $\qquad\square$

Corollary 2.3.62. *The Lie algebra of C^k-sections of the trivial bundle $\mathfrak{K} = M \times \mathfrak{k}$ is naturally isomorphic to $\mathrm{C}^k(M,\mathfrak{k})$, and $\Gamma^k\left(\mathrm{Aut}(M \times \mathfrak{k})\right)$ is naturally isomorphic to $\mathrm{C}^k(M, \mathrm{Aut}(\mathfrak{k}))$, and $\mathrm{Diff}^k(M)_{\mathfrak{K}} = \mathrm{Diff}^k(M)$ because $\sigma : \mathrm{Diff}^k(M) \to \mathrm{Aut}\left(\mathrm{C}^k(M,\mathfrak{k})\right)$ defined by $\sigma(\lambda)(f) := f \circ \lambda^{-1}$ lifts every diffeomorphism of M to an automorphism of $\Gamma^k(\mathfrak{K})$. Furthermore, σ is a section of the short exact sequence* (2.31), *thus there is an isomorphism*

$$\mathrm{Aut}(\mathrm{C}^k(M,\mathfrak{k})) \cong \mathrm{C}^k(M, \mathrm{Aut}(\mathfrak{k})) \rtimes \mathrm{Diff}^k(M).$$

We define the notion of equivalent Lie algebras of smooth sections analogously to the notion of equivalent bundles.

Definition 2.3.63. Let $\pi : \mathfrak{K} \to M$ and $\varpi : \mathfrak{G} \to M$ be two Lie algebra bundles with central simple fiber. An isomorphism of their section algebras $\mu : \Gamma^k(\mathfrak{K}) \to \Gamma^k(\mathfrak{G})$ is called an *equivalence*, if $\mathrm{id}_M = \Lambda(\mu)$ (cf. Corollary 2.3.59). In this case, the Lie algebras $\Gamma^k(\mathfrak{K})$ and $\Gamma^k(\mathfrak{G})$ are called *equivalent*.

With this notion we can reformulate Corollary 2.3.59.

Corollary 2.3.64. *Let $\pi : \mathfrak{K} \to M$ and $\varpi : \mathfrak{G} \to M$ be two Lie algebra bundles with central simple fiber. Then the Lie algebras $\Gamma^k(\mathfrak{K})$ and $\Gamma^k(\mathfrak{G})$ are equivalent if and only if the bundles $\pi : \mathfrak{K} \to M$ and $\varpi : \mathfrak{G} \to M$ are equivalent.*

Remark 2.3.65. If $\pi : \mathfrak{K} \to M$ and $\varpi : \mathfrak{G} \to M$ are two Lie algebra bundles with central simple fiber, then, by Proposition 2.3.61, an isomorphism $\mu : \Gamma^k(\mathfrak{K}) \to \Gamma^k(\mathfrak{G})$ is an equivalence if and only if it is $\mathrm{C}^k(M, \mathbb{K})$-linear.

3. Classification by Homotopy Theory

In this chapter we will classify Lie algebras of smooth sections by means of homotopy groups. For $n \in \mathbb{N}$ we will use the symbols $\mathcal{C}_n$, $\mathcal{S}_n$ and $\mathcal{D}_n$ for the cyclic group, the symmetric group and the dihedral group, respectively, of n elements. If G is a group, then

$$\mathrm{Conj}(G) := \left\{ \{ghg^{-1} : g \in G\} : h \in G \right\}$$

denotes the set of G-conjugacy classes. If not stated otherwise, the $\mathbb{K}$-Lie algebra $\mathfrak{g}$ is of finite dimension,

3.1. Classification of Bundles of Lie Algebras by Homotopy Groups

In the light of Corollary 2.3.64 it is appropriate to classify equivalence classes of Lie algebra bundles in order to classify equivalence classes of Lie algebras of smooth sections.

Remark 3.1.1.

1. There is a bijection between equivalence classes of $\mathrm{Aut}(\mathfrak{g})$-principal bundles $\rho : P \to M$ and those of Lie algebra bundles $\pi : \mathfrak{K} \to M$ with fiber $\mathfrak{g}$:

$$\mathrm{LABUN}(M, \mathfrak{g}) \cong \mathrm{PBUN}(M, \mathrm{Aut}(\mathfrak{g})).$$

2. The equivalence classes of G-principal bundles over B, where G is a Lie group modeled on a locally convex space and B is a paracompact manifold of finite dimension, in the smooth sense can be naturally mapped to the continuous equivalence classes of continuous G-principal bundles over B. In fact, this map is bijective (cf. Corollary II.13 of [MW09]):

$$\mathrm{PBUN}(B, G) \cong \mathrm{PBUN}_{\mathrm{ct}}(B, G).$$

3. The study of continuous G-principal bundles over paracompact base spaces, where G is a topological group, also leads to the notion of universal bundles: There are, up to homotopy equivalence, unique topological spaces EG, BG and a continuous G-principal bundle $pG : EG \to BG$, such that for any continuous G-principal bundle $p : E \to B$ there is, up to homotopy equivalence, a unique continuous map $f : B \to BG$ (called *classifying map*) such that the bundle $f^*(EG) \to B$ is isomorphic to $E \to B$. In addition, EG is contractible, so all its homotopy groups are trivial and BG is connected (cf. Chapter 4 of [Hus66]). Thus, there is a bijective correspondence between equivalence classes of continuous G-principal bundles $p : E \to B$ and homotopy classes of continuous maps $B \to BG$:

$$\mathrm{PBUN}_{\mathrm{ct}}(B, G) \cong [B; BG]_{\sim}.$$

4. There is a long exact homotopy sequence for locally trivial F-fiber bundles $p : E \to B$ with $b \in B$ and $x \in p^{-1}(b) \cong F$ as follows (cf. Theorem 4.41 and Proposition 4.48 of [Hat02]):

$$\cdots \to \pi_n(F, x) \to \pi_n(E, x) \to \pi_n(B, b) \to \pi_{n-1}(F, x) \to \pi_{n-1}(E, x) \to \cdots \to \pi_0(E, x) \to 1.$$

For the universal $\mathrm{Aut}(\mathfrak{g})$-bundle $p\,\mathrm{Aut}(\mathfrak{g}) : E\,\mathrm{Aut}(\mathfrak{g}) \to B\,\mathrm{Aut}(\mathfrak{g})$ and $n \in \mathbb{N}$ this leads to:

$$\pi_n(B\,\mathrm{Aut}(\mathfrak{g})) \cong \pi_{n-1}(\mathrm{Aut}(\mathfrak{g})).$$

5. It is well-known (cf. Theorem 18.5 of [Ste51]) that, for any topological group G and $n \in \mathbb{N}$, the equivalence classes of continuous G-principal bundles over $\mathbb{S}^n$ are in bijective correspondence with orbits of $\sigma : \pi_0(G) \times \pi_{n-1}(G) \to \pi_{n-1}(G)$, $(dG_0, [f]) \mapsto [d \cdot f \cdot d^{-1}]$:

$$\mathrm{PBUN}_{\mathrm{ct}}(\mathbb{S}^n, G) \cong \pi_{n-1}(G)/^\sigma \pi_0(G),$$

implying, e.g., for $n = 1$:

$$\mathrm{PBUN}_{\mathrm{ct}}(\mathbb{S}^1, G) \cong \mathrm{Conj}(\pi_0(G)).$$

6. If G is a Lie group with only finitely many connected components, then there is a diffeomorphism $G \cong K \times \mathbb{R}^d$ for a maximal compact subgroup $K \leq G$ and some $d \in \mathbb{N}_0$ (cf. Theorem 3.1 of [Hoc65]). Hence, $K \hookrightarrow G$ induces a group isomorphism $\pi_\ell(K) \cong \pi_\ell(G)$ for all $\ell \in \mathbb{N}_0$. Since, e.g., $\mathrm{Aut}(\mathfrak{g})$ is algebraic and algebraic groups only have finitely many connected components by Hassler Whitney's Theorem (cf. Theorem 3.5 of [Whi57]), we obtain $\pi_\ell(K) \cong \pi_\ell(\mathrm{Aut}(\mathfrak{g}))$ for any maximal compact $K \leq \mathrm{Aut}(\mathfrak{g})$.

The above remarks imply the following Classification[1] Theorems 3.1.2 and 3.1.3.

Theorem 3.1.2. *Lie algebra bundles $\pi : \mathfrak{K} \to M$ with fiber $\mathfrak{g}$ are classified by $[M; B\,\mathrm{Aut}(\mathfrak{g})]_\simeq$ and so are Lie algebras $\Gamma(\mathfrak{K})$ of smooth sections over M with fiber $\mathfrak{g}$, if $\mathfrak{g}$ is central simple (cf. Corollary 2.3.59).*

Theorem 3.1.3. *Lie algebra bundles $\pi : \mathfrak{K} \to \mathbb{S}^n$ with fiber $\mathfrak{g}$, for $n \in \mathbb{N}$, are classified by $\pi_{n-1}(\mathrm{Aut}(\mathfrak{g}))/^\sigma \pi_0(\mathrm{Aut}(\mathfrak{g}))$ and so are Lie algebras $\Gamma(\mathfrak{K})$ of smooth sections over $\mathbb{S}^n$ with fiber $\mathfrak{g}$, if $\mathfrak{g}$ is central simple.*

Corollary 3.1.4. *Equivalence classes of Lie algebras $\Gamma(\mathfrak{K})$ of smooth sections over $\mathbb{S}^1$ with central simple $\mathfrak{g}$ are parametrized by $\mathrm{Conj}(\pi_0(\mathrm{Aut}(\mathfrak{g})))$.*

Corollary 3.1.5. *Lie algebras $\Gamma(\mathfrak{K})$ of smooth sections over $\mathbb{S}^3$ with central simple $\mathfrak{g}$ are isomorphic to $\mathrm{C}^\infty(\mathbb{S}^3, \mathfrak{g})$.*

Proof. By Cartan's Theorem (cf. Proposition V.7.5 of [BtD85]) the group $\pi_2(K) \cong \pi_2(\mathrm{Aut}(\mathfrak{g}))$ is trivial for any maximal compact $K \leq \mathrm{Aut}(\mathfrak{g})$. So there is only one equivalence class of Lie algebra bundles $\pi : \mathfrak{K} \to \mathbb{S}^3$ with fiber $\mathfrak{g}$, thus $\Gamma(\mathfrak{K})$ is isomorphic to $\mathrm{C}^\infty(\mathbb{S}^3, \mathfrak{g})$. $\qquad\square$

---

[1] With respect to the notion of equivalent Lie algebras of smooth sections.

3.2. The Component Group of the Automorphism Group of a Simple Lie Algebra

For a simple Lie algebra $\mathfrak{g}$ we calculate $\pi_0(\mathrm{Aut}(\mathfrak{g})) = \mathrm{Aut}(\mathfrak{g})/\mathrm{Aut}(\mathfrak{g})_0 = \mathrm{Aut}(\mathfrak{g})/\mathrm{Inn}(\mathfrak{g})$ and the number of its conjugacy classes $\#\mathrm{Conj}(\pi_0(\mathrm{Aut}(\mathfrak{g})))$ and show that the corresponding short exact sequence $1 \to \mathrm{Aut}(\mathfrak{g})_0 \to \mathrm{Aut}(\mathfrak{g}) \to \pi_0(\mathrm{Aut}(\mathfrak{g})) \to 1$ is split or, equivalently, we have an isomorphism[2] $\mathrm{Aut}(\mathfrak{g}) \cong \mathrm{Aut}(\mathfrak{g})_0 \rtimes \pi_0(\mathrm{Aut}(\mathfrak{g}))$.

The Complex Simple Case

For complex simple Lie algebras, the statement is a classical result (cf., e.g., Theorem 10.6.10 of [Pro07]):

Theorem 3.2.1. *If $\mathfrak{g}$ is a complex simple Lie algebra, then $\pi_0(\mathrm{Aut}(\mathfrak{g}))$ is isomorphic to the symmetry group of $\mathfrak{g}$'s Dynkin diagram and there is an isomorphism $\mathrm{Aut}(\mathfrak{g}) \cong \mathrm{Aut}(\mathfrak{g})_0 \rtimes \pi_0(\mathrm{Aut}(\mathfrak{g}))$. The following table provides all possibilities[3] for $\pi_0(\mathrm{Aut}(\mathfrak{g}))$.*

complex simple $\mathfrak{g}$	$\pi_0(\mathrm{Aut}(\mathfrak{g}))$	$\#\mathrm{Conj}(\pi_0(\mathrm{Aut}(\mathfrak{g})))$
$\mathfrak{sl}(n+1, \mathbb{C})$ for $n \geq 2$	$\mathcal{C}_2$	2
$\mathfrak{so}(8, \mathbb{C})$	$\mathcal{S}_3$	3
$\mathfrak{so}(2n, \mathbb{C})$ for $n \geq 5$	$\mathcal{C}_2$	2
$\mathfrak{e}_6$	$\mathcal{C}_2$	2
all others	1	1

Table 3.1.: Component group $\pi_0(\mathrm{Aut}(\mathfrak{g}))$ for complex simple $\mathfrak{g}$

The Real Simple Case

In the real case, things are more complicated. From now on $\mathfrak{g}$ denotes a real simple Lie algebra.

Remark 3.2.2. It is a well-known fact (cf. Proposition X.1.5 of [Hel78] and Lemma 2.1.29) that each real simple Lie algebra fulfills exactly one of the two following conditions:

A. $\mathfrak{g}$ admits a complex structure J and the complex Lie algebra $\mathfrak{g}^{\mathbb{C}} = \mathfrak{g}^{\mathbb{C}}(J)$ (cf. Definition 2.1.30) is simple.[4] The complexification $\mathfrak{g}_{\mathbb{C}}$ is the direct sum of two simple isomorphic ideals, hence $\mathfrak{g}_{\mathbb{C}}$ is not a complex simple Lie algebra.

B. $\mathfrak{g}_{\mathbb{C}}$ is a complex simple Lie algebra. In the light of Remark 2.1.19, Lemma 2.1.29.2 and Definition 2.1.31, this is equivalent to the condition that the real simple Lie algebra $\mathfrak{g}$ is central simple.

---

[2] Since $\mathrm{Aut}(\mathfrak{g})_0$ is open in $\mathrm{Aut}(\mathfrak{g})$, the quotient group $\pi_0(\mathrm{Aut}(\mathfrak{g}))$ is discrete. Hence a section $\pi_0(\mathrm{Aut}(\mathfrak{g})) \to \mathrm{Aut}(\mathfrak{g})$ would automatically be continuous giving rise to an isomorphism of Lie groups $\mathrm{Aut}(\mathfrak{g}) \cong \mathrm{Aut}(\mathfrak{g})_0 \rtimes \pi_0(\mathrm{Aut}(\mathfrak{g}))$ (cf. Proposition 10.3.1. of [HN11]).

[3] The Lie algebras listed in this and the next tables exhaust all possibilities up to isomorphism.

[4] If $\mathfrak{g}$ is real simple with a complex structure J, then the only other complex structure is $-J$ and the corresponding simple complex Lie algebras $\mathfrak{g}^{\mathbb{C}}(J)$ and $\mathfrak{g}^{\mathbb{C}}(-J)$ are isomorphic.

The Real Central Simple Case

Now let $\mathfrak{g}$ be real central simple. We need the following lemma to determine $\pi_0(\mathrm{Aut}(\mathfrak{g}))$ in two distinct subcases: the compact and the non-compact case.

Lemma 3.2.3. *Let $\mathfrak{g}$ be a real semisimple Lie algebra of finite dimension with Cartan decomposition $\mathfrak{g} = \mathfrak{k} \oplus_\kappa^\tau \mathfrak{p}$ (cf. Theorem 2.1.42) and $\mathfrak{u} = \mathfrak{k} + i\mathfrak{p}$ the corresponding compact real Lie algebra with involution $\sigma = \mathrm{id}_\mathfrak{k} \oplus -\mathrm{id}_{i\mathfrak{p}}$. Then we have the following:* [5]

1. *The map $B_\tau : \mathfrak{g} \times \mathfrak{g} \to \mathbb{R}$, $(x, y) \mapsto -\kappa(x, \tau y)$ is a euclidean scalar product. Note that $B_\tau = -\kappa$ for compact $\mathfrak{g}$.*

2. *The operator $\mathrm{End}(\mathfrak{g}) \to \mathrm{End}(\mathfrak{g})$, $f \mapsto f^T$ defined by $B_\tau\left(f^T x, y\right) := B_\tau(x, fy)$ leaves $\mathrm{Aut}(\mathfrak{g})$ invariant. More precisely, $f^T = \tau f^{-1} \tau$ for all $f \in \mathrm{Aut}(\mathfrak{g})$. Also, the subgroup $\mathrm{Aut}(\mathfrak{g})^\tau = \{f \in \mathrm{Aut}(\mathfrak{g})|\, \tau f \tau = f\}$ is equal to $\mathrm{Aut}(\mathfrak{g}) \cap \mathrm{O}(\mathfrak{g}, B_\tau) = \{f \in \mathrm{Aut}(\mathfrak{g})|\, f^{-1} = f^T\}$ and smoothly isomorphic to $\mathrm{Aut}(\mathfrak{u})^\sigma$.*

3. *We have the inclusions*

$$\mathrm{ad}(\mathfrak{p}) \subseteq \mathrm{Sym}(\mathfrak{g}, B_\tau) = \left\{f \in \mathrm{End}(\mathfrak{g})|\, f = f^T\right\}$$

and

$$\mathrm{ad}(\mathfrak{k}) \subseteq \mathbf{L}(\mathrm{O}(\mathfrak{g}, B_\tau)) = \left\{f \in \mathrm{End}(\mathfrak{g})|\, f = -f^T\right\}.$$

Proof.

1. $+$ 3. These are the statements of Lemma 12.1.3 of [HN11].

2. Let $x, y \in \mathfrak{g}$ and $f \in \mathrm{Aut}(\mathfrak{g})$. Then:

$$B_\tau(f^T x, y) = B_\tau(x, fy) = -\kappa(x, \tau f \tau \tau(y)) = -\kappa\left(\tau f^{-1}\tau(x), \tau y\right) = B_\tau\left(\tau f^{-1}\tau(x), y\right).$$

Furthermore, we have: $f \in \mathrm{Aut}(\mathfrak{g})^\tau \iff \tau f \tau = f \iff \tau f^{-1}\tau = f^{-1} \iff f^T = f^{-1}$.
Now let $f \in \mathrm{Aut}(\mathfrak{g})^\tau$, $k \in \mathfrak{k}$ and $p \in \mathfrak{p}$. Then f preserves the Cartan decomposition:

$$f(k) = \tau f \tau(k) = \tau(f(k)) \implies f(k) \in \mathfrak{g}^\tau = \mathfrak{k},$$
$$f(p) = \tau f \tau(p) = -\tau(f(p)) \implies f(p) \in \mathfrak{g}^{-\tau} = \mathfrak{p}.$$

So we may define $\eta : \mathrm{Aut}(\mathfrak{g})^\tau \to \mathrm{Aut}(\mathfrak{u})$ by $\eta(f)(k + ip) := f(k) + if(p)$. Its image is in $\mathrm{Aut}(\mathfrak{u})^\sigma$:

$$\sigma\eta(f)\sigma(k + ip) = \sigma\eta(f)(k - ip) = \sigma(f(k) - if(p)) = f(k) + if(p) = \eta(f)(k + ip).$$

The map $\eta : \mathrm{Aut}(\mathfrak{g})^\tau \to \mathrm{Aut}(\mathfrak{u})^\sigma$ is a smooth group morphism and has an inverse defined by $\eta^{-1}(g)(k + ip) = g(k) + ig(p)$, thus is an isomorphism of Lie groups. □

---

[5] The statements are also true in the rather trivial case of $\mathfrak{g}$ being compact. Then $\mathfrak{g}$ has the Cartan decomposition $\mathfrak{g} = \mathfrak{g} \oplus_\kappa^1 \mathbf{0}$ and $(\mathfrak{u}, \sigma) = (\mathfrak{g}, \mathbf{1})$.

Remark 3.2.4. By a classical result (cf. Proposition 1.122 of [Kna96]), for any algebraic subgroup $H' \leq \mathrm{GL}(N, \mathbb{R})$ with $(H')^T = H'$, the map $(H' \cap \mathrm{O}(N, \mathbb{R})) \times (\mathbf{L}H' \cap \mathrm{Sym}(N, \mathbb{R})) \to H'$, $(k', x') \mapsto k'e^{x'}$ is a diffeomorphism. We can transfer this to the case of a euclidean scalar product B on a finite-dimensional vector space V and an algebraic subgroup $H \leq \mathrm{GL}(V)$ stable under B-transpositon and obtain a diffeomorphism $(H \cap \mathrm{O}(V, B)) \times (\mathbf{L}H \cap \mathrm{Sym}(V, B)) \to H$, $(k, x) \mapsto ke^x$.

If $\mathfrak{g}$ is a real semisimple Lie algebra with Cartan decomposition $\mathfrak{g} = \mathfrak{k} \oplus_\kappa^\tau \mathfrak{p}$, we have a euclidean scalar product B_τ on $\mathfrak{g}$ such that $\mathrm{Aut}(\mathfrak{g}) \leq \mathrm{GL}(\mathfrak{g})$ is stable under B_τ-transposition (cf. Lemma 3.2.3). We obtain, for algebraic and B_τ-transposition stable subgroup $H \leq \mathrm{Aut}(\mathfrak{g})$, the diffeomorphism

$$\Phi : (H \cap \mathrm{O}(\mathfrak{g}, B_\tau)) \times (\mathbf{L}(H) \cap \mathrm{Sym}(\mathfrak{g}, B_\tau)) \longrightarrow H,$$
$$(k, x) \longmapsto k \exp_{\mathrm{Aut}(\mathfrak{g})}(x).$$

Hence we obtain $H_0 \cap \mathrm{O}(\mathfrak{g}, B_\tau) = (H \cap \mathrm{O}(\mathfrak{g}, B_\tau))_0$ and, as an application of the Second Isomorphism Theorem, a group isomorphism

$$\pi_0 (H \cap \mathrm{O}(\mathfrak{g}, B_\tau)) \longrightarrow H_0 \cdot (H \cap \mathrm{O}(\mathfrak{g}, B_\tau)) / H_0 = H/H_0 = \pi_0(H),$$
$$h \cdot (H \cap \mathrm{O}(\mathfrak{g}, B_\tau))_0 \longmapsto h \cdot H_0.$$

The complexification of a real simple compact Lie algebra is always a complex simple Lie algebra, so any real simple compact Lie algebra is central simple. We want to show that $\pi_0(\mathrm{Aut}(\mathfrak{g}))$ is then isomorphic to $\pi_0(\mathrm{Aut}(\mathfrak{g}_\mathbb{C}))$.

Theorem 3.2.5. *If $\mathfrak{g}$ is a real simple compact Lie algebra, then there exists a group isomorphism $\pi_0(\mathrm{Aut}(\mathfrak{g})) \cong \pi_0(\mathrm{Aut}(\mathfrak{g}_\mathbb{C}))$. Since we know the latter group by Theorem 3.2.1, we can list all possibilities for $\pi_0(\mathrm{Aut}(\mathfrak{g}))$ in the following table.*

real simple compact $\mathfrak{g}$	$\pi_0(\mathrm{Aut}(\mathfrak{g}))$	$\# \mathrm{Conj}(\pi_0(\mathrm{Aut}(\mathfrak{g})))$
$\mathfrak{su}(n + 1, \mathbb{C})$ for $n \geq 2$	$\mathcal{C}_2$	2
$\mathfrak{so}(8, \mathbb{R})$	$\mathcal{S}_3$	3
$\mathfrak{so}(2n, \mathbb{R})$ for $n \geq 5$	$\mathcal{C}_2$	2
$\mathfrak{e}_{6(-78)}$	$\mathcal{C}_2$	2
all others	1	1

Table 3.2.: Component group $\pi_0 (\mathrm{Aut}(\mathfrak{g}))$ for real simple compact $\mathfrak{g}$

Furthemore, the embedding $\mathrm{Aut}(\mathfrak{g}) \hookrightarrow \mathrm{Aut}(\mathfrak{g}_\mathbb{C}, \mathfrak{g}) \cong \mathrm{Aut}(\mathfrak{g}_\mathbb{C})^\sigma$, where $\sigma : \mathfrak{g}_\mathbb{C} \to \mathfrak{g}_\mathbb{C}$ is the conjugation with respect to $\mathfrak{g}$ (cf. Proposition 2.1 of [Djo99]), leads to a group isomorphism $\mathrm{Aut}(\mathfrak{g}) \cong \mathrm{Aut}(\mathfrak{g})_0 \rtimes \pi_0(\mathrm{Aut}(\mathfrak{g}))$ (cf. Theorem 6.61.(vi) of [HM06]).

Proof. A Cartan decomposition of the real semisimple Lie algebra $(\mathfrak{g}_\mathbb{C})^\mathbb{R}$ is $\mathfrak{g} \oplus i\mathfrak{g}$ with the involution $\tau : \mathfrak{g}_\mathbb{C} \to \mathfrak{g}_\mathbb{C}$, $x + iy \mapsto x - iy$. The following calculation for $f \in \mathrm{Aut}(\mathfrak{g}_\mathbb{C})$ and $y \in \mathfrak{g}$ shows $\mathrm{Aut}(\mathfrak{g}_\mathbb{C})^T = \mathrm{Aut}(\mathfrak{g}_\mathbb{C})$:

$$f^T(iy) = \tau f^{-1} \tau(iy) = \tau f^{-1}(-iy) = -\tau i f^{-1}(y) = i\tau f^{-1}(y) = i\tau f^{-1}\tau(y) = if^T(y).$$

We apply Remark 3.2.4 to the algebraic subgroup $H = \mathrm{Aut}(\mathfrak{g}_{\mathbb{C}}) \leq \mathrm{Aut}((\mathfrak{g}_{\mathbb{C}})^{\mathbb{R}})$, obtaining the isomorphism $\pi_0(\mathrm{Aut}(\mathfrak{g}_{\mathbb{C}})) \cong \pi_0(\mathrm{Aut}(\mathfrak{g}_{\mathbb{C}}) \cap \mathrm{O}(\mathfrak{g}_{\mathbb{C}}, B_\tau))$. Finally, we note that $\mathrm{Aut}(\mathfrak{g}_{\mathbb{C}}) \cap \mathrm{O}(\mathfrak{g}_{\mathbb{C}}, B_\tau)$ and $\{f \in \mathrm{Aut}(\mathfrak{g}_{\mathbb{C}}) \mid \tau f \tau = f\}$ coincide and there is a smooth isomorphism of this group and $\mathrm{Aut}(\mathfrak{g})$ by $f \mapsto f_{|\mathfrak{g}}$, so $\pi_0(\mathrm{Aut}(\mathfrak{g}_{\mathbb{C}})) \cong \pi_0(\mathrm{Aut}(\mathfrak{g}))$. $\qquad\square$

We need the following statements for the calculation of $\pi_0(\mathrm{Aut}(\mathfrak{g}))$ for real central simple non-compact $\mathfrak{g}$. The first four lemmas are due to Groß and Heintze (cf. Appendix A of [HG09]), the fifth is Theorem IX.5.6 of [Hel78]. Lemma 3.2.12[6] is proven analogously to Proposition 1.1.9 of [HN11].

Lemma 3.2.6. *Let $\mathfrak{u}$ be a real simple compact Lie algebra and $\sigma \in \mathrm{Aut}(\mathfrak{u})$ a non-trivial involution. Then the group morphism $\omega : \pi_0(\mathrm{Aut}(\mathfrak{u})^\sigma) \to \pi_0(\mathrm{Aut}(\mathfrak{u}^\sigma))$ induced by restriction is injective for inner $\sigma \in \mathrm{Aut}(\mathfrak{u})$ and $\ker(\omega) = \{[\mathrm{id}_\mathfrak{u}], [\sigma]\}$ for outer $\sigma \in \mathrm{Aut}(\mathfrak{u})$.*

Lemma 3.2.7. *The so-called triality automorphism $\theta \in \mathrm{Aut}(\mathfrak{so}(8, \mathbb{R}))$ is outer, commutes with $\mathrm{Ad}\begin{pmatrix} -\mathbf{1}_4 & 0 \\ 0 & \mathbf{1}_4 \end{pmatrix}$ and is of order 3.*

Remark 3.2.8. We want to give an explicit description of (one realization of) the triality automorphism:

For the conjugation $\tau : \mathfrak{so}(8, \mathbb{C}) = \mathfrak{so}(8, \mathbb{R}) \oplus i\mathfrak{so}(8, \mathbb{R}) \to \mathfrak{so}(8, \mathbb{C})$, $x + iy \mapsto x - iy$, the map $\mathrm{Aut}(\mathfrak{so}(8, \mathbb{C}))^\tau \to \mathrm{Aut}(\mathfrak{so}(8, \mathbb{R}))$, $f \mapsto f_{|\mathfrak{so}(8, \mathbb{R})}$ is an isomorphism of Lie groups. We define an automorphism by means of the standard Chevalley generators[7], i.e., let

$$\mathfrak{h} := \mathbb{C}h_1 \oplus \mathbb{C}h_2 \oplus \mathbb{C}h_3 \oplus \mathbb{C}h_4$$

for $h_1 := E_{12} - E_{21}$, $h_2 := E_{34} - E_{43}$, $h_3 := E_{56} - E_{65}$, $h_4 := E_{78} - E_{87}$ be a Cartan subalgebra in $\mathfrak{so}(8, \mathbb{C})$, where the $E_{ij} - E_{ij}$'s denote the elementary skew-symmetric 8×8-matrices. With the dual elements $h_j^* : \mathfrak{h} \to \mathbb{C}$, $h_i \mapsto \delta_{ij}$ there is a root basis

$$\Pi := \Pi(\mathfrak{so}(8, \mathbb{C}), \mathfrak{h}) := \{\alpha_1 := h_1^* - h_2^*, \alpha_2 := h_2^* - h_3^*, \alpha_3 := h_3^* - h_4^*, \alpha_4 := h_3^* + h_4^*\}$$

with root spaces $\mathbb{C}x_i = \mathfrak{so}(8, \mathbb{C})^{\alpha_i}$, $\mathbb{C}y_i = \mathfrak{so}(8, \mathbb{C})^{-\alpha_i}$ for the matrices $x_1 := \begin{pmatrix} 0 & A & 0 & 0 \\ -A^T & 0 & 0 & 0 \\ 0 & 0 & 0 & 0 \\ 0 & 0 & 0 & 0 \end{pmatrix}$,

$$x_2 := \begin{pmatrix} 0 & 0 & 0 & 0 \\ 0 & 0 & A & 0 \\ 0 & -A^T & 0 & 0 \\ 0 & 0 & 0 & 0 \end{pmatrix}, \quad x_3 := \begin{pmatrix} 0 & 0 & 0 & 0 \\ 0 & 0 & 0 & 0 \\ 0 & 0 & 0 & A \\ 0 & 0 & -A^T & 0 \end{pmatrix} \text{ and } x_4 := \begin{pmatrix} 0 & 0 & 0 & 0 \\ 0 & 0 & 0 & 0 \\ 0 & 0 & 0 & 0 \\ 0 & 0 & -B^T & 0 \end{pmatrix}, \text{ where }$$

$A := \begin{pmatrix} 1 & -i \\ i & 1 \end{pmatrix}$, $B := \begin{pmatrix} 1 & i \\ i & -1 \end{pmatrix}$ and $y_i := \tau(x_i)$ for $i = 1, 2, 3, 4$. The triality automorphism corresponds to a cyclic permutation of the simple roots $\alpha_1, \alpha_4, \alpha_3$, which leads to the following

---

[6] Note that this lemma is proven for both, real and complex matrices. It is a generalization of Schur's Lemma applied to the canonical action of $\mathfrak{so}(n, \mathbb{C})$.

[7] By Theorem 3.2.1 of [Vin94], an automorphism of a simple complex Lie algebra is uniquely determined by its values on Chevalley generators.

definition of $\theta' \in \operatorname{Aut}(\mathfrak{so}(8,\mathbb{C}))$: Let $T := \frac{1}{2} \begin{pmatrix} 1 & 1 & 1 & 1 \\ 1 & 1 & -1 & -1 \\ 1 & -1 & 1 & -1 \\ -1 & 1 & 1 & -1 \end{pmatrix}$ be the map matrix of $\theta'_{|\mathfrak{h}}$

with respect to the basis (h_1, h_2, h_3, h_4) and $\theta'(x_i) := x_{\sigma(i)}$, $\theta'(y_i) := y_{\sigma(i)}$ for the permutation $\sigma = \begin{pmatrix} 1 & 2 & 3 & 4 \\ 3 & 2 & 4 & 1 \end{pmatrix}$ and $i = 1, \ldots, 4$. The entries of T and of the h_i's are real and for $z \in \mathbb{C}$, $i = 1, 2, 3, 4$ we have:

$$\tau\theta'\tau(zx_i) = \tau(\bar{z}\theta'(\tau(x_i))) = z\tau(\theta'(y_i)) = z\tau(y_{\sigma(i)}) = zx_{\sigma(i)} = \theta'(zx_i).$$

Analogously, we have $\tau\theta'\tau(zy_i) = \theta'(zy_i)$ for $z \in \mathbb{C}$, $i = 1, 2, 3, 4$. So $\theta' \in \operatorname{Aut}(\mathfrak{so}(8,\mathbb{C}))^\tau$. Set $\theta := \theta'_{|\mathfrak{so}(8,\mathbb{R})}$.

Lemma 3.2.9. *The four conjugacy classes of involutions of the real simple compact Lie algebra* $\mathfrak{e}_{6(-78)}$ *are represented by commuting elements. More precisely, if a maximal torus* $\mathfrak{t} \leq \mathfrak{e}_{6(-78)}$ *is fixed and* ρ_1 *is the corresponding Dynkin diagram involution, then there is an* $X \in \mathfrak{t}$ *such that* $\rho_2 := \rho_1 e^X$ *is a representative for the second conjugacy class of outer involutions. The two conjugacy class of inner involutions are represented by* $\rho_3 := e^Y$ *and* $\rho_4 := e^Z$ *for some elements* $Y, Z \in \mathfrak{t}^{\rho_1}$.

Lemma 3.2.10. *The non-trivial element of* $\pi_0\left(\operatorname{Aut}(\mathfrak{e}_{7(-133)})^{\sigma_{\mathfrak{e}_{7(7)}}}\right)$ *is represented by* $e^{\operatorname{ad}(X)}$ *for any non-zero element* $X \in \mathfrak{e}_{7(-133)}^{-\sigma_{\mathfrak{e}_{7(7)}}}$ *such that* $\operatorname{ad}(X)^3 = -\pi^2 \operatorname{ad}(X)$.

Lemma 3.2.11. *Let* $\mathfrak{u}$ *be a real simple compact Lie algebra and* $\sigma \in \operatorname{Aut}(\mathfrak{u})$ *a non-trivial involution. Then* σ *is inner if and only if the ranks of* $\mathfrak{u}^\sigma$ *and* $\mathfrak{u}$ *coincide. Here, the rank of a real compact Lie algebra is the dimension of any maximal abelian subalgebra or, equivalently, of any Cartan subalgebra.*

Lemma 3.2.12. *For* $n \geq 3$ *a matrix* $X \in \operatorname{M}(n,\mathbb{K})$ *commutes with all elementary skew-symmetric matrices* $E_{rs} - E_{sr}$, $1 \leq r < s \leq n$, *if and only if* $X \in \mathbb{K}\mathbf{1}_n$.

Proof. Let $E(m) := \{ E_{rs} - E_{sr} \,|\, 1 \leq r < s \leq m \}$ be the set of elementary skew-symmetric matrices in $\operatorname{M}(m,\mathbb{K})$ for $m \in \mathbb{N}$. For $\ell = 1, \ldots, n$ there is an embedding

$$j_\ell : E(n-1) \longrightarrow E(n)$$

$$y = (y_{i,j}) \longmapsto \begin{pmatrix} y_{1,1} & \cdots & y_{1,\ell-1} & 0 & y_{1,\ell} & \cdots & y_{1,n-1} \\ \vdots & \ddots & \vdots & \vdots & \vdots & \ddots & \vdots \\ y_{\ell-1,1} & \cdots & y_{\ell-1,\ell-1} & \vdots & y_{\ell-1,\ell} & \cdots & y_{\ell-1,n-1} \\ 0 & \cdots & \cdots & 0 & \cdots & \cdots & 0 \\ y_{\ell,1} & \cdots & y_{\ell,\ell-1} & \vdots & y_{\ell,\ell} & \cdots & y_{\ell,n-1} \\ \vdots & \ddots & \vdots & \vdots & \vdots & \ddots & \vdots \\ y_{n-1,1} & \cdots & y_{n-1,\ell-1} & 0 & y_{n-1,\ell} & \cdots & y_{n-1,n-1} \end{pmatrix}.$$

Note that $\bigcap_{y \in E(n-1)} \{ v \in \mathbb{K}^n | j_\ell(y) \cdot v = 0 \} = \mathbb{K}e_\ell$ for $(e_1, \ldots, e_n)$, the canonical basis of $\mathbb{K}^n$. So the condition $xX = Xx$ for all $x \in E(n)$ leads to $j_\ell(y)Xe_\ell = Xj_\ell(y)e_\ell = 0$ for all $y \in E(n-1)$, yielding $X(\mathbb{K}e_\ell) \subseteq \mathbb{K}e_\ell$. Thus X is diagonal. Furthermore, commuting with $E(n)$, the matrix X does not have distinct eigenvalues. Thus $X \in \mathbb{K}\mathbf{1}_n$. The other implication is trivial. $\square$

Corollary 3.2.13. *If $q \geq 2$ and $X \in \mathrm{O}(2q, \mathbb{R})$ with $\det(X) = -1$, then $\mathrm{Ad}(X) \in \mathrm{Aut}(\mathfrak{so}(2q, \mathbb{R}))$ is outer. In particular, if $q \geq 5$, then, since $\pi_0\left(\mathrm{Aut}(\mathfrak{so}(2q, \mathbb{R}))\right) \cong \mathcal{C}_2$ by Theorem 3.2.5, the map $\mathrm{Ad} : \mathrm{O}(2q, \mathbb{R}) \to \mathrm{Aut}(\mathfrak{so}(2q, \mathbb{R}))$ is surjective.*

Proof. The matrix $I := I_{1,2q-1} := \begin{pmatrix} -1 & \mathbf{0} \\ \mathbf{0} & \mathbf{1}_{2q-1} \end{pmatrix}$ is in $\mathrm{O}(2q, \mathbb{R}) \backslash \mathrm{SO}(2q, \mathbb{R})$. Furthermore, $\mathrm{Ad}(I) \neq \mathrm{Ad}(Q)$ as maps on $\mathfrak{so}(2q, \mathbb{R})$ for all $Q \in \mathrm{SO}(2q, \mathbb{R})$, since else, by Lemma 3.2.12 in the case $\mathbb{K} = \mathbb{R}$, there would exist $r \in \mathbb{R}^\times$ such that $Q^{-1}I = r\mathbf{1}_{2q}$ and

$$\det(I) = r^{2q}\det(Q) = r^{2q} \neq -1 = \det(I).$$

That is why the image of the map $\mathrm{Ad} : \mathrm{O}(2q, \mathbb{R}) \to \mathrm{Aut}(\mathfrak{so}(2q, \mathbb{R}))$ is strictly larger than $\mathrm{Ad}(\mathrm{SO}(2q, \mathbb{R})) = \mathrm{Aut}(\mathfrak{so}(2q, \mathbb{R}))_0$. Theorem 3.2.5 states $[\mathrm{Aut}(\mathfrak{so}(2q, \mathbb{R})) : \mathrm{Aut}(\mathfrak{so}(2q, \mathbb{R}))_0] = 2$, thus the surjectivity of Ad follows. $\qquad\square$

Let $\mathfrak{g}$ be real simple non-compact with Cartan involution τ and $\mathfrak{k}, \mathfrak{p}, \mathfrak{u}, \sigma \neq \mathrm{id}_\mathfrak{u}$ as in Theorem 2.1.42. We will calculate $\pi_0\left(\mathrm{Aut}(\mathfrak{u})^\sigma\right)$ and see that it is isomorphic to $\pi_0\left(\mathrm{Aut}(\mathfrak{g})\right)$ by using the classification of simply connected symmetric spaces.

Remark 3.2.14. $\mathrm{Aut}(\mathfrak{u})$ preserves the negative definite Cartan-Killing form $\kappa_\mathfrak{u}$, so it is compact and so are $U_{\mathrm{ad}} := \mathrm{Inn}(\mathfrak{u}) = \mathrm{Aut}(\mathfrak{u})_0$ and $U := \widetilde{U_{\mathrm{ad}}}$ because of Weyl's Theorem (cf. Theorem 11.1.17 of [HN11]) and the (semi)simplicity of $\mathrm{Der}(\mathfrak{u}) = \mathrm{ad}(\mathfrak{u}) \cong \mathfrak{u}$. We identify $\mathfrak{u}$ with $\mathrm{ad}(\mathfrak{u})$ and in this sense and since the exponential map of any connected Lie group with compact Lie algebra is surjective (cf. Proposition II.6.10 of [Hel78]), the map $\mathrm{Ad}_U : U \to U_{\mathrm{ad}}$, $\mathrm{Ad}_U(\exp_U(y))(x) := e^{\mathrm{ad}(y)}(x)$ is the universal cover of U_{ad}.

Let $\overline{\sigma} : U \to U$ be the unique Lie group morphism such that $\mathbf{L}\overline{\sigma} = \sigma$. The compact subgroup $K := U^{\overline{\sigma}} = \{g \in U \mid \overline{\sigma}(g) = g\}$ is connected with Lie algebra $\mathbf{L}K = \mathfrak{u}^\sigma = \mathfrak{k}$ and the homogeneous space $M := U/K$ is a simply connected compact Riemannian symmetric space (cf. Theorem VII.8.2 of [Hel78]) with compact group of displacements $U_{\mathrm{dis}} := U/\Gamma$ for $\Gamma := \mathrm{Z}(U) \cap K \trianglelefteq \mathrm{Z}(U)$, the maximal normal subgroup contained in K. So $\mathrm{Z}(U_{\mathrm{dis}}) \cong \mathrm{Z}(U)/\Gamma$. Furthermore, U_{dis} is, by Proposition IV.1.7 of [Loo69], isomorphic to the connected component of the isometry group of M and there is also an isomorphism of symmetric spaces $M \cong U_{\mathrm{dis}}/K'$, where $(U_{\mathrm{dis}}^\varsigma)_0 \subseteq K' \subseteq U_{\mathrm{dis}}^\varsigma$ for an involution $\varsigma \in \mathrm{Aut}(U_{\mathrm{dis}})$ such that $\mathbf{L}\varsigma = \mathbf{L}\overline{\sigma}$. The homotopy sequence $1 = \pi_1(M) \to \pi_0(K_2) \to \pi_0(U_{\mathrm{dis}}) = 1$ is exact, so $K' = (U_{\mathrm{dis}}^\varsigma)_0$ and $\mathbf{L}K = \mathbf{L}K'$.

Let $\rho_2 : U \to U_{\mathrm{dis}}$ be the universal covering morphism. Since $\Gamma = \ker(\rho_2) \subseteq \ker(\mathrm{Ad}_U) = \mathrm{Z}(U)$, the map $\mathrm{Ad}_{U_{\mathrm{dis}}} : U_{\mathrm{dis}} \to U_{\mathrm{ad}}$ is the unique morphism such that the diagram

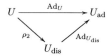

commutes, $\ker(\mathrm{Ad}_{U_{\mathrm{dis}}}) = \mathrm{Z}(U_{\mathrm{dis}}) \cong \mathrm{Z}(U)/\Gamma = \mathrm{Z}(U)/(\mathrm{Z}(U) \cap K)$ and $\mathrm{Ad}_{U_{\mathrm{dis}}}$ is a covering morphism, implying $\mathrm{Ad}_{U_{\mathrm{dis}}} \circ \exp_{U_{\mathrm{dis}}} = \exp_{U_{\mathrm{ad}}}$.

Tables 3.3 and 3.4 provide all possibilities for $\mathfrak{g}$, $\mathfrak{k}$ and $(\mathfrak{u}, \sigma)$ except for isomorphism of the Lie algebras and except for conjugation by automorphisms of the involutions. Furthermore, we make use of Loos' classification symbol X_r^*, where X_r is the root system of the corresponding complex simple Lie algebra $\mathfrak{g}_\mathbb{C} = \mathfrak{u}_\mathbb{C}$ and we list the groups U, U_{dis} with corresponding centers (cf. Chapter VII of [Loo69] and Table 4 of [Vin94]) and $Z(M)$, the center of M, an abelian discrete group acting on M (cf. Exercise X.C.4 of [Hel78]: The group $\pi_1(\mathrm{Inn}(\mathfrak{u})/\mathrm{Inn}(\mathfrak{u})_\theta)$ mentioned there is equal to $Z(M)$ by Proposition III.2.4 and the corollary of Theorem VI.3.6 of [Loo69]).

We also use the following symbols: cj is the complex conjugation, qj is the quaternionic conjugation and we have the matrices $I_{p,q} := \begin{pmatrix} -\mathbf{1}_p & \mathbf{0} \\ \mathbf{0} & \mathbf{1}_q \end{pmatrix}$ and $J_n := \begin{pmatrix} \mathbf{0} & \mathbf{1}_n \\ -\mathbf{1}_n & \mathbf{0} \end{pmatrix}$. If $\mathfrak{g}$ is exceptional, then $\sigma_\mathfrak{g}$ is the unique involution of $\mathfrak{u}$ such that $\mathfrak{g}_\mathbb{C} \to \mathfrak{g}_\mathbb{C}$, $x + iy \mapsto \sigma_\mathfrak{g}(x) + i\sigma_\mathfrak{g}(y)$ fixes $\mathfrak{g}$ pointwise, i.e., it is the complex conjugation with respect to the real form $\mathfrak{g}$. The unique simply connected real simple compact Lie groups with exceptional compact Lie algebras $\mathfrak{e}_{6(-78)}$, $\mathfrak{e}_{7(-133)}$, $\mathfrak{e}_{8(-248)}$, $\mathfrak{f}_{4(-52)}$, $\mathfrak{g}_{2(-14)}$ are denoted by $E_{6(-78)}$, $E_{7(-133)}$, $E_{8(-248)}$, $F_{4(-52)}$, $G_{2(-14)}$, respectively.

X_r^*	$\mathfrak{g}$	$\mathfrak{k}$	$(\mathfrak{u}, \sigma)$
$A_n^{\mathbb{R}},\ n \geq 1$	$\mathfrak{sl}(n+1, \mathbb{R})$	$\mathfrak{so}(n+1, \mathbb{R})$	$(\mathfrak{su}(n+1, \mathbb{C}), \mathrm{cj})$
$A_n^{\mathbb{C},q},\ q \leq \frac{n+1}{2} > 1$	$\mathfrak{su}(q, n+1-q, \mathbb{C})$	$\mathfrak{su}(q, \mathbb{C}) \oplus \mathfrak{u}(n+1-q, \mathbb{C})$	$(\mathfrak{su}(n+1, \mathbb{C}), \mathrm{Ad}(I_{n+1-q,q}))$
$A_{2n+1}^{\mathbb{H}},\ n \geq 1$	$\mathfrak{sl}(n+1, \mathbb{H}) \cong \mathfrak{su}^*(2n+2, \mathbb{C})$	$\mathfrak{sp}(n+1, \mathbb{H})$	$(\mathfrak{su}(2n+2, \mathbb{C}), \mathrm{cj} \circ \mathrm{Ad}(J_{n+1}))$
$B_n^{\mathbb{R},q},\ q \leq n \geq 2$	$\mathfrak{so}(q, 2n+1-q, \mathbb{R})$	$\mathfrak{so}(q, \mathbb{R}) \oplus \mathfrak{so}(2n+1-q, \mathbb{R})$	$(\mathfrak{so}(2n+1, \mathbb{R}), \mathrm{Ad}(I_{2n+1-q,q}))$
$C_n^{\mathbb{R}},\ n \geq 3$	$\mathfrak{sp}(2n, \mathbb{R})$	$\mathfrak{u}(n, \mathbb{C})$	$(\mathfrak{sp}(n, \mathbb{H}), \mathrm{cj})$
$C_n^{\mathbb{H},q},\ q \leq \frac{n}{2} > 1$	$\mathfrak{sp}(q, n-q, \mathbb{H})$	$\mathfrak{sp}(q, \mathbb{H}) \oplus \mathfrak{sp}(n-q, \mathbb{H})$	$(\mathfrak{sp}(n, \mathbb{H}), \mathrm{Ad}(I_{n-q,q}))$
$D_n^{\mathbb{R},q},\ q \leq n \geq 4$	$\mathfrak{so}(q, 2n-q, \mathbb{R})$	$\mathfrak{so}(q, \mathbb{R}) \oplus \mathfrak{so}(2n-q, \mathbb{R})$	$(\mathfrak{so}(2n, \mathbb{R}), \mathrm{Ad}(I_{2n-q,q}))$
$D_n^{\mathbb{H}},\ n \geq 5$	$\mathfrak{so}^*(2n, \mathbb{R})$	$\mathfrak{u}(n, \mathbb{C})$	$(\mathfrak{so}(2n, \mathbb{R}), \mathrm{Ad}(J_n))$
$E_{6(6)}$	$\mathfrak{e}_{6(6)}$	$\mathfrak{sp}(4, \mathbb{H})$	$(\mathfrak{e}_{6(-78)}, \sigma_{\mathfrak{e}_{6(6)}})$
$E_{6(2)}$	$\mathfrak{e}_{6(2)}$	$\mathfrak{su}(2, \mathbb{C}) \oplus \mathfrak{su}(6, \mathbb{C})$	$(\mathfrak{e}_{6(-78)}, \sigma_{\mathfrak{e}_{6(2)}})$
$E_{6(-14)}$	$\mathfrak{e}_{6(-14)}$	$\mathfrak{so}(10, \mathbb{C}) \oplus \mathbb{R}$	$(\mathfrak{e}_{6(-78)}, \sigma_{\mathfrak{e}_{6(-14)}})$
$E_{6(-26)}$	$\mathfrak{e}_{6(-26)}$	$\mathfrak{f}_{4(-52)}$	$(\mathfrak{e}_{6(-78)}, \sigma_{\mathfrak{e}_{6(-26)}})$
$E_{7(7)}$	$\mathfrak{e}_{7(7)}$	$\mathfrak{su}(8, \mathbb{C})$	$(\mathfrak{e}_{7(-133)}, \sigma_{\mathfrak{e}_{7(7)}})$
$E_{7(-5)}$	$\mathfrak{e}_{7(-5)}$	$\mathfrak{su}(2, \mathbb{C}) \oplus \mathfrak{so}(12, \mathbb{R})$	$(\mathfrak{e}_{7(-133)}, \sigma_{\mathfrak{e}_{7(-5)}})$
$E_{7(-25)}$	$\mathfrak{e}_{7(-25)}$	$\mathfrak{e}_{6(-78)} \oplus \mathbb{R}$	$(\mathfrak{e}_{7(-133)}, \sigma_{\mathfrak{e}_{7(-25)}})$
$E_{8(8)}$	$\mathfrak{e}_{8(8)}$	$\mathfrak{so}(16, \mathbb{R})$	$(\mathfrak{e}_{8(-248)}, \sigma_{\mathfrak{e}_{8(8)}})$
$E_{8(-24)}$	$\mathfrak{e}_{8(-24)}$	$\mathfrak{su}(2, \mathbb{C}) \oplus \mathfrak{e}_{7(-133)}$	$(\mathfrak{e}_{8(-248)}, \sigma_{\mathfrak{e}_{8(-24)}})$
$F_{4(4)}$	$\mathfrak{f}_{4(4)}$	$\mathfrak{su}(2, \mathbb{C}) \oplus \mathfrak{sp}(3, \mathbb{H})$	$(\mathfrak{f}_{4(-52)}, \sigma_{\mathfrak{f}_{4(4)}})$
$F_{4(-20)}$	$\mathfrak{f}_{4(-20)}$	$\mathfrak{so}(9, \mathbb{R})$	$(\mathfrak{f}_{4(-52)}, \sigma_{\mathfrak{f}_{4(-20)}})$
$G_{2(2)}$	$\mathfrak{g}_{2(2)}$	$\mathfrak{su}(2, \mathbb{C}) \oplus \mathfrak{su}(2, \mathbb{C})$	$(\mathfrak{g}_{2(-14)}, \sigma_{\mathfrak{g}_{2(2)}})$

Table 3.3.: Classification of real simple non-compact Lie algebras by symmetric spaces

X_r^*	U	$Z(U)$	U_{dis}	$Z(U_{\mathrm{dis}})$	$Z(M)$
$A_n^{\mathbb{R}}$, odd $n \geq 1$	$SU(n+1,\mathbb{C})$	C_{n+1}	$SU(n+1,\mathbb{C})/C_21$	$C_{\frac{n+1}{2}}$	C_{n+1}
$A_n^{\mathbb{R}}$, even $n \geq 2$	$SU(n+1,\mathbb{C})$	C_{n+1}	$SU(n+1,\mathbb{C})$	C_{n+1}	C_{n+1}
$A_n^{C,q}$, $q < \frac{n+1}{2} > 1$	$SU(n+1,\mathbb{C})$	C_{n+1}	$SU(n+1,\mathbb{C})/C_{n+1}1$	1	1
$A_{2q-1}^{C,q}$, $q > 1$	$SU(2q,\mathbb{C})$	C_{2q}	$SU(2q,\mathbb{C})/C_{2q}1$	1	C_2
$A_{2n+1}^{\mathbb{H}}$, $n \geq 1$	$SU(2n+2,\mathbb{C})$	C_{2n+2}	$SU(2n+2,\mathbb{C})/C_21$	C_{n+1}	C_{n+1}
$B_n^{\mathbb{R},q}$, $q \leq n \geq 2$	$Spin(2n+1,\mathbb{R})$	C_2	$SO(2n+1,\mathbb{R})$	1	C_2
$C_n^{\mathbb{R}}$, $n \geq 3$	$Sp(n,\mathbb{H})$	C_2	$Sp(n,\mathbb{H})/C_21$	1	C_2
$C_n^{\mathbb{H},q}$, $q < \frac{n}{2} > 1$	$Sp(n,\mathbb{H})$	C_2	$Sp(n,\mathbb{H})/C_21$	1	1
$C_{2q}^{\mathbb{H},q}$, $q \geq 2$	$Sp(2q,\mathbb{H})$	C_2	$Sp(2q,\mathbb{H})/C_21$	1	C_2
$D_n^{\mathbb{R},q}$, even $n \geq 4$, odd $q < n$	$Spin(2n,\mathbb{R})$	$C_2 \times C_2$	$SO(2n,\mathbb{R})$	C_2	C_2
$D_n^{\mathbb{R},q}$, even $n \geq 4$, even $q < n$	$Spin(2n,\mathbb{R})$	$C_2 \times C_2$	$SO(2n,\mathbb{R})/C_21$	1	C_2
$D_q^{\mathbb{R},q}$, even $q \geq 4$	$Spin(2q,\mathbb{R})$	$C_2 \times C_2$	$SO(2q,\mathbb{R})/C_21$	1	$C_2 \times C_2$
$D_n^{\mathbb{R},q}$, odd $n \geq 5$, odd $q < n$	$Spin(2n,\mathbb{R})$	C_4	$SO(2n,\mathbb{R})$	C_2	C_2
$D_q^{\mathbb{R},q}$, odd $q \geq 5$	$Spin(2q,\mathbb{R})$	C_4	$SO(2q,\mathbb{R})$	C_2	C_4
$D_n^{\mathbb{R},q}$, odd $n \geq 5$, even $q < n$	$Spin(2n,\mathbb{R})$	C_4	$SO(2n,\mathbb{R})/C_21$	1	C_2
$D_n^{\mathbb{H}}$, odd $n \geq 5$	$Spin(2n,\mathbb{R})$	C_4	$SO(2n,\mathbb{R})/C_21$	1	1
$D_n^{\mathbb{H}}$, even $n \geq 6$	$Spin(2n,\mathbb{R})$	$C_2 \times C_2$	$SO(2n,\mathbb{R})/C_21$	1	C_2
$E_{6(6)}$	$E_{6(-78)}$	C_3	$E_{6(-78)}$	C_3	C_3
$E_{6(2)}$	$E_{6(-78)}$	C_3	$E_{6(-78)}/C_3$	1	1
$E_{6(-14)}$	$E_{6(-78)}$	C_3	$E_{6(-78)}/C_3$	1	1
$E_{6(-26)}$	$E_{6(-78)}$	C_3	$E_{6(-78)}$	C_3	C_3
$E_{7(7)}$	$E_{7(-133)}$	C_2	$E_{7(-133)}/C_2$	1	C_2
$E_{7(-5)}$	$E_{7(-133)}$	C_2	$E_{7(-133)}/C_2$	1	1
$E_{7(-25)}$	$E_{7(-133)}$	C_2	$E_{7(-133)}/C_2$	1	C_2
$E_{8(8)}$	$E_{8(-248)}$	1	$E_{8(-248)}$	1	1
$E_{8(-24)}$	$E_{8(-248)}$	1	$E_{8(-248)}$	1	1
$F_{4(4)}$	$F_{4(-52)}$	1	$F_{4(-52)}$	1	1
$F_{4(-20)}$	$F_{4(-52)}$	1	$F_{4(-52)}$	1	1
$G_{2(2)}$	$G_{2(-14)}$	1	$G_{2(-14)}$	1	1

Table 3.4.: Classification of real simple non-compact Lie algebras by symmetric spaces (continued)

3. Classification by Homotopy Theory

We prove our key proposition by slightly modifying Loos' proof of Theorem VII.4.4 of [Loo69].

Proposition 3.2.15. *Let $\mathfrak{u}$ be a real simple compact Lie algebra, $\sigma \in \mathrm{Aut}(\mathfrak{u})$ a non-trivial involution, i.e., $\sigma \neq \mathrm{id}_{\mathfrak{u}}$ and $\sigma^2 = \mathrm{id}_{\mathfrak{u}}$, set $\mathfrak{k} := \mathfrak{u}^\sigma$ and $\mathfrak{p} := \mathfrak{u}^{-\sigma}$ and let $\mathfrak{g} = \mathfrak{k} + i\mathfrak{p}$ be the corresponding real simple non-compact Lie algebra with Cartan involution $\tau = \mathrm{id}_{\mathfrak{k}} \oplus -\mathrm{id}_{i\mathfrak{p}}$. Then:*

1. *The group $\pi_0((\mathrm{Aut}(\mathfrak{u})_0)^\sigma)$ is elementary abelian, i.e., isomorphic to $\mathcal{C}_2^k$ for some $k \in \mathbb{N}_0$.*

2. *The identity components of $\mathrm{Aut}(\mathfrak{u})^\sigma$ and $(\mathrm{Aut}(\mathfrak{u})_0)^\sigma$ coincide.*

3. *The inclusion $\mathrm{Aut}(\mathfrak{u})^\sigma \cong \mathrm{Aut}(\mathfrak{g})^\tau \hookrightarrow \mathrm{Aut}(\mathfrak{g})$ induces an isomorphism between $\pi_0(\mathrm{Aut}(\mathfrak{u})^\sigma)$ and $\pi_0(\mathrm{Aut}(\mathfrak{g}))$. There is an isomorphism $\mathrm{Aut}(\mathfrak{u})^\sigma \cong (\mathrm{Aut}(\mathfrak{u})^\sigma)_0 \rtimes \pi_0(\mathrm{Aut}(\mathfrak{u})^\sigma)$ if and only if $\mathrm{Aut}(\mathfrak{g}) \cong \mathrm{Aut}(\mathfrak{g})_0 \rtimes \pi_0(\mathrm{Aut}(\mathfrak{g}))$.*

4. *There is a finite group F such that $\mathrm{Aut}(\mathfrak{u})^\sigma \cong (\mathrm{Aut}(\mathfrak{u})_0)^\sigma \rtimes F$ and $\pi_0(\mathrm{Aut}(\mathfrak{u})^\sigma) \cong \mathcal{C}_2^k \rtimes F$.*

5. *The following table provides all possibilities for 2^k, F and $\pi_0(\mathrm{Aut}(\mathfrak{u})^\sigma)$.*

$(\mathfrak{u}, \sigma)$	2^k	F	$\pi_0(\mathrm{Aut}(\mathfrak{u})^\sigma)$
$(\mathfrak{su}(2,\mathbb{C}), \mathrm{cj})$	2	1	$\mathcal{C}_2$
$(\mathfrak{su}(n+1,\mathbb{C}), \mathrm{cj})$ for even $n \geq 2$	1	$\mathcal{C}_2$	$\mathcal{C}_2$
$(\mathfrak{su}(n+1,\mathbb{C}), \mathrm{cj})$ for odd $n \geq 3$	2	$\mathcal{C}_2$	$\mathcal{C}_2 \times \mathcal{C}_2$
$(\mathfrak{su}(n+1,\mathbb{C}), \mathrm{Ad}(I_{n+1-q,q}))$ for $q < \frac{n+1}{2} > 1$	1	$\mathcal{C}_2$	$\mathcal{C}_2$
$(\mathfrak{su}(2n+2,\mathbb{C}), \mathrm{Ad}(I_{n+1,n+1}))$ for $n \geq 1$	2	$\mathcal{C}_2$	$\mathcal{C}_2 \times \mathcal{C}_2$
$(\mathfrak{su}(2n+2,\mathbb{C}), \mathrm{cj} \circ \mathrm{Ad}(J_{n+1}))$ for $n \geq 1$	1	$\mathcal{C}_2$	$\mathcal{C}_2$
$(\mathfrak{so}(2n+1,\mathbb{R}), \mathrm{Ad}(I_{2n+1-q,q}))$ for $q \leq n \geq 2$	2	1	$\mathcal{C}_2$
$(\mathfrak{sp}(n,\mathbb{H}), \mathrm{qj})$ for $n \geq 3$	2	1	$\mathcal{C}_2$
$(\mathfrak{sp}(2n,\mathbb{H}), \mathrm{Ad}(I_{n,n}))$ for $n \geq 1$	2	1	$\mathcal{C}_2$
$(\mathfrak{so}(2n,\mathbb{R}), \mathrm{Ad}(I_{2n-q,q}))$ for odd $q < n \geq 4$	1	$\mathcal{C}_2$	$\mathcal{C}_2$
$(\mathfrak{so}(2n,\mathbb{R}), \mathrm{Ad}(I_{2n-q,q}))$ for even $q < n \geq 4$	2	$\mathcal{C}_2$	$\mathcal{C}_2 \times \mathcal{C}_2$
$(\mathfrak{so}(2n,\mathbb{R}), \mathrm{Ad}(I_{n,n}))$ for odd $n \geq 5$	2	$\mathcal{C}_2$	$\mathcal{C}_2 \times \mathcal{C}_2$
$(\mathfrak{so}(2n,\mathbb{R}), \mathrm{Ad}(I_{n,n}))$ for even $n \geq 6$	4	$\mathcal{C}_2$	$\mathcal{D}_4$
$(\mathfrak{so}(8,\mathbb{R}), \mathrm{Ad}(I_{4,4}))$	4	$\mathcal{S}_3$	$\mathcal{S}_4$
$(\mathfrak{so}(4n,\mathbb{R}), \mathrm{Ad}(J_{2n}))$ for $n \geq 3$	2	1	$\mathcal{C}_2$
$(\mathfrak{so}(4n+2,\mathbb{R}), \mathrm{Ad}(J_{2n+1}))$ for $n \geq 2$	1	$\mathcal{C}_2$	$\mathcal{C}_2$
$\left(\mathfrak{e}_{6(-78)}, \sigma_{\mathfrak{e}_{6(j)}}\right)$ for $j = 6, 2, -14, -26$	1	$\mathcal{C}_2$	$\mathcal{C}_2$
$\left(\mathfrak{e}_{7(-133)}, \sigma_{\mathfrak{e}_{7(j)}}\right)$ for $j = 7, -25$	2	1	$\mathcal{C}_2$
all others	1	1	1

Proof. We use the notation of Remark 3.2.14.

1. The closed subgroup $U_{\mathrm{ad}}^\sigma = \{f \in U_{\mathrm{ad}} | \sigma f \sigma = f\}$ is compact and so is $\pi_0(U_{\mathrm{ad}}^\sigma)$. In addition, the latter group is discrete, hence finite. We will show that every element in $\pi_0(U_{\mathrm{ad}}^\sigma)$ is self-inverse: The Lie algebra of U_{ad}^σ is $LU_{\mathrm{ad}}^\sigma = LK = \mathfrak{k}$, thus the identity component $(U_{\mathrm{ad}}^\sigma)_0$ is $\exp_{U_{\mathrm{ad}}^\sigma}(\mathfrak{k}) = \exp_{U_{\mathrm{ad}}}(\mathfrak{k})$, so by using the surjectivity of

$$\Phi : \exp_{U_{\mathrm{ad}}}(\mathfrak{k}) \times \mathfrak{p} \longrightarrow \exp_{U_{\mathrm{ad}}}(\mathfrak{u}) = U_{\mathrm{ad}}$$
$$(f, x) \longmapsto f \cdot \exp_{U_{\mathrm{ad}}}(x),$$

(cf. Remark 3.2.4), we have $U_{\mathrm{ad}}^\sigma = (U_{\mathrm{ad}}^\sigma)_0 \cdot (U_{\mathrm{ad}}^\sigma \cap \exp_{U_{\mathrm{ad}}}(\mathfrak{p}))$ and every element of $\pi_0(U_{\mathrm{ad}}^\sigma)$ takes the form $\exp_{U_{\mathrm{ad}}}(x) \cdot (U_{\mathrm{ad}}^\sigma)_0 = \big[\exp_{U_{\mathrm{ad}}}(x)\big]$ for some $x \in \mathfrak{p}$. Thus the calculation

$$\big[\exp_{U_{\mathrm{ad}}}(x)\big]^{-1} = \big[\exp_{U_{\mathrm{ad}}}(-x)\big] = \big[\exp_{U_{\mathrm{ad}}}(\sigma(x))\big] = \big[\sigma \exp_{U_{\mathrm{ad}}}(x)\sigma\big] = \big[\exp_{U_{\mathrm{ad}}}(x)\big]$$

shows that every element of $\pi_0(U_{\mathrm{ad}}^\sigma)$ is self-inverse, hence $\pi_0(U_{\mathrm{ad}}^\sigma) \cong \mathcal{C}_2^k$ for some $k \in \mathbb{N}_0$.

In order to derive a formula for k, we consider the following commutative diagram of pointed spaces:

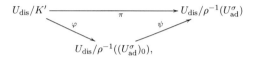

where π, φ, ψ are induced by the identity on U_{dis} and $\rho := \mathrm{Ad}_{U_{\mathrm{dis}}}$. As a first step, we calculate:

$$\begin{aligned}
\rho^{-1}\left((U_{\mathrm{ad}}^\sigma)_0\right) &= \big\{ g \in U_{\mathrm{dis}} \,|\, (\exists x \in \mathfrak{k}) : \rho(g) = \exp_{U_{\mathrm{ad}}}(x) \big\} \\
&= \big\{ g \in U_{\mathrm{dis}} \,|\, (\exists x \in \mathfrak{k}) : \rho(g) = \rho \circ \exp_{U_{\mathrm{dis}}}(x) \big\} \\
&= \big\{ g \in U_{\mathrm{dis}} \,|\, (\exists h \in K') : gh^{-1} \in \mathrm{Z}(U_{\mathrm{dis}}) \big\} = \mathrm{Z}(U_{\mathrm{dis}})K'.
\end{aligned}$$

So we have $\ker(\varphi) = \mathrm{Z}(U_{\mathrm{dis}})K'/K' \cong \mathrm{Z}(U_{\mathrm{dis}})/(\mathrm{Z}(U_{\mathrm{dis}}) \cap K') \cong \mathrm{Z}(U_{\mathrm{dis}})$, where the last isomorphism follows from the fact that $\mathrm{Z}(U_{\mathrm{dis}}) \cap K'$ acts trivially on $M = U_{\mathrm{dis}}/K'$.

There exists a group isomorphism $\rho^{-1}(U_{\mathrm{ad}}^\sigma)/\rho^{-1}\left((U_{\mathrm{ad}}^\sigma)_0\right) \to \pi_0(U_{\mathrm{ad}}^\sigma)$ by $[g] \mapsto [\rho(g)]$: The map is well-defined and its kernel is $\big\{ [g] |\, \rho(g) \in (U_{\mathrm{ad}}^\sigma)_0 \big\} = \mathbf{1} \in \rho^{-1}(U_{\mathrm{ad}}^\sigma)/\rho^{-1}\left((U_{\mathrm{ad}}^\sigma)_0\right)$. Its surjectivity follows from the surjectivity of $\rho^{U_{\mathrm{ad}}^\sigma}_{|\rho^{-1}(U_{\mathrm{ad}}^\sigma)}$. This shows $\ker(\psi) \cong \pi_0(U_{\mathrm{ad}}^\sigma)$.

In order to calculate $\ker(\pi)$ we use the bijection $U_{\mathrm{dis}}/\rho^{-1}(U_{\mathrm{ad}}^\sigma) \to U_{\mathrm{ad}}/U_{\mathrm{ad}}^\sigma$, $[g] \mapsto [\rho(g)]$, leading to $\widetilde{\pi} : U_{\mathrm{dis}}/K' \to U_{\mathrm{ad}}/U_{\mathrm{ad}}^\sigma$, $[g] \mapsto [\rho(g)]$ and, by Proposition III.2.4 and the corollary of Theorem VI.3.6 of [Loo69], the [1]-fiber of this map is $\mathrm{Z}(M)$.

We thus obtain the following formula for the order of the group $\pi_0(U_{\mathrm{ad}}^\sigma)$:

$$2^k = \frac{\#\,\mathrm{Z}(M)}{\#\,\mathrm{Z}(U_{\mathrm{dis}})},$$

leading to the following preliminary result:

X_r^*	2^k
$A_n^{\mathbb{R}}$, odd $n \geq 1$	2
$A_{2q-1}^{\mathbb{C},q}$, $q > 1$	2
$B_n^{\mathbb{R},q}$, $q \leq n \geq 2$	2
$C_n^{\mathbb{R}}$, $n \geq 3$	2
$C_{2q}^{\mathbb{H},q}$, $q \geq 2$	2
$D_n^{\mathbb{R},q}$, $n \geq 4$, even $q < n$	2
$D_q^{\mathbb{R},q}$, even $q \geq 4$	4
$D_q^{\mathbb{R},q}$, odd $q \geq 5$	2
$D_n^{\mathbb{H}}$, even $n \geq 6$	2
$E_{7(7)}$	2
$E_{7(-25)}$	2
all others	1

2. Any element in U_{ad}^σ is of the form $\exp_{U_{\mathrm{ad}}}(x)$ for some $x \in \mathfrak{k}$. If $f \in \mathrm{Aut}(\mathfrak{u})^\sigma$ is arbitrary, then $f \exp_{U_{\mathrm{ad}}}(x) f^{-1} = \exp_{U_{\mathrm{ad}}}(f(x))$ is again in U_{ad}, so in U_{ad}^σ, hence $U_{\mathrm{ad}}^\sigma \trianglelefteq \mathrm{Aut}(\mathfrak{u})^\sigma$. The Lie algebra of $\mathrm{Aut}(\mathfrak{u})^\sigma$ is $\mathbf{L}\,\mathrm{Aut}(\mathfrak{u})^\sigma = LU_{\mathrm{ad}}^\sigma = \mathfrak{k}$ and $\mathrm{Aut}(\mathfrak{u})^\sigma$, U_{ad}^σ are closed subgroups of $\mathrm{Aut}(\mathfrak{u})$, so $(\mathrm{Aut}(\mathfrak{u})^\sigma)_0 = (U_{\mathrm{ad}}^\sigma)_0$. The short exact sequence

$$1 \longrightarrow U_{\mathrm{ad}}^\sigma \xrightarrow{\text{incl.}} \mathrm{Aut}(\mathfrak{u})^\sigma \xrightarrow{\text{quot.}} F \longrightarrow 1$$

induces the short exact sequence

$$1 \longrightarrow \pi_0\left(U_{\mathrm{ad}}^\sigma\right) \xrightarrow{\text{incl.}} \pi_0\left(\mathrm{Aut}(\mathfrak{u})^\sigma\right) \xrightarrow{\text{quot.}} F \longrightarrow 1$$

and if the former sequence is split, then the latter is split, too.

3. We apply Remark 3.2.4 to $H = \mathrm{Aut}(\mathfrak{g})$. So we obtain $\mathrm{Aut}(\mathfrak{g})_0 \cap \mathrm{Aut}(\mathfrak{g})^\tau = (\mathrm{Aut}(\mathfrak{g})^\tau)_0$ and $\mathrm{Aut}(\mathfrak{g})_0 \cdot \mathrm{Aut}(\mathfrak{g})^\tau = \mathrm{Aut}(\mathfrak{g})$ and $\pi_0\left(\mathrm{Aut}(\mathfrak{g})^\tau\right) \to \pi_0\left(\mathrm{Aut}(\mathfrak{g})\right)$, $[\omega] \mapsto [\omega]$ is a group isomorphism.

Furthermore, the isomorphism $\eta : \mathrm{Aut}(\mathfrak{g})^\tau \to \mathrm{Aut}(\mathfrak{u})^\sigma$ from Lemma 3.2.3.2 induces the isomorphism

$$\pi_0(\eta) : \pi_0\left(\mathrm{Aut}(\mathfrak{g})^\tau\right) \longrightarrow \pi_0\left(\mathrm{Aut}(\mathfrak{u})^\sigma\right), [f] \longmapsto [\eta(f)],$$

yielding $\pi_0(\mathrm{Aut}(\mathfrak{g})) \cong \pi_0\left(\mathrm{Aut}(\mathfrak{u})^\sigma\right)$.

If there are representatives of the elements in $\pi_0\left(\mathrm{Aut}(\mathfrak{u})^\sigma\right)$ forming a subgroup of $\mathrm{Aut}(\mathfrak{u})^\sigma$ isomorphic to $\pi_0\left(\mathrm{Aut}(\mathfrak{u})^\sigma\right)$, we may apply η^{-1} to them and they turn into representatives in $\pi_0\left(\mathrm{Aut}(\mathfrak{g})^\tau\right)$ and thus in $\pi_0\left(\mathrm{Aut}(\mathfrak{g})\right)$, showing that $\pi_0\left(\mathrm{Aut}(\mathfrak{g})\right)$ can be embedded into $\mathrm{Aut}(\mathfrak{g})$ with its image intersecting $\mathrm{Aut}(\mathfrak{g})_0$ trivially, i.e., $\mathrm{Aut}(\mathfrak{g}) \cong \mathrm{Aut}(\mathfrak{g})_0 \rtimes \pi_0(\mathrm{Aut}(\mathfrak{g}))$.

This argumentation works conversely by applying η to the elements in $\pi_0\left(\mathrm{Aut}(\mathfrak{g})^\tau\right)$ forming a subgroup of $\mathrm{Aut}(\mathfrak{g})^\tau$ isomorphic to $\pi_0\left(\mathrm{Aut}(\mathfrak{g})^\tau\right)$.

4. + 5. We will proof $\mathrm{Aut}(\mathfrak{u})^\sigma \cong U_{\mathrm{ad}}^\sigma \rtimes F$, where F is some finite group, by a case-by-case discussion:

Case 1: If $\mathrm{Aut}(\mathfrak{u})$ is connected, i.e.

$$X_r^\star \in \left\{ A_1^{\mathbb{R}};\ B_n^{\mathbb{R},q}, q \leq n \geq 2;\ C_n^{\mathbb{R}}, n \geq 3;\ C_n^{\mathbb{H},q}, q \leq \frac{n}{2} > 1; \right.$$
$$\left. E_{7(7)};\ E_{7(-5)};\ E_{7(-25)};\ E_{8(8)};\ E_{8(-24)};\ F_{4(4)};\ F_{4(-20)};\ G_{2(2)} \right\}$$

by Theorem 3.2.5, then $\mathrm{Aut}(\mathfrak{u})^\sigma \cong U_{\mathrm{ad}}^\sigma \rtimes F$ follows for trivial F. In these cases, $\mathrm{Z}\,(U_{\mathrm{dis}})$ is trivial, so $\pi_0\,(\mathrm{Aut}(\mathfrak{u})^\sigma)$ has order $\# \mathrm{Z}(M) \in \{1, 2\}$, thus $\pi_0\,(\mathrm{Aut}(\mathfrak{u})^\sigma) \in \{\mathbf{1}, \mathcal{C}_2\}$ (cf. Table 3.4).

Case 2: Let the automorphism $\sigma \in \mathrm{Aut}(\mathfrak{u})$ be outer and $\mathfrak{u} \neq \mathfrak{so}(8, \mathbb{R})$. The former condition is fulfilled if and only if the ranks of $\mathfrak{u}$ and $\mathfrak{k}$ do not coincide by Lemma 3.2.11, so we obtain:

$$X_r^\star \in \left\{ A_n^{\mathbb{R}}, n > 1;\ A_{2n+1}^{\mathbb{H}}, n \geq 1;\ D_n^{\mathbb{R},q}, \text{odd } q \leq n > 4;\ E_{6(6)};\ E_{6(-26)} \right\}.$$

In all these cases, we have $k = 0$, except for the case $X_r^\star = A_n^{\mathbb{R}}$, odd $n > 1$, and $D_q^{\mathbb{R},q}$, odd $q > 4$, where $k = 1$. Then $\sigma \notin \mathrm{Aut}(\mathfrak{u})_0 = U_{\mathrm{ad}} \supseteq U_{\mathrm{ad}}^\sigma \supseteq (U_{\mathrm{ad}}^\sigma)_0 = (\mathrm{Aut}(\mathfrak{u})^\sigma)_0$, so $[\sigma] \in \pi_0\,(\mathrm{Aut}(\mathfrak{u})^\sigma)$ and a map $p : \mathrm{Aut}(\mathfrak{u})^\sigma \to F$ by $p(\theta) := [\mathrm{id}_\mathfrak{u}]$ if and only if $\theta \in U_{\mathrm{ad}}^\sigma$ and $p(\theta) := [\sigma]$ otherwise. If $\theta, \theta' \in \mathrm{Aut}(\mathfrak{u})^\sigma \backslash U_{\mathrm{ad}}^\sigma$, then the equivalence classes of θ^{-1}, θ' in $\pi_0\,(\mathrm{Aut}(\mathfrak{u}))$ are different from $[\mathrm{id}_\mathfrak{u}]$, so, by Theorem 3.2.5, the equivalence classes coincide and $\theta\theta' \in \mathrm{Aut}(\mathfrak{u})_0 \cap \mathrm{Aut}(\mathfrak{u})^\sigma = U_{\mathrm{ad}}^\sigma$. So $p(\theta\theta') = [\mathrm{id}_\mathfrak{u}] = [\sigma]^2 = p(\theta)p(\theta')$ and hence p is a morphism of groups. This morphism has a natural section $s : F \to \mathrm{Aut}(\mathfrak{u})^\sigma$, thus $\mathrm{Aut}(\mathfrak{u})^\sigma \cong U_{\mathrm{ad}}^\sigma \rtimes_\alpha F$ with $\alpha([\sigma])(f) := \sigma f \sigma = f$ for all $f \in U_{\mathrm{ad}}^\sigma$, so $\mathrm{Aut}(\mathfrak{u})^\sigma \cong U_{\mathrm{ad}}^\sigma \times \mathcal{C}_2$.

Case 3: $X_r^\star \in \left\{ D_4^{\mathbb{R},1}, D_4^{\mathbb{R},3} \right\}$, i.e., $(\mathfrak{u}, \sigma) \in \{(\mathfrak{so}(8, \mathbb{R}), \mathrm{Ad}(I_{7,1})), (\mathfrak{so}(8, \mathbb{R}), \mathrm{Ad}(I_{5,3}))\}$:
The automorphism group of $\mathfrak{k} = \mathfrak{u}^\sigma$ is $\mathrm{Aut}(\mathfrak{so}(1, \mathbb{R})) \times \mathrm{Aut}(\mathfrak{so}(7, \mathbb{R})) \cong \mathrm{Aut}(\mathfrak{so}(7, \mathbb{R}))$ or $\mathrm{Aut}(\mathfrak{so}(3, \mathbb{R})) \times \mathrm{Aut}(\mathfrak{so}(5, \mathbb{R}))$, but in both cases $\mathrm{Aut}(\mathfrak{k})$ is connected, because the Dynkin diagrams A_1, B_2 and B_3 have no non-trivial symmetries. Since the automorphism σ is outer, Lemma 3.2.6 implies that $\pi_0\,(\mathrm{Aut}(\mathfrak{u})^\sigma) = \{[\mathrm{id}_\mathfrak{u}], [\sigma]\}$ and, since U_{ad}^σ is connected in both cases, we have a short exact sequence as follows:

$$1 \longrightarrow U_{\mathrm{ad}}^\sigma \overset{\mathrm{incl.}}{\longrightarrow} \mathrm{Aut}(\mathfrak{u})^\sigma \overset{p}{\longrightarrow} \{[\mathrm{id}_\mathfrak{u}], [\sigma]\} \longrightarrow 1,$$

where $p(\theta) := [\mathrm{id}_\mathfrak{u}]$ if $\theta \in U_{\mathrm{ad}}^\sigma = (U_{\mathrm{ad}}^\sigma)_0 = (\mathrm{Aut}(\mathfrak{u})^\sigma)_0$ and $p(\theta) := [\sigma]$ otherwise. There is a natural section $s : \{[\mathrm{id}_\mathfrak{u}], [\sigma]\} \to \mathrm{Aut}(\mathfrak{u})^\sigma$ of the sequence, leading to the isomorphism $\mathrm{Aut}(\mathfrak{u})^\sigma \cong U_{\mathrm{ad}}^\sigma \rtimes_\alpha \{[\mathrm{id}_\mathfrak{u}], [\sigma]\}$ with $\alpha([\sigma])(f) := \sigma f \sigma = f$ for all $f \in U_{\mathrm{ad}}^\sigma$, so $\mathrm{Aut}(\mathfrak{u})^\sigma \cong U_{\mathrm{ad}}^\sigma \times \mathcal{C}_2$.

Case 4: It remains to consider the case where $\mathrm{Aut}(\mathfrak{u})$ is disconnected, but σ is inner. We have the following subcases:

    i. $X_r^\star = A_n^{\mathbb{C},q}$, i.e., $(\mathfrak{u}, \sigma) = (\mathfrak{su}(n + 1, \mathbb{C}), \mathrm{Ad}(I_{n+1-q,q}))$, for $q \leq \frac{n+1}{2} > 1$:
We consider the complex conjugation $\mathrm{cj} : \mathfrak{su}(n+1, \mathbb{C}) \to \mathfrak{su}(n+1, \mathbb{C})$. It commutes with σ because $I_{n+1-q,q}$ is real. The rank of $\mathfrak{u}^{\mathrm{cj}} = \mathfrak{so}(n + 1, \mathbb{R})$ is $\frac{n}{2}$ for even n and $\frac{n+1}{2}$ for odd n, so never equal to n, the rank of $\mathfrak{u}$. Lemma 3.2.11 implies that the group element $[\mathrm{cj}] \in \pi_0\,(\mathrm{Aut}(\mathfrak{u})^\sigma) \backslash \pi_0\,(U_{\mathrm{ad}}^\sigma)$ has order two. We define $F := \{[\mathrm{id}_\mathfrak{u}], [\mathrm{cj}]\} \leq \pi_0\,(\mathrm{Aut}(\mathfrak{u})^\sigma)$ and a map $p : \mathrm{Aut}(\mathfrak{u})^\sigma \to F$ by $p(\theta) := [\mathrm{id}_\mathfrak{u}]$ if

$\theta \in U_{\mathrm{ad}}^{\sigma}$ and $p(\theta) := [\mathrm{cj}]$ otherwise. If $\theta, \theta' \in \mathrm{Aut}(\mathfrak{u})^{\sigma} \backslash U_{\mathrm{ad}}^{\sigma}$, then the equivalence classes of θ^{-1}, θ' in $\pi_0(\mathrm{Aut}(\mathfrak{u}))$ are different from $[\mathrm{id}_{\mathfrak{u}}]$, so, by Theorem 3.2.5, the equivalence classes coincide and $\theta\theta' \in \mathrm{Aut}(\mathfrak{u})_0 \cap \mathrm{Aut}(\mathfrak{u})^{\sigma} = U_{\mathrm{ad}}^{\sigma}$. We obtain $p(\theta\theta') = [\mathrm{id}_{\mathfrak{u}}] = [\mathrm{cj}]^2 = p(\theta)p(\theta')$ and hence p is a morphism of groups. This morphism has a natural section $s : F \rightarrow \mathrm{Aut}(\mathfrak{u})^{\sigma}$, thus $\mathrm{Aut}(\mathfrak{u})^{\sigma} \cong U_{\mathrm{ad}}^{\sigma} \rtimes F$.

ii. $X_r^* = D_q^{\mathbb{R},q}$, i.e., $(\mathfrak{u}, \sigma) = (\mathfrak{so}(2q, \mathbb{R}), \mathrm{Ad}(I_{q,q}))$, for even $q \geq 6$:
By Corollary 3.2.13 we have:

$$\mathrm{Aut}(\mathfrak{u})^{\sigma} = \{\, \mathrm{Ad}(X) | \, X \in O(2q, \mathbb{R}) \text{ and } I_{q,q} X I_{q,q} = \pm X \}$$
$$= \left\{ \mathrm{Ad}(X) \, \middle| \, X = \begin{pmatrix} A & 0 \\ 0 & B \end{pmatrix} \text{ or } X = \begin{pmatrix} 0 & A \\ B & 0 \end{pmatrix} \text{ with } A, B \in O(q, \mathbb{R}) \right\}.$$

Since $\mathrm{Inn}(\mathfrak{u}) = U_{\mathrm{ad}} = \{\mathrm{Ad}(X) | X \in SO(2q, \mathbb{R})\}$, it follows that

$$U_{\mathrm{ad}}^{\sigma} = \left\{ \mathrm{Ad}(X) \, \middle| \, X \in \left\{ \begin{pmatrix} A & 0 \\ 0 & B \end{pmatrix}, \begin{pmatrix} 0 & A \\ B & 0 \end{pmatrix} \right\} \text{ with } A, B \in O(q, \mathbb{R}), \tfrac{\det(A)}{\det(B)} = 1 \right\}.$$

By the surjectivity of the exponential maps $\mathfrak{u}^{\sigma} \rightarrow (U_{\mathrm{ad}}^{\sigma})_0$, $\mathfrak{so}(q, \mathbb{R}) \rightarrow SO(q, \mathbb{R})$ and by $\mathfrak{u}^{\sigma} = \mathfrak{so}(q, \mathbb{R}) \oplus \mathfrak{so}(q, \mathbb{R})$ we have:

$$(U_{\mathrm{ad}}^{\sigma})_0 = (\mathrm{Aut}(\mathfrak{u})^{\sigma})_0 = \left\{ \mathrm{Ad}(X) \, \middle| \, X = \begin{pmatrix} A & 0 \\ 0 & B \end{pmatrix} \text{ with } A, B \in SO(q, \mathbb{R}) \right\}.$$

The short exact sequence

$$1 \rightarrow (U_{\mathrm{ad}}^{\sigma})_0 \rightarrow U_{\mathrm{ad}}^{\sigma} \rightarrow \pi_0(U_{\mathrm{ad}}^{\sigma}) = \{[\mathrm{Ad}(X_1)], \dots, [\mathrm{Ad}(X_4)]\} \cong \mathcal{C}_2^2 \rightarrow 1$$

has the section $\pi_0(\mathrm{Aut}(\mathfrak{u})^{\sigma}) \rightarrow \mathrm{Aut}(\mathfrak{u})^{\sigma}$, $[\mathrm{Ad}(X_i)] \mapsto \mathrm{Ad}(X_i)$, where we use the matrices $X_1 = \begin{pmatrix} 1_q & 0 \\ 0 & 1_q \end{pmatrix}$, $X_2 = \begin{pmatrix} I_{1,q-1} & 0 \\ 0 & I_{1,q-1} \end{pmatrix}$, $X_3 = \begin{pmatrix} 0 & 1_q \\ -1_q & 0 \end{pmatrix}$ and $X_4 = \begin{pmatrix} 0 & I_{1,q-1} \\ -I_{1,q-1} & 0 \end{pmatrix}$. The eight components of $\mathrm{Aut}(\mathfrak{u})^{\sigma}$ are represented by $\mathrm{Ad}(X_i)$, $i = 1, \dots, 8$ with $X_1, \dots, X_4$ as above and $X_5 = \begin{pmatrix} I_{1,q-1} & 0 \\ 0 & 1_q \end{pmatrix}$, $X_6 = \begin{pmatrix} 1_q & 0 \\ 0 & -I_{1,q-1} \end{pmatrix}$, $X_7 = \begin{pmatrix} 0 & I_{1,q-1} \\ 1_q & 0 \end{pmatrix}$, $X_8 = \begin{pmatrix} 0 & 1_q \\ -I_{1,q-1} & 0 \end{pmatrix}$.

The group $\pi_0(\mathrm{Aut}(\mathfrak{u})^{\sigma})$ of eight elements has exactly two elements of order four, $[\mathrm{Ad}(X_7)]$ and $[\mathrm{Ad}(X_8)]$, and, by

$$[\mathrm{Ad}(X_3 X_5)] = [\mathrm{Ad}(X_8)] \neq [\mathrm{Ad}(X_7)] = [\mathrm{Ad}(X_5 X_3)],$$

it is not abelian, hence is isomorphic to $\mathcal{D}_4$. We obtain the split short exact sequence

$$1 \rightarrow U_{\mathrm{ad}}^{\sigma} \rightarrow \mathrm{Aut}(\mathfrak{u})^{\sigma} \rightarrow \mathrm{Aut}(\mathfrak{u})^{\sigma}/U_{\mathrm{ad}}^{\sigma} = \{[\mathrm{Ad}(X_1)], [\mathrm{Ad}(X_5)]\} \cong \mathcal{C}_2 \rightarrow 1.$$

iii. $X_r^* = D_q^{\mathbb{R},n}$, i.e., $(\mathfrak{u}, \sigma) = (\mathfrak{so}(2q, \mathbb{R}), \mathrm{Ad}(I_{2n-q,q}))$, for even $q < n \geq 5$:

In analogy with the previous subcase we have:

$$\mathrm{Aut}(\mathfrak{u})^\sigma = \{\, \mathrm{Ad}(X) \,|\, X \in \mathrm{O}(2n, \mathbb{R}) \text{ and } I_{2n-q,q} X I_{2n-q,q} = \pm X \,\}$$

$$= \left\{ \mathrm{Ad}(X) \,\middle|\, X = \begin{pmatrix} A & 0 \\ 0 & B \end{pmatrix} \text{ with } A \in \mathrm{O}(2n-q, \mathbb{R}), B \in \mathrm{O}(q, \mathbb{R}) \right\}$$

because, for matrices $A \in M_{2n-q,q}(\mathbb{R})$, $B \in M_{q,2n-q}(\mathbb{R})$, a matrix $X = \begin{pmatrix} 0 & A \\ B & 0 \end{pmatrix}$ is singular if $q < n$. By $U_{\mathrm{ad}} = \{\mathrm{Ad}(X) | X \in \mathrm{SO}(2n, \mathbb{R})\}$, it follows that

$$U_{\mathrm{ad}}^\sigma = \left\{ \mathrm{Ad}(X) \,\middle|\, X = \begin{pmatrix} A & 0 \\ 0 & B \end{pmatrix}, A \in \mathrm{O}(2q-n, \mathbb{R}), B \in \mathrm{O}(q, \mathbb{R}), \tfrac{\det(A)}{\det(B)} = 1 \right\}.$$

By $\mathfrak{u}^\sigma = \mathfrak{so}(2n-q, \mathbb{R}) \oplus \mathfrak{so}(q, \mathbb{R})$ and the surjectivity of the exponential maps $\mathfrak{u}^\sigma \to (U_{\mathrm{ad}}^\sigma)_0$, $\mathfrak{so}(2n-q, \mathbb{R}) \to \mathrm{SO}(2n-q, \mathbb{R})$ and $\mathfrak{so}(q, \mathbb{R}) \to \mathrm{SO}(q, \mathbb{R})$ we have:

$$(U_{\mathrm{ad}}^\sigma)_0 = (\mathrm{Aut}(\mathfrak{u})^\sigma)_0$$

$$= \left\{ \mathrm{Ad}(X) \,\middle|\, X = \begin{pmatrix} A & 0 \\ 0 & B \end{pmatrix} \text{ with } A \in \mathrm{SO}(2n-q, \mathbb{R}), B \in \mathrm{SO}(q, \mathbb{R}) \right\}.$$

The short exact sequence

$$1 \to (U_{\mathrm{ad}}^\sigma)_0 \to U_{\mathrm{ad}}^\sigma \to \pi_0\,(U_{\mathrm{ad}}^\sigma) = \{[\mathrm{Ad}(X_1)], [\mathrm{Ad}(X_2)]\} \cong \mathcal{C}_2 \to 1$$

with the representatives $X_1 = \begin{pmatrix} \mathbf{1}_{2n-q} & 0 \\ 0 & \mathbf{1}_q \end{pmatrix}$, $X_2 = \begin{pmatrix} I_{1,2n-q-1} & 0 \\ 0 & I_{1,q-1} \end{pmatrix}$ has the section $\pi_0\,(\mathrm{Aut}(\mathfrak{u})^\sigma) \to \mathrm{Aut}(\mathfrak{u})^\sigma$, $[\mathrm{Ad}(X_i)] \mapsto \mathrm{Ad}(X_i)$. The four components of $\mathrm{Aut}(\mathfrak{u})^\sigma$ are represented by $\mathrm{Ad}(X_i)$, $i = 1, \dots, 4$ with X_1, X_2 as above and $X_3 = \begin{pmatrix} I_{1,2n-q-1} & 0 \\ 0 & \mathbf{1}_q \end{pmatrix}$ and $X_4 = \begin{pmatrix} \mathbf{1}_{2n-q} & 0 \\ 0 & I_{1,q-1} \end{pmatrix}$. The group $\pi_0\,(\mathrm{Aut}(\mathfrak{u})^\sigma)$ of four elements has no element of order four, hence is isomorphic to $\mathcal{C}_2 \times \mathcal{C}_2$. We obtain the split short exact sequence

$$1 \to U_{\mathrm{ad}}^\sigma \to \mathrm{Aut}(\mathfrak{u})^\sigma \to \mathrm{Aut}(\mathfrak{u})^\sigma / U_{\mathrm{ad}}^\sigma = \{[\mathrm{Ad}(X_1)], [\mathrm{Ad}(X_3)]\} \cong \mathcal{C}_2 \to 1.$$

iv. $X_r^* = D_n^{\mathbb{H}}$, i.e., $(\mathfrak{u}, \sigma) = (\mathfrak{so}(2n, \mathbb{R}), \mathrm{Ad}(J_n))$, for $n \geq 5$:

In analogy with the two previous subcases we have, for $J_n = \begin{pmatrix} 0 & \mathbf{1}_n \\ -\mathbf{1}_n & 0 \end{pmatrix}$:

$$\mathrm{Aut}(\mathfrak{u})^\sigma = \{\, \mathrm{Ad}(X) \,|\, X \in \mathrm{O}(2n, \mathbb{R}) \text{ and } J_n X J_n^{-1} = \pm X \,\}$$

$$= \{\, \mathrm{Ad}(X) \,|\, X \in \mathrm{O}(2n, \mathbb{R})^{J_n} = \mathrm{U}(n, \mathbb{C}) \,\}$$

$$\cup \{\, \mathrm{Ad}(X) \,|\, X \in \mathrm{O}(2n, \mathbb{R})^{-J_n} = I_{n,n}\, \mathrm{U}(n, \mathbb{C}) \,\},$$

where we consider $\mathrm{U}(n, \mathbb{C})$ as the subgroup of $\mathrm{O}(2n, \mathbb{R})$ fixed by the conjugation with J_n. The relation $\det_{\mathbb{R}}(X) = |\det_{\mathbb{C}}(X)|^2$ tells us that the matrices in $\mathrm{U}(n, \mathbb{C})$

have real determinant 1, whereas the matrices in $I_{n,n} \mathrm{U}(n,\mathbb{C})$ have real determinant $(-1)^n$, so we have the following decomposition into closed open subsets:

$$\mathrm{Aut}(\mathfrak{u})^\sigma = \{\mathrm{Ad}(X)\,|\,X \in \mathrm{U}(n,\mathbb{C})\} \cup \{\mathrm{Ad}(X)\,|\,X \in I_{n,n}\,\mathrm{U}(n,\mathbb{C})\}$$

Since $U_{\mathrm{ad}} = \{\mathrm{Ad}(X)|X \in \mathrm{SO}(2n,\mathbb{R})\}$, it follows that

$$U_{\mathrm{ad}}^\sigma = \underbrace{\{\mathrm{Ad}(X)\,|\,X \in \mathrm{U}(n,\mathbb{C})\}}_{=(U_{\mathrm{ad}}^\sigma)_0 = (\mathrm{Aut}(\mathfrak{u})^\sigma)_0} \cup \left\{\mathrm{Ad}(X)\,\Big|\,X \in I_{n,n}\,\mathrm{U}(n,\mathbb{C}), \det_{\mathbb{R}}(X) = 1\right\}.$$

If n is odd, then $k = 0$ and $U_{\mathrm{ad}}^\sigma = (U_{\mathrm{ad}}^\sigma)_0$. We obtain the split short exact sequence

$$\mathbf{1} \to U_{\mathrm{ad}}^\sigma \to \mathrm{Aut}(\mathfrak{u})^\sigma \to \mathrm{Aut}(\mathfrak{u})^\sigma/U_{\mathrm{ad}}^\sigma = \{[\mathrm{id}_\mathfrak{u}], [\mathrm{Ad}(I_{n,n})]\} \cong \mathcal{C}_2 \to \mathbf{1}.$$

If n is even, then $k = 1$ and $\mathrm{Aut}(\mathfrak{u})^\sigma = U_{\mathrm{ad}}^\sigma$.

v. $X_r^* = D_4^{\mathbb{R},2}$, i.e., $(\mathfrak{u},\sigma) = (\mathfrak{so}(8,\mathbb{R}), \mathrm{Ad}(I_{6,2}))$:

The automorphism group of the compact subalgebra $\mathfrak{k} = \mathfrak{u}^\sigma = \mathfrak{so}(2,\mathbb{R}) \oplus \mathfrak{so}(6,\mathbb{R})$ is $\mathrm{Aut}(\mathbb{R} \oplus \mathfrak{su}(4,\mathbb{C}))$, so $\pi_0(\mathrm{Aut}(\mathfrak{k})) \cong \mathcal{C}_2 \times \mathcal{C}_2$, since the Dynkin diagram A_3 has one non-trivial symmetry only. Since $\sigma \in \mathrm{Aut}(\mathfrak{u})$ is inner, Lemma 3.2.6 implies that $\pi_0(\mathrm{Aut}(\mathfrak{u})^\sigma)$ is isomorphic to a subgroup of $\mathcal{C}_2 \times \mathcal{C}_2$ and since $k = 1$, we either have $\pi_0(\mathrm{Aut}(\mathfrak{u})^\sigma) \cong \mathcal{C}_2$ or $\pi_0(\mathrm{Aut}(\mathfrak{u})^\sigma) \cong \mathcal{C}_2 \times \mathcal{C}_2$.

By $\mathrm{Aut}(\mathfrak{so}(8,\mathbb{C})) \supseteq \mathrm{Ad}(\mathrm{O}(8,\mathbb{C}))$, we have at least:

$$\mathrm{Aut}(\mathfrak{u})^\sigma \supseteq \left\{\mathrm{Ad}(X)\,\Big|\,X = \begin{pmatrix} A & \mathbf{0} \\ \mathbf{0} & B \end{pmatrix} \text{ with } A \in \mathrm{O}(6,\mathbb{R}), B \in \mathrm{O}(2,\mathbb{R})\right\},$$

Since $U_{\mathrm{ad}} = \{\mathrm{Ad}(X)|X \in \mathrm{SO}(8,\mathbb{R})\}$, it follows that

$$U_{\mathrm{ad}}^\sigma = \left\{\mathrm{Ad}(X)\,\Big|\,X = \begin{pmatrix} A & \mathbf{0} \\ \mathbf{0} & B \end{pmatrix} \text{ with } A \in \mathrm{O}(6,\mathbb{R}), B \in \mathrm{O}(2,\mathbb{R}), \tfrac{\det(A)}{\det(B)} = 1\right\}.$$

By the surjectivity of the exponential maps $\mathfrak{so}(2,\mathbb{R}) \oplus \mathfrak{so}(6,\mathbb{R}) = \mathfrak{u}^\sigma \to (U_{\mathrm{ad}}^\sigma)_0$, $\mathfrak{so}(6,\mathbb{R}) \to \mathrm{SO}(6,\mathbb{R})$ and $\mathfrak{so}(2,\mathbb{R}) \cong \mathbb{R} \to \mathrm{SO}(2,\mathbb{R})$ we have:

$$(U_{\mathrm{ad}}^\sigma)_0 = (\mathrm{Aut}(\mathfrak{u})^\sigma)_0 = \left\{\mathrm{Ad}(X)\,\Big|\,X = \begin{pmatrix} A & \mathbf{0} \\ \mathbf{0} & B \end{pmatrix}, A \in \mathrm{SO}(6,\mathbb{R}), B \in \mathrm{SO}(2,\mathbb{R})\right\}.$$

But we already know that $\pi_0(U_{\mathrm{ad}}^\sigma) \cong \mathcal{C}_2$, implying the equality

$$U_{\mathrm{ad}}^\sigma = \left\{\mathrm{Ad}(X)\,\Big|\,X = \begin{pmatrix} A & \mathbf{0} \\ \mathbf{0} & B \end{pmatrix} \text{ with } A \in \mathrm{O}(6,\mathbb{R}), B \in \mathrm{O}(2,\mathbb{R}), \tfrac{\det(A)}{\det(B)} = 1\right\}.$$

We consider $X' = \begin{pmatrix} 1_6 & \mathbf{0} & \mathbf{0} \\ \mathbf{0} & 0 & 1 \\ \mathbf{0} & 1 & 0 \end{pmatrix}$. Hence $\mathrm{Ad}(X') \in \mathrm{Aut}(\mathfrak{u}) \backslash U_{\mathrm{ad}}^\sigma$, and we calculate:

$$\begin{aligned}
\sigma\,\mathrm{Ad}(X')\sigma(x) &= \mathrm{Ad}(I_{6,2}X'I_{6,2})(x) \\
&= \mathrm{Ad}\left(\begin{pmatrix} -1_6 & \mathbf{0} & \mathbf{0} \\ \mathbf{0} & 1 & 0 \\ \mathbf{0} & 0 & 1 \end{pmatrix}\begin{pmatrix} 1_6 & \mathbf{0} & \mathbf{0} \\ \mathbf{0} & 0 & 1 \\ \mathbf{0} & 1 & 0 \end{pmatrix}\begin{pmatrix} -1_6 & \mathbf{0} & \mathbf{0} \\ \mathbf{0} & 1 & 0 \\ \mathbf{0} & 0 & 1 \end{pmatrix}\right)(x) \\
&= \mathrm{Ad}(X')(x).
\end{aligned}$$

So we see that $[\mathrm{Aut}(\mathfrak{u})^\sigma : U_{\mathrm{ad}}^\sigma] > 1$, so $[\mathrm{Aut}(\mathfrak{u})^\sigma : U_{\mathrm{ad}}^\sigma] = 2$, yielding the equality

$$\mathrm{Aut}(\mathfrak{u})^\sigma = \left\{ \mathrm{Ad}(X) \,\middle|\, X = \begin{pmatrix} A & \mathbf{0} \\ \mathbf{0} & B \end{pmatrix} \text{ with } A \in \mathrm{O}(6,\mathbb{R}), B \in \mathrm{O}(2,\mathbb{R}) \right\}.$$

We obtain the split short exact sequence

$$1 \to U_{\mathrm{ad}}^\sigma \to \mathrm{Aut}(\mathfrak{u})^\sigma \to \mathrm{Aut}(\mathfrak{u})^\sigma/U_{\mathrm{ad}}^\sigma = \left\{ [\mathrm{id}_\mathfrak{u}] , \left[\mathrm{Ad}(X') \right] \right\} \cong \mathcal{C}_2 \to 1.$$

vi. $X_r^* = D_4^{\mathbb{R},2}$, i.e., $(\mathfrak{u}, \sigma) = (\mathfrak{so}(8,\mathbb{R}), \mathrm{Ad}(I_{4,4}))$:
The automorphism group of the compact subalgebra $\mathfrak{k} = \mathfrak{u}^\sigma$ is

$$\mathrm{Aut}(\mathfrak{so}(4,\mathbb{R}) \oplus \mathfrak{so}(4,\mathbb{R})) \cong \mathrm{Aut}(\mathfrak{su}(2,\mathbb{C}) \oplus \mathfrak{su}(2,\mathbb{C}) \oplus \mathfrak{su}(2,\mathbb{C}) \oplus \mathfrak{su}(2,\mathbb{C})),$$

so $\pi_0(\mathrm{Aut}(\mathfrak{k}))$ is isomorphic the symmetry group of the Dynkin diagram with four disconnected nodes, hence isomorphic to $\mathcal{S}_4$. Since $\sigma \in \mathrm{Aut}(\mathfrak{u})$ is inner, Lemma 3.2.6 implies that $\pi_0\left(\mathrm{Aut}(\mathfrak{u})^\sigma\right)$ is isomorphic to a subgroup of $\mathcal{S}_4$. Furthermore, we know that $\mathcal{C}_2^2 \cong \pi_0\left(U_{\mathrm{ad}}^\sigma\right) \trianglelefteq \pi_0\left(\mathrm{Aut}(\mathfrak{u})^\sigma\right)$, thus the Third Isomorphism Theorem yields that the order of $\mathrm{Aut}(\mathfrak{u})^\sigma/U_{\mathrm{ad}}^\sigma \cong \pi_0\left(\mathrm{Aut}(\mathfrak{u})^\sigma\right)/\pi_0\left(U_{\mathrm{ad}}^\sigma\right)$ is at most six.

By Lemma 3.2.7, there is an element $[\theta] \in \mathrm{Aut}(\mathfrak{u})^\sigma/U_{\mathrm{ad}}^\sigma \cong \pi_0\left(\mathrm{Aut}(\mathfrak{u})^\sigma\right)/\pi_0\left(U_{\mathrm{ad}}^\sigma\right)$ of order three. The automorphism $\mathrm{Ad}(Y)$ with $Y = \begin{pmatrix} \mathbf{1}_4 & \mathbf{0} \\ \mathbf{0} & -I_{3,1} \end{pmatrix}$ commutes with $\sigma = \mathrm{Ad}(I_{4,4})$ and the rank of $\mathfrak{u}^{\mathrm{Ad}(Y)} = \mathfrak{so}(7,\mathbb{R})$ is different from the rank of $\mathfrak{u}$, thus, by Lemma 3.2.11, the order of $[\mathrm{Ad}(Y)] \in \mathrm{Aut}(\mathfrak{u})^\sigma/U_{\mathrm{ad}}^\sigma$ is two. Hence $\mathrm{Aut}(\mathfrak{u})^\sigma/U_{\mathrm{ad}}^\sigma$ is a group of six elements and $\pi_0\left(\mathrm{Aut}(\mathfrak{u})^\sigma\right)$ has 24 elements, so $\pi_0\left(\mathrm{Aut}(\mathfrak{u})^\sigma\right) \cong \mathcal{S}_4$.

The Second Isomorphism Theorem yields

$$\mathrm{Aut}(\mathfrak{u})^\sigma/U_{\mathrm{ad}}^\sigma = \mathrm{Aut}(\mathfrak{u})^\sigma/\left(U_{\mathrm{ad}} \cap \mathrm{Aut}(\mathfrak{u})^\sigma\right) \cong \left(U_{\mathrm{ad}} \cdot \mathrm{Aut}(\mathfrak{u})^\sigma\right)/U_{\mathrm{ad}} \leq \pi_0\left(\mathrm{Aut}(\mathfrak{u})\right).$$

Since $\pi_0\left(\mathrm{Aut}(\mathfrak{u})\right) \cong \mathcal{S}_3$ by Theorem 3.2.5, we have $\mathrm{Aut}(\mathfrak{u})^\sigma/U_{\mathrm{ad}}^\sigma \cong \mathcal{S}_3$. We obtain the short exact sequence

$$1 \to U_{\mathrm{ad}}^\sigma \to \mathrm{Aut}(\mathfrak{u})^\sigma \to \mathrm{Aut}(\mathfrak{u})^\sigma/U_{\mathrm{ad}}^\sigma$$
$$= \underbrace{\left\{ [\mathrm{id}_\mathfrak{u}] , [\theta] , \left[\theta^2\right] , [\mathrm{Ad}(Y)] , [\theta\,\mathrm{Ad}(Y)] , \left[\theta^2\,\mathrm{Ad}(Y)\right] \right\}}_{\cong \mathcal{S}_3} \to 1,$$

which is split: We take $\theta' \in \mathrm{Aut}(\mathfrak{so}(8,\mathbb{C}))^\tau$ and $\theta \in \mathrm{Aut}(\mathfrak{so}(8,\mathbb{R}))^\sigma$ as described in and with the notation of Remark 3.2.8. The subgroup $\Gamma \leq \mathrm{Aut}(\mathfrak{u})^\sigma$ generated by $\mathrm{Ad}(Y)$ and θ is mapped onto $\mathcal{S}_3$ by a group morphism. So the First Isomorphism Theorem yields that $\mathcal{S}_3$ is isomorphic to a quotient of Γ, thus it suffices to show that the order of Γ is at most six.

We calculate, for the standard Chevalley generators h_i, x_i, y_i, where $i = 1, 2, 3, 4$, of $\mathfrak{so}(8, \mathbb{C})$ and $\mathfrak{r} := \mathbb{C}x_1 \oplus \mathbb{C}x_2 \oplus \mathbb{C}x_3 \oplus \mathbb{C}x_4$:

$$T_1 := \left[\theta'_{|\mathfrak{h}}\right]_{(h_1,\ldots,h_4)} = \frac{1}{2}\begin{pmatrix} 1 & 1 & 1 & 1 \\ 1 & 1 & -1 & -1 \\ 1 & -1 & 1 & -1 \\ -1 & 1 & 1 & -1 \end{pmatrix},$$

$$T_2 := \left[\theta'_{|\mathfrak{r}}\right]_{(x_1,\ldots,x_4)} = \begin{pmatrix} 0 & 0 & 0 & 1 \\ 0 & 1 & 0 & 0 \\ 1 & 0 & 0 & 0 \\ 0 & 0 & 1 & 0 \end{pmatrix},$$

$$Y_1 := \left[\mathrm{Ad}(Y)_{|\mathfrak{h}}\right]_{(h_1,\ldots,h_4)} = \begin{pmatrix} 1 & 0 & 0 & 0 \\ 0 & 1 & 0 & 0 \\ 0 & 0 & 1 & 0 \\ 0 & 0 & 0 & -1 \end{pmatrix},$$

$$Y_2 := \left[\mathrm{Ad}(Y)_{|\mathfrak{r}}\right]_{(x_1,\ldots,x_4)} = \begin{pmatrix} 1 & 0 & 0 & 0 \\ 0 & 1 & 0 & 0 \\ 0 & 0 & 0 & 1 \\ 0 & 0 & 1 & 0 \end{pmatrix}.$$

The identities $T_1 Y_1 T_1 = Y_1$ and $T_2 Y_2 T_2 = Y_2$ show[8] that $\theta' \mathrm{Ad}(Y)\theta' = \mathrm{Ad}(Y)$ on $\mathfrak{u}_{\mathbb{C}}$, so also $\theta \mathrm{Ad}(Y)\theta = \mathrm{Ad}(Y)$ on $\mathfrak{u}$, hence $\mathrm{Ad}(Y)\{1, \theta, \theta^2\}\mathrm{Ad}(Y) = \{1, \theta, \theta^2\}$, thus $\{1, \theta, \theta^2\} \trianglelefteq \Gamma$ and $\{1, \mathrm{Ad}(Y)\} \cdot \{1, \theta, \theta^2\}$ is a subgroup of Γ, yielding $\{1, \mathrm{Ad}(Y)\} \cdot \{1, \theta, \theta^2\} = \Gamma$, showing $\#\Gamma \leq 6$ and hence $\Gamma \cong \mathcal{S}_3$.

vii. $X_r^* \in \left\{E_{6(2)}, E_{6(-14)}\right\}$:
We use the the notation of Lemma 3.2.9. The involution σ is inner, so we have $\sigma = e^Y = \exp_{\mathrm{Aut}(\mathfrak{u})}(\mathrm{ad}_{\mathfrak{u}}(Y))$ or $\sigma = e^Z = \exp_{\mathrm{Aut}(\mathfrak{u})}(\mathrm{ad}_{\mathfrak{u}}(Z))$, respectively. Since $\mathrm{Aut}(\mathfrak{u})^{\sigma}/U_{\mathrm{ad}}^{\sigma}$ is isomorphic to a subgroup of $\pi_0\left(\mathrm{Aut}(\mathfrak{u})\right)$, which is isomorphic to $\mathcal{C}_2$ by Theorem 3.2.5, and the non-trivial Dynkin diagram involution ρ_1 is outer and commutes with e^Y and e^Z, we obtain the split short exact sequence

$$1 \to U_{\mathrm{ad}}^{\sigma} \to \mathrm{Aut}(\mathfrak{u})^{\sigma} \to \mathrm{Aut}(\mathfrak{u})^{\sigma}/U_{\mathrm{ad}}^{\sigma} = \{[\mathrm{id}_{\mathfrak{u}}], [\rho_1]\} \cong \mathcal{C}_2 \to 1.$$

$\square$

Corollary 3.2.16. *If $\mathfrak{g}$ is a real central simple non-compact Lie algebra, then the following table provides all possibilities for $\pi_0(\mathrm{Aut}(\mathfrak{g}))$.*

---

[8]The automorphisms are uniquely determined by their values on the h_i's and x_i's and by commuting with the conjugation $\tau : \mathfrak{so}(8, \mathbb{R}) + i\mathfrak{so}(8, \mathbb{R}) = \mathfrak{so}(8, \mathbb{C}) \to \mathfrak{so}(8, \mathbb{C})$.

$\mathfrak{g}$	$\pi_0(\mathrm{Aut}(\mathfrak{g}))$	$\#\,\mathrm{Conj}(\pi_0(\mathrm{Aut}(\mathfrak{g})))$
$\mathfrak{sl}(2,\mathbb{R})$	$\mathcal{C}_2$	2
$\mathfrak{sl}(n+1,\mathbb{R})$ for even $n \geq 2$	$\mathcal{C}_2$	2
$\mathfrak{sl}(n+1,\mathbb{R})$ for odd $n \geq 3$	$\mathcal{C}_2 \times \mathcal{C}_2$	4
$\mathfrak{su}(q,n+1-q,\mathbb{C})$ for $q < \frac{n+1}{2} > 1$	$\mathcal{C}_2$	2
$\mathfrak{su}(n+1,n+1,\mathbb{C})$ for $n \geq 1$	$\mathcal{C}_2 \times \mathcal{C}_2$	4
$\mathfrak{sl}(n+1,\mathbb{H}) = \mathfrak{su}^*(2n+2,\mathbb{C})$ for $n \geq 1$	$\mathcal{C}_2$	2
$\mathfrak{so}(q,2n+1-q,\mathbb{R})$ for $q \leq n \geq 2$	$\mathcal{C}_2$	2
$\mathfrak{sp}(2n,\mathbb{R})$ for $n \geq 3$	$\mathcal{C}_2$	2
$\mathfrak{sp}(n,n,\mathbb{H})$ for $n \geq 1$	$\mathcal{C}_2$	2
$\mathfrak{so}(q,2n-q,\mathbb{R})$ for odd $q < n \geq 4$	$\mathcal{C}_2$	2
$\mathfrak{so}(q,2n-q,\mathbb{R})$ for even $q < n \geq 4$	$\mathcal{C}_2 \times \mathcal{C}_2$	4
$\mathfrak{so}(n,n,\mathbb{R})$ for odd $n \geq 5$	$\mathcal{C}_2 \times \mathcal{C}_2$	4
$\mathfrak{so}(n,n,\mathbb{R})$ for even $n \geq 6$	$\mathcal{D}_4$	5
$\mathfrak{so}(4,4,\mathbb{R})$	$\mathcal{S}_4$	5
$\mathfrak{so}^*(4n,\mathbb{R})$ for $n \geq 3$	$\mathcal{C}_2$	2
$\mathfrak{so}^*(4n+2,\mathbb{R})$ for $n \geq 2$	$\mathcal{C}_2$	2
$\mathfrak{e}_{6(j)}$ for $j = 6, 2, -14, -26$	$\mathcal{C}_2$	2
$\mathfrak{e}_{7(j)}$ for $j = 7, -25$	$\mathcal{C}_2$	2
all others	1	1

Corollary 3.2.17. *If $\mathfrak{g}$ is a real central simple non-compact Lie algebra such that $\pi_0(\mathrm{Aut}(\mathfrak{g})) = 1$ or $\mathfrak{g}$ is isomorphic to one of the following:*

1. *$\mathfrak{sl}(n+1,\mathbb{R})$ for even $n \geq 2$,*

2. *$\mathfrak{su}(q,n+1-q,\mathbb{C})$ for $q < \frac{n+1}{2} > 1$,*

3. *$\mathfrak{sl}(n+1,\mathbb{H}) = \mathfrak{su}^*(2n+2,\mathbb{C})$ for $n \geq 1$,*

4. *$\mathfrak{so}(q,2n-q,\mathbb{R})$ for odd $q < n \geq 4$,*

5. *$\mathfrak{so}^*(4n+2,\mathbb{R})$ for $n \geq 2$,*

6. *$\mathfrak{e}_{6(j)}$ for $j = 6, 2, -14, -26$,*

then there is an isomorphism $\mathrm{Aut}(\mathfrak{g}) \cong \mathrm{Aut}(\mathfrak{g})_0 \rtimes \pi_0(\mathrm{Aut}(\mathfrak{g}))$.

Proof. By Proposition 3.2.15, we have $(\mathrm{Aut}(\mathfrak{u})_0)^\sigma = ((\mathrm{Aut}(\mathfrak{u})_0)^\sigma)_0 = (\mathrm{Aut}(\mathfrak{u})^\sigma)_0$ in the mentioned cases, yielding $\mathrm{Aut}(\mathfrak{g}) \cong \mathrm{Aut}(\mathfrak{g})_0 \rtimes \pi_0(\mathrm{Aut}(\mathfrak{g}))$. $\square$

Corollary 3.2.18. *If $\mathfrak{g}$ is a real central simple non-compact Lie algebra isomorphic to one of the following:*

1. *$\mathfrak{so}(q,2n+1-q,\mathbb{R})$ for $q \leq n \geq 2$,*

2. *$\mathfrak{so}(q,2n-q,\mathbb{R})$ for even $q < n \geq 4$,*

3. *$\mathfrak{so}(n,n,\mathbb{R})$ for odd $n \geq 5$,*

4. $\mathfrak{so}(n,n,\mathbb{R})$ for even $n \geq 6$,

5. $\mathfrak{so}(4,4,\mathbb{R})$,

6. $\mathfrak{so}^*(4n,\mathbb{R})$ for $n \geq 3$,

then $\mathrm{Aut}(\mathfrak{g}) \cong \mathrm{Aut}(\mathfrak{g})_0 \rtimes \pi_0(\mathrm{Aut}(\mathfrak{g}))$.

Proof. We will see that, in the mentioned cases and with the notation of Remark 3.2.14, there are representatives of the elements in $\pi_0\left(\mathrm{Aut}(\mathfrak{u})^\sigma\right)$ which form a subgroup of $\mathrm{Aut}(\mathfrak{u})^\sigma$ isomorphic to $\pi_0\left(\mathrm{Aut}(\mathfrak{u})^\sigma\right)$. Hence, by Proposition 3.2.15.3, we have $\mathrm{Aut}(\mathfrak{g}) \cong \mathrm{Aut}(\mathfrak{g})_0 \rtimes \pi_0(\mathrm{Aut}(\mathfrak{g}))$.

1. $\mathfrak{g} \cong \mathfrak{so}(q, 2n+1-q, \mathbb{R})$ for $q \leq n \geq 2$:
   $\mathrm{Aut}(\mathfrak{so}(2n+1, \mathbb{R}))$ is connected by Theorem 3.2.5, hence equal to $\mathrm{Ad}\left(\mathrm{SO}(2n+1, \mathbb{R})\right)$. Thus, in analogy with the proof of Proposition 3.2.15.4, Case 4, iii, we know that

$$\pi_0\left(\mathrm{Aut}(\mathfrak{so}(2n+1, \mathbb{R}))^\sigma\right) = \{\mathbf{1}, [\mathrm{Ad}(X_2)]\} \cong \mathcal{C}_2$$

   for the matrix $X_2 = \begin{pmatrix} I_{1,2n-q} & \mathbf{0} \\ \mathbf{0} & I_{1,q-1} \end{pmatrix}$.

2. $\mathfrak{g} \cong \mathfrak{so}(q, 2n-q, \mathbb{R})$ for even $q < n \geq 4$:
   By the proof of Proposition 3.2.15.4, Case 4, iii and v, we know that

$$\pi_0\left(\mathrm{Aut}(\mathfrak{so}(2n, \mathbb{R}))^\sigma\right) = \{\mathbf{1}, [\mathrm{Ad}(X_2)], [\mathrm{Ad}(X_3)], [\mathrm{Ad}(X_4)]\} \cong \mathcal{C}_2 \times \mathcal{C}_2$$

   for $X_2 = \begin{pmatrix} I_{1,2n-q-1} & \mathbf{0} \\ \mathbf{0} & I_{1,q-1} \end{pmatrix}$, $X_3 = \begin{pmatrix} I_{1,2n-q-1} & \mathbf{0} \\ \mathbf{0} & 1_q \end{pmatrix}$, $X_4 = \begin{pmatrix} 1_{2n-q} & \mathbf{0} \\ \mathbf{0} & I_{1,q-1} \end{pmatrix}$.

3. $\mathfrak{g} \cong \mathfrak{so}(n,n,\mathbb{R})$ for odd $n \geq 5$:
   In analogy with the proof of Proposition 3.2.15.4, Case 4, ii, we have:

$$\mathrm{Aut}(\mathfrak{so}(2n, \mathbb{R}))^\sigma = \{\,\mathrm{Ad}(X) \mid X \in \mathrm{O}(2n, \mathbb{R}) \text{ and } I_{n,n} X I_{n,n} = \pm X\}$$

$$= \left\{\mathrm{Ad}(X) \,\middle|\, X \in \left\{\begin{pmatrix} A & \mathbf{0} \\ \mathbf{0} & B \end{pmatrix}, \begin{pmatrix} \mathbf{0} & A \\ B & \mathbf{0} \end{pmatrix}\right\}, A, B \in \mathrm{O}(n, \mathbb{R})\right\},$$

$$(\mathrm{Aut}(\mathfrak{so}(2n, \mathbb{R}))_0)^\sigma = \left\{\mathrm{Ad}(X) \,\middle|\, X = \begin{pmatrix} A & \mathbf{0} \\ \mathbf{0} & B \end{pmatrix}, A, B \in \mathrm{O}(n, \mathbb{R}), \tfrac{\det(A)}{\det(B)} = 1\right\}$$

$$\cup \left\{\mathrm{Ad}(X) \,\middle|\, X = \begin{pmatrix} \mathbf{0} & A \\ B & \mathbf{0} \end{pmatrix} \text{ with } A, B \in \mathrm{O}(n, \mathbb{R}), \tfrac{\det(A)}{\det(B)} = -1\right\},$$

$$((\mathrm{Aut}(\mathfrak{so}(2n, \mathbb{R}))_0)^\sigma)_0 = (\mathrm{Aut}(\mathfrak{so}(2n, \mathbb{R}))^\sigma)_0$$

$$= \left\{\mathrm{Ad}(X) \,\middle|\, X = \begin{pmatrix} A & \mathbf{0} \\ \mathbf{0} & B \end{pmatrix} \text{ with } A, B \in \mathrm{SO}(n, \mathbb{R})\right\}.$$

We know that n is odd, $\mathrm{Ad}(A) = \mathrm{Ad}(-A)$ for each orthogonal matrix $A \in \mathrm{O}(n, \mathbb{R})$, and $\mathrm{SO}(n, \mathbb{R}) \to \mathrm{O}(n, \mathbb{R}) \backslash \mathrm{SO}(n, \mathbb{R})$, $A \mapsto -A$ is a bijection, so we can also write:

$$(\mathrm{Aut}(\mathfrak{so}(2n, \mathbb{R}))_0)^\sigma = \left\{\mathrm{Ad}(X) \,\middle|\, X \in \left\{\begin{pmatrix} A & \mathbf{0} \\ \mathbf{0} & B \end{pmatrix}, \begin{pmatrix} \mathbf{0} & A \\ -B & \mathbf{0} \end{pmatrix}\right\} \text{ with } A, B \in \mathrm{SO}(n, \mathbb{R})\right\}.$$

Thus we have

$$\pi_0\left(\mathrm{Aut}(\mathfrak{so}(2n,\mathbb{R}))^\sigma\right) = \{\mathbf{1}, [\mathrm{Ad}(X_3)], [\mathrm{Ad}(X_5)], [\mathrm{Ad}(X_8)]\} \cong \mathcal{C}_2 \times \mathcal{C}_2$$

for the matrices $X_3 = \begin{pmatrix} 0 & 1_n \\ -1_n & 0 \end{pmatrix}$, $X_5 = \begin{pmatrix} I_{1,n-1} & 0 \\ 0 & 1_n \end{pmatrix}$, $X_8 = \begin{pmatrix} 0 & 1_n \\ -I_{1,n-1} & 0 \end{pmatrix}$.

4. $\mathfrak{g} \cong \mathfrak{so}(n,n,\mathbb{R})$ for even $n \geq 6$:
By the proof of Proposition 3.2.15.4, Case 4, ii, we know that

$$\pi_0\left(\mathrm{Aut}(\mathfrak{so}(2n,\mathbb{R}))^\sigma\right) = \{\mathbf{1}, [\mathrm{Ad}(X_2)], \ldots, [\mathrm{Ad}(X_8)]\} \cong \mathcal{D}_4$$

for the matrices $X_2 = \begin{pmatrix} I_{1,n-1} & 0 \\ 0 & I_{1,n-1} \end{pmatrix}$, $X_3 = \begin{pmatrix} 0 & 1_n \\ -1_n & 0 \end{pmatrix}$, $X_4 = \begin{pmatrix} 0 & I_{1,n-1} \\ -I_{1,n-1} & 0 \end{pmatrix}$,
$X_5 = \begin{pmatrix} I_{1,n-1} & 0 \\ 0 & 1_n \end{pmatrix}$, $X_6 = \begin{pmatrix} 1_n & 0 \\ 0 & -I_{1,n-1} \end{pmatrix}$, $X_7 = \begin{pmatrix} 0 & I_{1,n-1} \\ 1_n & 0 \end{pmatrix}$, $X_8 = \begin{pmatrix} 0 & 1_n \\ -I_{1,n-1} & 0 \end{pmatrix}$.

5. $\mathfrak{g} \cong \mathfrak{so}(4,4,\mathbb{R})$:
In analogy with the proof of Proposition 3.2.15.4, Case 4, ii, we have:

$$\mathrm{Aut}(\mathfrak{so}(8,\mathbb{R}))^\sigma \supseteq \{\mathrm{Ad}(X)\,|\, X \in \mathrm{O}(8,\mathbb{R}) \text{ and } I_{4,4}XI_{4,4} = \pm X\}$$
$$= \left\{\mathrm{Ad}(X)\,\Big|\, X = \begin{pmatrix} A & 0 \\ 0 & B \end{pmatrix} \text{ or } X = \begin{pmatrix} 0 & A \\ B & 0 \end{pmatrix} \text{ with } A, B \in \mathrm{O}(4,\mathbb{R}) \right\},$$
$$(\mathrm{Aut}(\mathfrak{so}(8,\mathbb{R}))_0)^\sigma = \left\{\mathrm{Ad}(X)\,\Big|\, X = \begin{pmatrix} A & 0 \\ 0 & B \end{pmatrix} \text{ or } X = \begin{pmatrix} 0 & A \\ B & 0 \end{pmatrix} \text{ with } A, B \in \mathrm{O}(4,\mathbb{R}), \right.$$
$$\left. \det(A) = \det(B)\right\},$$
$$((\mathrm{Aut}(\mathfrak{so}(8,\mathbb{R}))_0)^\sigma)_0 = (\mathrm{Aut}(\mathfrak{so}(8,\mathbb{R}))^\sigma)_0$$
$$= \left\{\mathrm{Ad}(X)\,\Big|\, X = \begin{pmatrix} A & 0 \\ 0 & B \end{pmatrix} \text{ with } A, B \in \mathrm{SO}(4,\mathbb{R}) \right\},$$
$$\pi_0\left((\mathrm{Aut}(\mathfrak{so}(8,\mathbb{R}))_0)^\sigma\right) = \{\mathbf{1}, [\mathrm{Ad}(X_2)], [\mathrm{Ad}(X_3)], [\mathrm{Ad}(X_4)]\} \cong \mathcal{C}_2 \times \mathcal{C}_2$$

for the matrices $X_2 = \begin{pmatrix} I_{3,1} & 0 \\ 0 & I_{3,1} \end{pmatrix}$, $X_3 = \begin{pmatrix} 0 & 1_4 \\ -1_4 & 0 \end{pmatrix}$, $X_4 = \begin{pmatrix} 0 & I_{3,1} \\ -I_{3,1} & 0 \end{pmatrix}$. By the proof
of Proposition 3.2.15.4, Case 4, vi, we know that

$$\mathrm{Aut}(\mathfrak{so}(8,\mathbb{R}))^\sigma / (\mathrm{Aut}(\mathfrak{so}(8,\mathbb{R}))_0)^\sigma \cong \left\{\mathbf{1}, [\theta], [\theta^2], [\mathrm{Ad}(Y)], [\theta\,\mathrm{Ad}(Y)], [\theta^2\,\mathrm{Ad}(Y)]\right\} \cong \mathcal{S}_3$$

for the matrix $Y = \begin{pmatrix} 1_4 & 0 \\ 0 & -I_{3,1} \end{pmatrix}$ and the triality automorphism θ as described in and with
the notation of Remark 3.2.8 and that the corresponding short exact sequence is split.
Furthermore, $\pi_0\left(\mathrm{Aut}(\mathfrak{so}(8,\mathbb{R}))^\sigma\right) \cong \mathcal{S}_4$.

We will show that also the short exact sequence

$$1 \to (\mathrm{Aut}(\mathfrak{so}(8,\mathbb{R}))^\sigma)_0 \to \mathrm{Aut}(\mathfrak{so}(8,\mathbb{R}))^\sigma \to \pi_0\left(\mathrm{Aut}(\mathfrak{so}(8,\mathbb{R}))^\sigma\right) \cong \mathcal{S}_4 \to 1$$

is split by showing that $\{\mathrm{Ad}(X_2), \mathrm{Ad}(X_3), \mathrm{Ad}(Y), \theta\} \subseteq \mathrm{Aut}(\mathfrak{so}(8,\mathbb{R}))^\sigma$ generates a subgroup Γ isomorphic to $\mathcal{S}_4$. We already know that Γ is mapped onto $\mathcal{S}_4$ by a group morphism. So the First Isomorphism Theorem yields that $\mathcal{S}_4$ is isomorphic to a quotient of Γ, thus it suffices to show that the order of Γ is at most 24.

First consider the subgroup $S \leq \Gamma$ generated by the set $\{\mathrm{Ad}(X_2), \mathrm{Ad}(X_3), \mathrm{Ad}(Y)\}$. With $a := \mathrm{Ad}(X_2)$, $b := \mathrm{Ad}(X_3)$, $c := \mathrm{Ad}(Y)$ we see that $ab = ba$ and $ac = ca$ and hence each product of three different elements is equal to abc or acb. We calculate:

$$cb = \mathrm{Ad}\left(\begin{pmatrix} \mathbf{1}_4 & \mathbf{0} \\ \mathbf{0} & -I_{3,1} \end{pmatrix}\begin{pmatrix} \mathbf{0} & \mathbf{1}_4 \\ -\mathbf{1}_4 & \mathbf{0} \end{pmatrix}\right) = \mathrm{Ad}\begin{pmatrix} \mathbf{0} & \mathbf{1}_4 \\ I_{3,1} & \mathbf{0} \end{pmatrix} = \mathrm{Ad}\left(\begin{pmatrix} \mathbf{0} & I_{3,1} \\ \mathbf{1}_4 & \mathbf{0} \end{pmatrix}\begin{pmatrix} I_{3,1} & \mathbf{0} \\ \mathbf{0} & I_{3,1} \end{pmatrix}\right)$$

$$= bca = abc.$$

Multiplying by a from the left yields: $acb = bc$. So the fact that also b, c are of order two yields: $S = \{1, a, b, c, ab, ac, bc, cb\}$. In S, the elements $d := ab$ and $e := bc$ are of order two and four, respectively, and fulfill the relation

$$de = (ab)(bc) = ac = ca = (cb)(ba) = (cabbccab)(ab) = e^3 d,$$

hence $S \cong \mathcal{D}_4$ is of order eight.

We calculate, for the standard Chevalley generators h_i, x_i, y_i, where $i = 1, 2, 3, 4$, of $\mathfrak{so}(8, \mathbb{C})$ and $x_5 := \frac{1}{4}[y_1, [y_2, y_4]]$, $x_6 := \frac{1}{16}[[x_1, x_2], [x_3, [x_4, x_2]]]$, $x_7 := \frac{1}{4}[y_3, [y_2, y_1]]$, $x_8 := \frac{1}{4}[y_4, [y_2, y_3]]$ and $\mathfrak{x} := \mathbb{C}x_1 \oplus \ldots \oplus \mathbb{C}x_8$:

$$T_1 = \frac{1}{2}\begin{pmatrix} 1 & 1 & 1 & 1 \\ 1 & 1 & -1 & -1 \\ 1 & -1 & 1 & -1 \\ -1 & 1 & 1 & -1 \end{pmatrix}, \quad D_1 = \begin{pmatrix} 0 & 0 & -1 & 0 \\ 0 & 0 & 0 & 1 \\ -1 & 0 & 0 & 0 \\ 0 & 1 & 0 & 0 \end{pmatrix},$$

$$E_1 = \begin{pmatrix} 0 & 0 & 1 & 0 \\ 0 & 0 & 0 & -1 \\ 1 & 0 & 0 & 0 \\ 0 & 1 & 0 & 0 \end{pmatrix}, \quad T_2 = \begin{pmatrix} 0 & 0 & 0 & 1 & 0 & 0 & 0 & 0 \\ 0 & 1 & 0 & 0 & 0 & 0 & 0 & 0 \\ 1 & 0 & 0 & 0 & 0 & 0 & 0 & 0 \\ 0 & 0 & 1 & 0 & 0 & 0 & 0 & 0 \\ 0 & 0 & 0 & 0 & 0 & 0 & 0 & 1 \\ 0 & 0 & 0 & 0 & 0 & 1 & 0 & 0 \\ 0 & 0 & 0 & 0 & 1 & 0 & 0 & 0 \\ 0 & 0 & 0 & 0 & 0 & 0 & 1 & 0 \end{pmatrix},$$

$$D_2 = \begin{pmatrix} 0 & 0 & 0 & 1 & 0 & 0 & 0 & 0 \\ 0 & 0 & 0 & 0 & 1 & 0 & 0 & 0 \\ 0 & 0 & 0 & 0 & 0 & 1 & 0 & 0 \\ 1 & 0 & 0 & 0 & 0 & 0 & 0 & 0 \\ 0 & 1 & 0 & 0 & 0 & 0 & 0 & 0 \\ 0 & 0 & 1 & 0 & 0 & 0 & 0 & 0 \\ 0 & 0 & 0 & 0 & 0 & 0 & 0 & 1 \\ 0 & 0 & 0 & 0 & 0 & 0 & 1 & 0 \end{pmatrix}, \quad E_2 = \begin{pmatrix} 0 & 0 & 0 & 1 & 0 & 0 & 0 & 0 \\ 0 & 0 & 0 & 0 & 1 & 0 & 0 & 0 \\ 1 & 0 & 0 & 0 & 0 & 0 & 0 & 0 \\ 0 & 0 & 0 & 0 & 0 & 1 & 0 & 0 \\ 0 & 0 & 0 & 0 & 0 & 0 & 0 & 1 \\ 0 & 0 & 1 & 0 & 0 & 0 & 0 & 0 \\ 0 & 1 & 0 & 0 & 0 & 0 & 0 & 0 \\ 0 & 0 & 0 & 0 & 0 & 0 & 1 & 0 \end{pmatrix},$$

where $T_1 := \left[\theta'_{|\mathfrak{h}}\right]_{(h_1,\ldots,h_4)}$, $D_1 := \left[d_{|\mathfrak{h}}\right]_{(h_1,\ldots,h_4)}$, $E_1 := \left[e_{|\mathfrak{h}}\right]_{(h_1,\ldots,h_4)}$, $T_2 := \left[\theta'_{|\mathfrak{x}}\right]_{(x_1,\ldots,x_8)}$, $D_2 := \left[\theta'_{|\mathfrak{x}}\right]_{(x_1,\ldots,x_8)}$, $E_2 := \left[\theta'_{|\mathfrak{x}}\right]_{(x_1,\ldots,x_8)}$. Since $T_\ell D_\ell = D_\ell E_\ell^2 T_\ell$ and $T_\ell E_\ell = D_\ell E_\ell T_\ell^2$ for

$\ell = 1, 2$, we have $\theta d = de^2 \theta$ and $\theta e = de\theta^2$ on $\mathfrak{so}(8, \mathbb{R})$, hence:

$$\Gamma = \left\{ d^i e^j \theta^k : i \in \{0, 1\}, j \in \{0, 1, 2, 3\}, k \in \{0, 1, 2\} \right\},$$

showing $\#\Gamma \leq 24$ and thus $\Gamma \cong \mathcal{S}_4$.

6. $\mathfrak{g} \cong \mathfrak{so}^*(4n, \mathbb{R})$ for $n \geq 3$:

By the proof of Proposition 3.2.15.4, Case 4, iv, we have:

$$(\mathrm{Aut}(\mathfrak{so}(4n, \mathbb{R}))^\sigma)_0 = \{\mathrm{Ad}(X) \,|\, X \in \mathrm{U}(2n, \mathbb{C})\},$$
$$\mathrm{Aut}(\mathfrak{so}(4n, \mathbb{R}))^\sigma = (\mathrm{Aut}(\mathfrak{so}(4n, \mathbb{R}))_0)^\sigma$$
$$= (\mathrm{Aut}(\mathfrak{so}(4n, \mathbb{R}))^\sigma)_0 \cup \left\{ \mathrm{Ad}(X) \,\middle|\, X \in I_{2n,2n}\, \mathrm{U}(2n, \mathbb{C}), \det_{\mathbb{R}}(X) = 1 \right\}.$$

Thus we have:

$$\pi_0 \left(\mathrm{Aut}(\mathfrak{so}(4n, \mathbb{R}))^\sigma \right) = \{\mathbf{1}, [\mathrm{Ad}(I_{2n,2n})]\} \cong \mathcal{C}_2.$$

$\square$

The split real forms of complex simple Lie algebras and the hermitian Lie algebras both form important classes of real central simple Lie algebras. We treat them in the next two theorems.

Theorem 3.2.19. *If $\mathfrak{g}$ is a hermitian Lie algebra, i.e., isomorphic to one of the algebras $\mathfrak{sl}(2, \mathbb{R})$, $\mathfrak{su}(q, n + 1 - q, \mathbb{C})$ for $q \leq \frac{n+1}{2} > 1$, $\mathfrak{so}^*(2n + 6, \mathbb{R})$, $\mathfrak{so}(2, n + 2, \mathbb{R})$ or $\mathfrak{sp}(2n + 4, \mathbb{R})$ for some $n \in \mathbb{N}$ or isomorphic to $\mathfrak{e}_{6(-14)}$ or $\mathfrak{e}_{7(-25)}$, then $\mathrm{Aut}(\mathfrak{g}) \cong \mathrm{Aut}(\mathfrak{g})_0 \rtimes \pi_0(\mathrm{Aut}(\mathfrak{g}))$.*

Proof. Let $\mathfrak{g} = \mathfrak{k} \oplus_\kappa^\tau \mathfrak{p}$ be a Cartan decomposition, G a Lie simply connected Lie group with Lie algebra $\mathfrak{g}$ and $K \leq G$ a connected compact subalgebra with Lie algebra $\mathfrak{k}$. Since $\mathfrak{g}$ is hermitian, $\mathfrak{z}(\mathfrak{k}) = \mathbb{R} \cdot z$ for some $0 \neq z \in \mathfrak{k}$. Kobayashi has shown (cf. Lemma 2.4 of [Kob07]) the existence of an involutive Lie group automorphism of G such that its derivative in $\mathbf{1} \in G$ is a Lie algebra automorphism $\omega : \mathfrak{g} \to \mathfrak{g}$ with $\omega(z) = -z$ and $\omega\tau = \tau\omega$.

By Corollary 3.2.17 and Corollary 3.2.18, we have to prove the statement of the theorem only if $\mathfrak{g}$ is isomorphic to one of $\mathfrak{sl}(2, \mathbb{R})$, $\mathfrak{e}_{7(-25)}$ or $\mathfrak{sp}(2n + 4, \mathbb{R})$ for some $n \in \mathbb{N}$. By $\pi_0(\mathrm{Aut}(\mathfrak{g})) \cong \mathcal{C}_2$, it suffices to show that $\omega \in \mathrm{Aut}(\mathfrak{g})$ is outer.

Supposing ω inner, we can write $\omega = e^{\mathrm{ad}(p)} e^{\mathrm{ad}_\mathfrak{g}(k_n)} \cdots e^{\mathrm{ad}_\mathfrak{g}(k_1)}$ for some $n \in \mathbb{N}$, $k_1, \ldots, k_n \in \mathfrak{k}$ and $p \in \mathfrak{p}$ (cf. Cartan Decomposition Theorem 12.1.7 of [HN11]), thus $\omega(z) = e^{\mathrm{ad}(p)}(z)$.

Let $B_\tau : \mathfrak{g} \times \mathfrak{g} \to \mathbb{R}$ be the scalar product from Lemma 3.2.3. Then $(\mathrm{ad}(p))^{2\ell+1}(z) \in \mathfrak{p} = \mathfrak{k}^{\perp_{B_\tau}}$ for all $\ell \in \mathbb{N}_0$, so $B_\tau(z, \omega(z)) = B_\tau(z, \omega'(z))$ for the map

$$\omega' : \mathfrak{g} \longrightarrow \mathfrak{g}, \quad x \longmapsto \sum_{\ell=0}^{\infty} \frac{1}{(2\ell)!} (\mathrm{ad}_\mathfrak{g}(p))^{2\ell}(x) = \cosh(\mathrm{ad}(p))(x).$$

The B_τ-symmetry of $\mathrm{ad}(p)$ yields:

$$B_\tau \left(x, (\mathrm{ad}(p))^{2\ell}(x) \right) = B_\tau \left((\mathrm{ad}(p))^\ell(x), (\mathrm{ad}(p))^\ell(x) \right) \geq 0$$

for all $x \in \mathfrak{g}$ and $\ell \in \mathbb{N}_0$. So we know that $B_\tau(z, \omega(z)) = B_\tau(z, \omega'(z)) \geq B_\tau(z, z) > 0$, contradicting the fact that $\omega(z) = -z$. So ω is outer. $\square$

Theorem 3.2.20. *If $\mathfrak{g}$ is a split real form of its complexification, i.e., isomorphic to one of $\mathfrak{sl}(n+1,\mathbb{R})$, $\mathfrak{so}(n+1,n+2,\mathbb{R})$, $\mathfrak{sp}(2n+4,\mathbb{R})$, $\mathfrak{so}(n+3,n+3,\mathbb{R})$ for some $n \in \mathbb{N}$ or isomorphic to $\mathfrak{e}_{6(6)}$, $\mathfrak{e}_{7(7)}$, $\mathfrak{e}_{8(8)}$, $\mathfrak{f}_{4(4)}$ or $\mathfrak{g}_{2(2)}$, then $\mathrm{Aut}(\mathfrak{g}) \cong \mathrm{Aut}(\mathfrak{g})_0 \rtimes \pi_0(\mathrm{Aut}(\mathfrak{g}))$.*

Proof. By Corollary 3.2.17, Corollary 3.2.18 and Theorem 3.2.19, we have to prove the statement of the theorem only if $\mathfrak{g}$ is isomorphic to $\mathfrak{sl}(n+1,\mathbb{R})$ for some odd $n \geq 3$ or $\mathfrak{e}_{7(7)}$. We will show this by finding, with the notation of Remark 3.2.14, representatives of the elements in $\pi_0\left(\mathrm{Aut}(\mathfrak{u})^{\sigma}\right)$ forming a subgroup of $\mathrm{Aut}(\mathfrak{u})^{\sigma}$ isomorphic to $\pi_0\left(\mathrm{Aut}(\mathfrak{u})^{\sigma}\right)$.

1. $\mathfrak{g} \cong \mathfrak{sl}(n+1,\mathbb{R})$ for odd $n \geq 3$:

    By the proof of Proposition 3.2.15.1 and 3.2.15.4, Case 2, the group $(\mathrm{Aut}(\mathfrak{su}(n+1,\mathbb{C}))_0)^{\mathrm{cj}}$ has two connected components and there is a short exact sequence

    $$1 \longrightarrow (\mathrm{Aut}(\mathfrak{su}(n+1,\mathbb{C}))_0)^{\mathrm{cj}} \longrightarrow \mathrm{Aut}(\mathfrak{su}(n+1,\mathbb{C}))^{\mathrm{cj}} \longrightarrow \{1, [\mathrm{cj}]\} \longrightarrow 1$$

    which is split because cj is of order two in $\mathrm{Aut}(\mathfrak{su}(n+1,\mathbb{C}))^{\mathrm{cj}}$. Since $\mathrm{SU}(n+1,\mathbb{C})$ is compact and connected, we know that $\mathrm{Aut}(\mathfrak{su}(n+1,\mathbb{C}))_0 = \{\mathrm{Ad}(X) | X \in \mathrm{SU}(n+1,\mathbb{C})\}$, so we can write:

    $$(\mathrm{Aut}(\mathfrak{su}(n+1,\mathbb{C}))_0)^{\mathrm{cj}} = \{\mathrm{Ad}(X) | X \in \mathrm{SU}(n+1,\mathbb{C})\}^{\mathrm{cj}} = \{\mathrm{Ad}(X) | X \in \mathrm{U}(n+1,\mathbb{C})\}^{\mathrm{cj}}$$
    $$= \{\mathrm{Ad}(X) | X \in \mathrm{U}(n+1,\mathbb{C}) \text{ and } X^{-1}\overline{X}x = x X^{-1}\overline{X}$$
    $$\text{for all } x \in \mathfrak{su}(n+1,\mathbb{C}) \supseteq \mathfrak{so}(n+1,\mathbb{R})\}.$$

    By Lemma 3.2.12 in the case $\mathbb{K} = \mathbb{C}$, we may write:

    $$(\mathrm{Aut}(\mathfrak{su}(n+1,\mathbb{C}))_0)^{\mathrm{cj}} = \{\mathrm{Ad}(X) | X \in \mathrm{U}(n+1,\mathbb{C}) \text{ and } X^{-1}\overline{X} \in \mathbb{C}\mathbf{1}_{n+1}\}$$
    $$= \{\mathrm{Ad}(X) | X \in \mathrm{U}(n+1,\mathbb{C}) \text{ and } \overline{X} = \lambda X \text{ for some } \lambda \in \mathbb{C}\}.$$

    For $X \in \mathrm{U}(n+1,\mathbb{C})$ the condition $\overline{X} = -X$ implies $X = iR$ and hence $\mathrm{Ad}(X) = \mathrm{Ad}(R)$ for some real-valued matrix R which is then orthogonal. The condition $\overline{X} = \lambda X$ for some $\lambda \in \mathbb{C} \setminus \{-1\}$ yields that $\mathrm{Ad}(X) = \mathrm{Ad}(|\mu| Y)$ for $\mu := \frac{2}{1+\lambda}$ and the real-valued matrix $Y := \frac{X+\overline{X}}{2} = \mu^{-1}X$. But also $|\mu| Y$ is orthogonal:

    $$(|\mu| Y)^T (|\mu| Y) = |\mu|^2 Y^T Y = \overline{(\mu Y)^T} (\mu Y) = \overline{(X)^T} X = \mathbf{1}_{n+1}.$$

    So we have:

    $$(\mathrm{Aut}(\mathfrak{su}(n+1,\mathbb{C}))_0)^{\mathrm{cj}} = \{\mathrm{Ad}(X) | X \in \mathrm{O}(n+1,\mathbb{R})\}.$$

    By Corollary 3.2.13, the involution $\mathrm{Ad}(I_{1,n}) \in (\mathrm{Aut}(\mathfrak{su}(n+1,\mathbb{C}))_0)^{\mathrm{cj}}$ is not in the connected component of $(\mathrm{Aut}(\mathfrak{su}(n+1,\mathbb{C}))_0)^{\mathrm{cj}}$. Since cj and $\mathrm{Ad}(I_{1,n})$ commute, they form an elementary abelian subgroup of order four in $\mathrm{Aut}(\mathfrak{su}(n+1,\mathbb{C}))^{\mathrm{cj}}$ and we have a split short exact sequence

    $$1 \longrightarrow \left(\mathrm{Aut}(\mathfrak{su}(n+1,\mathbb{C}))^{\mathrm{cj}}\right)_0 = \left((\mathrm{Aut}(\mathfrak{su}(n+1,\mathbb{C}))_0)^{\mathrm{cj}}\right)_0 \longrightarrow \mathrm{Aut}(\mathfrak{su}(n+1,\mathbb{C}))^{\mathrm{cj}}$$
    $$\longrightarrow \{\mathbf{1}, [\mathrm{cj}], [\mathrm{Ad}(I_{1,n})], [\mathrm{cj}\,\mathrm{Ad}(I_{1,n})]\} \longrightarrow 1.$$

2. $\mathfrak{g} \cong \mathfrak{e}_{7(7)}$:

By the proof of Proposition 3.2.15.1, and 3.2.15.4, Case 2, we know that the group

$$\left(\mathrm{Aut}(\mathfrak{e}_{7(-133)})_0\right)^{\sigma_{\mathfrak{e}_{7(7)}}} = \mathrm{Aut}(\mathfrak{e}_{7(-133)})^{\sigma_{\mathfrak{e}_{7(7)}}}$$

has two connected components. By Lemma 3.2.10, the non-trivial element of the group $\pi_0\left(\mathrm{Aut}(\mathfrak{e}_{7(-133)})^{\sigma_{\mathfrak{e}_{7(7)}}}\right)$ is represented by $e^{\mathrm{ad}(X)}$ for any non-zero element $X \in \mathfrak{e}_{7(-133)}^{-\sigma_{\mathfrak{e}_{7(7)}}}$ such that $\mathrm{ad}(X)^3 = -\pi^2 \mathrm{ad}(X)$. So the minimal polynomial of $\mathrm{ad}(X)$ is $p(\lambda) = \lambda(\lambda^2 + \pi^2)$, yielding the existence of a basis of $\mathfrak{g}$ with respect to which $\mathrm{ad}(X)$ is represented by a block diagonal matrix $\mathrm{diag}\left(0,\ldots,0, \begin{pmatrix} 0 & \pi \\ -\pi & 0 \end{pmatrix}, \ldots, \begin{pmatrix} 0 & \pi \\ -\pi & 0 \end{pmatrix}\right)$, hence the block diagonal matrix $\mathrm{diag}\left(1,\ldots,1, \begin{pmatrix} -1 & 0 \\ 0 & -1 \end{pmatrix}, \ldots, \begin{pmatrix} -1 & 0 \\ 0 & -1 \end{pmatrix}\right)$ represents $e^{\mathrm{ad}(X)}$, which is of order two in $\mathrm{Aut}(\mathfrak{e}_{7(-133)})^{\sigma_{\mathfrak{e}_{7(7)}}}$.

$\square$

There is only one isomorphism class of real central simple Lie algebras $\mathfrak{g}$ left, for which we have not proven yet the isomorphism $\mathrm{Aut}(\mathfrak{g}) \cong \mathrm{Aut}(\mathfrak{g})_0 \rtimes \pi_0(\mathrm{Aut}(\mathfrak{g}))$.

Theorem 3.2.21. *If $\mathfrak{g}$ is a real central simple Lie algebra isomorphic to $\mathfrak{sp}(n,n,\mathbb{H})$ for some $n \in \mathbb{N}$, then $\mathrm{Aut}(\mathfrak{g}) \cong \mathrm{Aut}(\mathfrak{g})_0 \rtimes \pi_0(\mathrm{Aut}(\mathfrak{g}))$.*

Proof. We know that $(\mathrm{Aut}(\mathfrak{sp}(2n,\mathbb{H}))_0)^{\mathrm{Ad}(I_{n,n})} = \mathrm{Aut}(\mathfrak{sp}(2n,\mathbb{H}))^{\mathrm{Ad}(I_{n,n})}$ has two connected components by the proof of Proposition 3.2.15.1, and 3.2.15.4, Case 2. Since $\mathrm{Sp}(2n,\mathbb{H})$ is compact and connected, we know that $\mathrm{Aut}(\mathfrak{sp}(2n,\mathbb{H}))_0 = \{\mathrm{Ad}(X) | X \in \mathrm{Sp}(2n,\mathbb{H})\}$, so we can write:

$$(\mathrm{Aut}(\mathfrak{sp}(2n,\mathbb{H}))_0)^{\mathrm{Ad}(I_{n,n})} = \{\mathrm{Ad}(X) | X \in \mathrm{Sp}(2n,\mathbb{H})\}^{\mathrm{Ad}(I_{n,n})}.$$

Note that $X \in \mathrm{Sp}(2n,\mathbb{H})$ means that X preserves the hermitian form $\langle (x_i)_i, (y_i)_i \rangle := \sum_i \mathrm{qj}(x_i) y_i$ on $\mathbb{H}^{2n}$. Then also $I_{n,n} \in \mathrm{Sp}(2n,\mathbb{H})$, thus also $X I_{n,n} X^{-1} I_{n,n} \in \mathrm{Sp}(2n,\mathbb{H})$ and we can write:

$$\begin{aligned}
(\mathrm{Aut}(\mathfrak{sp}(2n,\mathbb{H}))_0)^{\mathrm{Ad}(I_{n,n})} &= \left\{\mathrm{Ad}(X) | X \in \mathrm{Sp}(2n,\mathbb{H}), \mathrm{Ad}(X I_{n,n} X^{-1} I_{n,n}) \in \ker\left(\mathrm{Ad}_{\mathrm{Sp}(2n,\mathbb{H})}\right)\right\} \\
&= \left\{\mathrm{Ad}(X) | X \in \mathrm{Sp}(2n,\mathbb{H}) \text{ and } \mathrm{Ad}(X I_{n,n} X^{-1} I_{n,n}) \in \mathrm{Z}\left(\mathrm{Sp}(2n,\mathbb{H})\right)\right\} \\
&= \left\{\mathrm{Ad}(X) | X \in \mathrm{Sp}(2n,\mathbb{H}) \text{ and } \mathrm{Ad}\left(X I_{n,n} X^{-1} I_{n,n}\right) = \pm \mathbf{1}_{2n}\right\}.
\end{aligned}$$

So, for $J_n := \begin{pmatrix} \mathbf{0} & \mathbf{1}_n \\ -\mathbf{1}_n & \mathbf{0} \end{pmatrix}$, we have $\mathrm{Ad}(J_n) \in (\mathrm{Aut}(\mathfrak{sp}(2n,\mathbb{H}))_0)^{\mathrm{Ad}(I_{n,n})}$, which is of order two.

It remains to show that $\mathrm{Ad}(J_n) \notin \left((\mathrm{Aut}(\mathfrak{sp}(2n,\mathbb{H}))_0)^{\mathrm{Ad}(I_{n,n})}\right)_0 = \left(\mathrm{Aut}(\mathfrak{sp}(2n,\mathbb{H}))^{\mathrm{Ad}(I_{n,n})}\right)_0$. By Lemma 3.2.6, the map

$$\pi_0\left(\mathrm{Aut}(\mathfrak{sp}(2n,\mathbb{H}))^{\mathrm{Ad}(I_{n,n})}\right) \to \pi_0\left(\mathrm{Aut}(\mathfrak{sp}(n,\mathbb{H}) \oplus \mathfrak{sp}(n,\mathbb{H}))\right)$$

$$\cong \pi_0\left(\mathrm{Aut}(\mathfrak{sp}(n,\mathbb{H})) \times \mathrm{Aut}(\mathfrak{sp}(n,\mathbb{H}))\right) \cong \mathcal{C}_2$$

induced by restriction is injective. But $\mathrm{Ad}(J_n)_{|\mathfrak{sp}(n,\mathbb{H}) \oplus \mathfrak{sp}(n,\mathbb{H})}$ represents the only non-trivial class in $\pi_0\left(\mathrm{Aut}(\mathfrak{sp}(n,\mathbb{H}) \oplus \mathfrak{sp}(n,\mathbb{H}))\right)$ because $\mathrm{Ad}(J_n)_{|\mathfrak{sp}(n,\mathbb{H}) \oplus \mathfrak{sp}(n,\mathbb{H})}(x,y) = (y,x)$ for $x,y \in \mathfrak{sp}(n,\mathbb{H})$. So $\mathrm{Ad}(J_n) \in \mathrm{Aut}(\mathfrak{sp}(2n,\mathbb{H}))^{\mathrm{Ad}(I_{n,n})}$ is outer. $\square$

The Real Non-Central Simple Case

We now turn to Case A of Lemma 3.2.2, i.e., the real non-central simple case. Here, Theorem 3.2.22 due to Djoković (cf. Proposition 4.1 and 7.1 of [Djo99]), helps us to connect the group $\pi_0(\mathrm{Aut}(\mathfrak{g}))$ to $\pi_0(\mathrm{Aut}(\mathfrak{g}^{\mathbb{C}}))$.

Theorem 3.2.22. *If $\mathfrak{g}$ is a real simple Lie algebra admitting a complex structure with conjugation σ, then there is the Lie group isomorphism $\phi = \phi_\sigma : \mathrm{Aut}(\mathfrak{g}^{\mathbb{C}}) \rtimes C_2 \to \mathrm{Aut}(\mathfrak{g})$, $(f, (-1)^\ell) \mapsto \sigma^\ell \circ f$ and the induced isomorphism $\eta = \eta_\sigma : \pi_0(\mathrm{Aut}(\mathfrak{g})) \to \pi_0(\mathrm{Aut}(\mathfrak{g}^{\mathbb{C}})) \times C_2$.*

Corollary 3.2.23. *If $\mathfrak{g}$ is a real simple Lie algebra admitting a complex structure, then $\mathrm{Aut}(\mathfrak{g})$ is isomorphic to $\mathrm{Aut}(\mathfrak{g})_0 \rtimes \pi_0(\mathrm{Aut}(\mathfrak{g}))$ and $\#\mathrm{Conj}(\pi_0(\mathrm{Aut}(\mathfrak{g}))) = \#\mathrm{Conj}(\pi_0(\mathrm{Aut}(\mathfrak{g}^{\mathbb{C}}))) \cdot 2$.*

Proof. The subgroups $\mathrm{Aut}(\mathfrak{g}^{\mathbb{C}})_0$ and $\mathrm{Aut}(\mathfrak{g})_0$, generated by the sets $\left\{ e^x : x \in \mathbf{L}\left(\mathrm{Aut}(\mathfrak{g}^{\mathbb{C}})\right) \right\}$ and $\{ e^x : x \in \mathbf{L}\left(\mathrm{Aut}(\mathfrak{g})\right) \}$, respectively, coincide, since $\mathbf{L}\left(\mathrm{Aut}(\mathfrak{g}^{\mathbb{C}})\right) = \mathrm{Der}(\mathfrak{g}) = \mathbf{L}\left(\mathrm{Aut}(\mathfrak{g})\right)$ by the surjectivity of $\mathrm{ad} : \mathfrak{g} \to \mathrm{Der}(\mathfrak{g})$.

We fix a split real form $\mathfrak{s}$ of the complex simple Lie algebra $\mathfrak{g}^{\mathbb{C}}$ and a corresponding conjugation σ, i.e., $\sigma : \mathfrak{g}^{\mathbb{C}} = \mathfrak{s}_{\mathbb{C}} = \mathfrak{s} + i\mathfrak{s} \to \mathfrak{g}^{\mathbb{C}}$, $r + is \mapsto r - is$. Let then ϕ and η be given by Theorem 3.2.22 and $\omega : \pi_0\left(\mathrm{Aut}(\mathfrak{s})\right) = \pi_0(\mathrm{Aut}(\mathfrak{g}^{\mathbb{C}})) \to \mathrm{Aut}(\mathfrak{s})$ be a section of the short exact sequence

$$1 \to \mathrm{Aut}(\mathfrak{s})_0 \to \mathrm{Aut}(\mathfrak{s}) \to \pi_0\left(\mathrm{Aut}(\mathfrak{s})\right) \to 1$$

given by Theorem 3.2.20. We define a morphism $\mathrm{ex}_{\mathfrak{s}} : \mathrm{Aut}(\mathfrak{s}) \to \mathrm{Aut}(\mathfrak{g}^{\mathbb{C}})^\sigma \subseteq \mathrm{Aut}(\mathfrak{g}^{\mathbb{C}})$ by $\mathrm{ex}_{\mathfrak{s}}(g)(r + is) := g(r) + ig(s)$ for $g \in \mathrm{Aut}(\mathfrak{s})$ and $r, s \in \mathfrak{s}$. We will now show that for the short exact sequence

$$1 \to \mathrm{Aut}(\mathfrak{g})_0 \to \mathrm{Aut}(\mathfrak{g}) \to \pi_0\left(\mathrm{Aut}(\mathfrak{g})\right) \to 1$$

the map $\gamma := \phi \circ (\mathrm{ex}_{\mathfrak{s}} \times \mathrm{id}_{C_2}) \circ (\omega \times \mathrm{id}_{C_2}) \circ \eta$ is a section:

Let $f \in \mathrm{Aut}(\mathfrak{g})$. Then $((\omega \times \mathrm{id}_{C_2}) \circ \eta)([f]) \in \mathrm{Aut}(\mathfrak{s}) \times C_2$ is a homomorphic image of the class $[f] \in \pi_0(\mathrm{Aut}(\mathfrak{g}))$ and can be written as $(g, (-1)^\ell)$ for some $g \in \mathrm{Aut}(\mathfrak{s})$ and $\ell \in \{0, 1\}$. Let, for $f' \in \mathrm{Aut}(\mathfrak{g})$, be $(g', (-1)^{\ell'})$ a corresponding pair. Then

$$\gamma([f]) = (\phi \circ (\mathrm{ex}_{\mathfrak{s}} \times \mathrm{id}_{C_2}))\,(g, (-1)^\ell) = \phi(\mathrm{ex}_{\mathfrak{s}}(g), (-1)^\ell)$$

and $\gamma([f']) = \phi(\mathrm{ex}_{\mathfrak{s}}(g'), (-1)^{\ell'})$ and so:

$$\gamma([f]) \circ \gamma([f']) = \phi(\mathrm{ex}_{\mathfrak{s}}(g), (-1)^\ell) \circ \phi(\mathrm{ex}_{\mathfrak{s}}(g'), (-1)^{\ell'}) = \sigma^\ell \circ \mathrm{ex}_{\mathfrak{s}}(g) \circ \sigma^{\ell'} \circ \mathrm{ex}_{\mathfrak{s}}(g')$$
$$= \sigma^{\ell + \ell'} \circ \mathrm{ex}_{\mathfrak{s}}(gg') = \gamma([f]\,[f']).$$

The second statement of the corollary immediately follows from Theorem 3.2.22. $\square$

3.3. Classification for Two-Dimensional Closed Base Manifolds

We want to give a classification of (equivalence classes of) section algebras of Lie algebra bundles $\pi : \mathfrak{K} \to M$ with certain central simple fibers $\mathfrak{k}$, for closed base manifolds M of dimension 2. By Theorem 3.1.2, the homotopy classes in $[M; B\,\mathrm{Aut}(\mathfrak{k})]_\sim$ are in bijective correspondence to the equivalence classes of algebras $\Gamma(\mathfrak{K})$. Our main tool in this section is the following theorem due to Bredon (Corollary VII.13.16 of [Bre93]).

Theorem 3.3.1. *Let Y be an $(n-1)$-connected topological space for some $n \in \mathbb{N}$. Let X be a CW-complex such that the $\pi_i(Y)$-valued (simplicial) cohomology groups $\mathrm{H}^{i+1}(X, \pi_i(Y))$ and $\mathrm{H}^i(X, \pi_i(Y))$ vanish for all $i > n$. Using $\mathrm{H}^n(Y, \pi_n(Y)) \cong \mathrm{End}(\pi_n(Y))$, there is a bijection*

$$[X; Y]_{\simeq} \longrightarrow \mathrm{H}^n(X, \pi_n(Y))$$
$$[f] \longmapsto f^*\left[\mathrm{id}_{\pi_n(Y)}\right].$$

We want to use this theorem for $n = 2$ and $Y = B\,\mathrm{Aut}(\mathfrak{k})$ (cf Remark 3.1.1.2). So we have to assure that the higher cohomology groups vanish. Therefore, we need the following short exact sequence of abelian groups (cf. Theorem 3.2 of [Hat02]) for $k \in \mathbb{N}$ and any abelian group Λ:

$$0 \to \mathrm{Ext}(\mathrm{H}_{k-1}(M), \Lambda) \to \mathrm{H}^k(M, \Lambda) \to \mathrm{Hom}(\mathrm{H}_k(M), \Lambda) \to \mathbf{0}, \tag{3.1}$$

where $\mathrm{Ext}(A, B)$ is, for abelian groups A and B, the group of equivalence classes of abelian group extensions of A by B. In particular, $\mathrm{Ext}(\mathbf{0}, B) = \mathbf{0}$ and $\mathrm{End}(A, \mathbf{0}) = \mathbf{0}$. Also, if $A = \mathbb{Z}^n$ for some $n \in \mathbb{N}$, then $\mathrm{Ext}(A, B) = \mathbf{0}$.

Lemma 3.3.2. *If A is a free abelian group, i.e., $A = \mathbb{Z}^n$ for some $n \in \mathbb{N}$, then every abelian group extension of A by any abelian group B is trivial.*

Proof. Let

$$0 \longrightarrow B \overset{j}{\longrightarrow} C \overset{q}{\longrightarrow} \mathbb{Z}^n \longrightarrow \mathbf{0} \tag{3.2}$$

be a short exact sequence of abelian groups. If $e_1, \ldots, e_n$ are the canonical generators of $\mathbb{Z}^n$ and, for each $i = 1, \ldots, n$, the element $c_i \in C$ is chosen such that $q(c_i) = e_i$, then the map $\sigma : \mathbb{Z}^n \to C$, $\sum_{i=1}^n \lambda_i e_i \mapsto \sum_{i=1}^n \lambda_i c_i$ is a set-theoretical section of the short exact sequence (3.2), but it is also a group morphism, since the c_i's commute. Hence C is isomorphic to $B \oplus \mathbb{Z}^n$ via

$$B \oplus \mathbb{Z}^n \longrightarrow C$$
$$(b, z) \longmapsto j(b) + \sigma(z). \qquad \square$$

In the context of this thesis, a manifold is closed if and only if it is compact.

Definition 3.3.3. A manifold is called *closed* if it is compact, connected and without boundary.

The following is Theorem 3.26 of [Hat02].

Theorem 3.3.4. *Let M be a closed manifold of dimension $m \in \mathbb{N}$.*

1. *The group $\mathrm{H}_i(M)$ vanishs for all $i > m$.*

2. *If M is orientable, then $\mathrm{H}_m(M) \cong \mathbb{Z}$.*

3. *If M is non-orientable, then $\mathrm{H}_m(M) = \mathbf{0}$.*

Corollary 3.3.5. *Let M be a closed manifold of dimension $m \in \mathbb{N}$ and Λ an abelian group. Then the group $\mathrm{H}^{i+1}(M, \Lambda)$ vanishs for $i \geq m$.*

Proof. This follows from the exact sequence (3.1), Lemma 3.3.2 and Theorem 3.3.4. $\square$

For a closed manifold of dimension 2, there is the notion of its genus.

Definition 3.3.6. Let M be a closed manifold of dimension 2. We say that M has *genus* 0 if it is diffeomorphic to the sphere $\mathbb{S}^2$, and genus $n \in \mathbb{N}$ if it is diffeomorphic to the connected sum of n copies of the torus $\mathbb{T}^2$ or diffeomorphic to the connected sum of n copies of the projective plane $\mathbb{RP}^2$.

The construction of the connected sum of two manifolds is compatible with the calculation of the cohomology group in the following sense (cf. Exercise 3.3.6 of [Hat02]).

Lemma 3.3.7. *Let M_1, M_2 be closed manifolds with connected sum $M = M_1 \# M_2$ and dimension n. Then $\mathrm{H}_i(M) \cong \mathrm{H}_i(M_1) \oplus \mathrm{H}_i(M_2)$ for all $i = 1, \ldots, n-1$, except for the case that M_1, M_2 are both non-orientable and $i = n-1$, where $\mathrm{H}_{n-1}(M_1) \oplus \mathrm{H}_{n-1}(M_2) \cong \mathcal{C}_2 \oplus \mathcal{C}_2 \oplus \mathbb{Z}^k$ for some $k \in \mathbb{N}_0$ and $\mathrm{H}_{n-1}(M) \cong \mathcal{C}_2 \oplus \mathbb{Z}^{k+1}$.*

We now cite the Classification Theorem for closed 2-manifolds (cf. Theorem I.8.2 of [Mas91]).

Theorem 3.3.8. *Two closed 2-manifolds are diffeomorphic if and only if their genera coincide and they are both orientable or they are both non-orientable.*

Corollary 3.3.9. *Let M be a closed manifold of dimension 2 and genus $n \in \mathbb{N}_0$.*

1. If M is non-orientable, then $\mathrm{H}_1(M) \cong \mathcal{C}_2 \oplus \mathbb{Z}^{n-1}$.

2. If M is orientable, then $\mathrm{H}_1(M) \cong \mathbb{Z}^{2n}$.

Proof. The first statement follows from $\mathrm{H}_1(\mathbb{RP}^2) \cong \mathcal{C}_2$ and Lemma 3.3.7. The second statement is Example 2.36 of [Hat02]. □

If M is a closed manifold of dimension 1, then it is diffeomorphic to $\mathbb{S}^1$ and we know, by Remark 3.1.1, how to classify $\Gamma(\mathfrak{K})$ in this case. We now discuss the case of $\dim(M) = 2$.

Proposition 3.3.10. *Let M be a closed 2-manifold of genus n and Y be a 1-connected space.*

1. If M is non-orientable, then there is a bijection

$$[M; Y]_{\simeq} \longrightarrow \mathrm{H}^2\left(M, \pi_2(Y)\right) \cong \mathrm{Ext}(\mathrm{H}_1(M), \pi_2(Y))$$
$$[f] \longmapsto f^*\left[\mathrm{id}_{\pi_2(Y)}\right].$$

2. If M is orientable, then there is a bijection

$$[M; Y]_{\simeq} \longrightarrow \mathrm{H}^2\left(M, \pi_2(Y)\right) \cong \mathrm{Hom}(\mathrm{H}_2(M), \pi_2(Y))$$
$$[f] \longmapsto f^*\left[\mathrm{id}_{\pi_2(Y)}\right].$$

Proof. These statements follow from Theorem 3.3.1, the short exact sequence (3.1), Corollary 3.3.5, the fact that $\mathrm{H}_2(M) = \mathbf{0}$ for non-orientable M and the fact that $\mathrm{H}_1(M) = \mathbb{Z}^{2n}$, and hence $\mathrm{Ext}(\mathrm{H}_1(M), \pi_2(Y)) = \mathbf{0}$, for orientable M by Corollary 3.3.9.2. □

Corollary 3.3.11. *Let M be a closed 2-manifold of genus n and Y be a 1-connected space.*

1. If M is non-orientable, then there is a bijection $[M; Y]_{\simeq} \to \pi_2(Y)/2 \cdot \pi_2(Y)$ for the subgroup $2 \cdot \pi_2(Y) := \{g + g : g \in \pi_2(Y)\}$.

2. *If M is orientable, then there is a bijection $[M; Y]_{\simeq} \to \pi_2(Y)$.*

Proof.

1. We have $\mathrm{H}_1(M) \cong \mathcal{C}_2 \oplus \mathbb{Z}^{n-1}$ and thus

$$\mathrm{Ext}(\mathrm{H}_1(M), \pi_2(Y)) \cong \mathrm{Ext}(\mathcal{C}_2, \pi_2(Y)) \times \prod_{i=1}^{n-1} \mathrm{Ext}(\mathbb{Z}, \pi_2(Y)) \cong \mathrm{Ext}(\mathcal{C}_2, \pi_2(Y)).$$

By Proposition 3F.11 of [Hat02], for any abelian group G and any abelian extension of abelian groups $0 \to A \to B \to C \to 0$, there is the following short exact sequence:

$$0 \to \mathrm{Hom}(C, G) \to \mathrm{Hom}(B, G) \to \mathrm{Hom}(A, G)$$
$$\to \mathrm{Ext}(C, G) \to \mathrm{Ext}(B, G) \to \mathrm{Ext}(A, G) \to 0.$$

The extension $0 \to \mathbb{Z} \xrightarrow{f} \mathbb{Z} \xrightarrow{q} \mathcal{C}_2 \to 0$, where $f(\ell) := 2\ell$ for $\ell \in \mathbb{Z}$, thus induces the sequence

$$0 \to \mathrm{Hom}(\mathcal{C}_2, \pi_2(Y)) \xrightarrow{q^*} \mathrm{Hom}(\mathbb{Z}, \pi_2(Y)) \cong \pi_2(Y) \xrightarrow{f^*} \mathrm{Hom}(\mathbb{Z}, \pi_2(Y)) \cong \pi_2(Y)$$
$$\to \mathrm{Ext}(\mathcal{C}_2, \pi_2(Y)) \to 0.$$

We denote the isomorphism $\mathrm{Hom}(\mathbb{Z}, \pi_2(Y)) \cong \pi_2(Y), \varphi \mapsto \varphi(1)$ by Ψ. Then the image of $\Psi \circ q^*$ is $2 \cdot \pi_2(Y)$, proving

$$\pi_2(Y)/2 \cdot \pi_2(Y) \cong \mathrm{Ext}(\mathcal{C}_2, \pi_2(Y)) \cong \mathrm{Ext}(\mathrm{H}_1(M), \pi_2(Y)) \cong [M; Y]_{\simeq}.$$

2. Since $\mathrm{H}_2(M) = \mathbb{Z}$ by Theorem 3.3.4.2, we have $\mathrm{Hom}(\mathrm{H}_2(M), \pi_2(Y)) \cong \pi_2(Y)$. $\square$

Remark 3.3.12. By Remark 3.1.1.4, we know that $\pi_n(Y) \cong \pi_{n-1}(\mathrm{Aut}(\mathfrak{k}))$ for the CW-complex $Y := B\,\mathrm{Aut}(\mathfrak{k})$ for any Lie algebra $\mathfrak{k}$. We recall the tables of $\pi_0(\mathrm{Aut}(\mathfrak{k}))$ for simple $\mathfrak{k}$ from the previous section, study lists of homotopy groups of Lie groups (cf. Section 19.3 of [Jam95]) and consider the fact that, for real simple compact $\mathfrak{k}$, the group $\pi_1(\mathrm{Aut}(\mathfrak{k}_\mathbb{C}))$ is trivial if $\pi_1(\mathrm{Aut}(\mathfrak{k}))$ is so (cf. Theorem 14.1.4.1 of [HN11]) and summarize these results to the following.

1. $B\,\mathrm{Aut}(\mathfrak{k})$ is 1-connected, if and only if $\mathrm{Aut}(\mathfrak{k})$ is connected, if and only if one of the following conditions is satisfied:

   - $\mathfrak{k}$ is a complex simple Lie algebra and isomorphic to $\mathfrak{sl}(2, \mathbb{C})$, $\mathfrak{so}(2n + 1, \mathbb{C})$ for $n \geq 2$, $\mathfrak{sp}(2n, \mathbb{C})$ for $n \geq 3$, $\mathfrak{e}_7$, $\mathfrak{e}_8$, $\mathfrak{f}_4$ or $\mathfrak{g}_2$,
   - $\mathfrak{k}$ is a real central simple Lie algebra, compact and isomorphic to $\mathfrak{su}(2, \mathbb{C})$, $\mathfrak{so}(2n+1, \mathbb{R})$ for $n \geq 2$, $\mathfrak{sp}(n, \mathbb{H})$ for $n \geq 3$, $\mathfrak{e}_{7(-133)}$, $\mathfrak{e}_{8(-248)}$, $\mathfrak{f}_{4(-52)}$ or $\mathfrak{g}_{2(-14)}$,
   - $\mathfrak{k}$ is a real central simple Lie algebra, non-compact and isomorphic to $\mathfrak{sp}(q, n - q, \mathbb{H})$ for $1 \leq q < \frac{n}{2}$, $\mathfrak{e}_{7(-5)}$, $\mathfrak{e}_{8(8)}$, $\mathfrak{e}_{8(-24)}$, $\mathfrak{f}_{4(4)}$, $\mathfrak{f}_{4(-20)}$ or $\mathfrak{g}_{2(2)}$.

2. $B\,\mathrm{Aut}(\mathfrak{k})$ is 2-connected, if and only if $\mathrm{Aut}(\mathfrak{k})$ is 1-connected, if and only if one the following conditions is satisfied:

   - $\mathfrak{k}$ is a complex simple Lie algebra and isomorphic to $\mathfrak{sl}(2, \mathbb{C})$, $\mathfrak{sp}(2n, \mathbb{C})$ for $n \geq 3$, $\mathfrak{e}_7$, $\mathfrak{e}_8$, $\mathfrak{f}_4$ or $\mathfrak{g}_2$,

- $\mathfrak{k}$ is a real central simple Lie algebra, compact and isomorphic to $\mathfrak{su}(2,\mathbb{C})$, $\mathfrak{sp}(n,\mathbb{H})$ for $n \geq 3$, $\mathfrak{e}_{7(-133)}$, $\mathfrak{e}_{8(-248)}$, $\mathfrak{f}_{4(-52)}$ or $\mathfrak{g}_{2(-14)}$.

We know that Lie algebra bundles $\mathfrak{K} \to \mathbb{S}^1$ with a fiber $\mathfrak{k}$ such that $\mathrm{Aut}(\mathfrak{k})$ is connected and Lie algebra bundles $\mathfrak{K} \to \mathbb{S}^2$ with a fiber $\mathfrak{k}$ such that $\mathrm{Aut}(\mathfrak{k})$ is 1-connected are trivial (cf. Theorem 3.1.3). The latter statement is still true, when replacing $\mathbb{S}^2$ by any other closed manifold of dimension 2.

Theorem 3.3.13. *Let M be a closed manifold of dimension 2 and $\pi : \mathfrak{K} \to M$ be a Lie algebra bundle with fiber $\mathfrak{k}$ such that $\mathrm{Aut}(\mathfrak{k})$ is connected.*

1. *If M is non-orientable, then the equivalence classes of $\Gamma(\mathfrak{K})$ are parametrized by the quotient $\pi_1(\mathrm{Aut}(\mathfrak{k}))/2 \cdot \pi_1(\mathrm{Aut}(\mathfrak{k}))$. In particular, if $\mathrm{Aut}(\mathfrak{k})$ is even 1-connected, then $\mathfrak{K}$ is trivial and $\Gamma(\mathfrak{K}) \cong C^\infty(M,\mathfrak{k})$.*

2. *If M is orientable, then the equivalence classes of $\Gamma(\mathfrak{K})$ are parametrized by $\pi_1(\mathrm{Aut}(\mathfrak{k}))$. If $\mathrm{Aut}(\mathfrak{k})$ is even 1-connected, then $\mathfrak{K}$ is trivial and $\Gamma(\mathfrak{K}) \cong C^\infty(M,\mathfrak{k})$.*

Proof. These statements directly follow from Corollary 3.3.11. $\qquad\square$

4. Universal Invariant Symmetric Bilinear Forms

In this chapter we want to determine a universal invariant symmetric bilinear form on the Lie algebra of smooth sections $\Gamma(\mathfrak{K})$ of a Lie algebra bundle $\pi : \mathfrak{K} \to M$ with fiber $\mathfrak{k}$.

4.1. Universal Invariant Symmetric Bilinear Forms

In this section, we will present, based on the paper [MN03] of Maier and Neeb, the algebraic theory of universal invariant symmetric bilinear forms on Lie algebras. In the following, $\mathfrak{g}$ is a (possibly infinite-dimensional) $\mathbb{K}$-Lie algebra and V a vector space considered as a trivial Lie algebra and a trivial $\mathfrak{g}$-module.

Definition 4.1.1. Let B be a commutative algebra and $\mathfrak{h}$ a B-module, i.e., there is a bilinear map $B \times \mathfrak{h} \to \mathfrak{h}$, $(b, h) \mapsto b.h$ such that $(a \cdot b).h = a.(b.h)$ for $a, b \in B$, $h \in \mathfrak{h}$. Then there is a B-module $S_B^2(\mathfrak{h})$ and a symmetric B-bilinear map $\vee_B : \mathfrak{h} \times \mathfrak{h} \to S_B^2(\mathfrak{h})$ with the following universal property:

For each B-module W and each symmetric B-bilinear map $\beta : \mathfrak{h} \times \mathfrak{h} \to W$ there is a unique B-linear map $\check{\beta} : S_B^2(\mathfrak{h}) \to W$ such that $\beta = \check{\beta} \circ \vee_B$.

The B-module $S_B^2(\mathfrak{h})$ is unique up to isomorphy. It is called the *symmetric square* of $\mathfrak{h}$ with respect to B.

Remark 4.1.2. Let B be a commutative algebra and $\mathfrak{h}$ a B-module.

1. The symmetric square of $\mathfrak{h}$ with respect to B can be realized as the quotient of the B-tensor square $\bigotimes_B^2(\mathfrak{h})$ of $\mathfrak{h}$ modulo the subspace generated by elements of the form $h_1 \otimes h_2 - h_2 \otimes h_1$ for $h_1, h_2 \in \mathfrak{h}$. If $q_B : \bigotimes_B^2(\mathfrak{h}) \to S_B^2(\mathfrak{h})$ is the corresponding quotient map, then $q_B(h_1 \otimes_B h_2) = q_B(h_1) \vee_B q_B(h_2)$ for $h_1, h_2 \in \mathfrak{h}$ and the set $\{h_1 \vee_B h_2 : h_1, h_2 \in \mathfrak{h}\}$ generates $S_B^2(\mathfrak{h})$.

2. The B-module structure of $S_B^2(\mathfrak{h})$ can be described by $b.(x \vee_B y) := b(x) \vee_B y$, which is well-defined, since, for fixed $b \in B$ and for the symmetric B-bilinear map $\mathfrak{g} \times \mathfrak{g} \to S_B^2(\mathfrak{g})$, $(x, y) \mapsto b(x) \vee_B y$, there is a unique linear map $f_b : S_B^2(\mathfrak{g}) \to S_B^2(\mathfrak{g})$ such that, for $x, y \in \mathfrak{g}$, we have $f_b(x \vee_B y) = b(x) \vee_B y$.

3. If $\mathfrak{h} = \mathfrak{g}$ is a Lie algebra and $B \subseteq \text{Cent}(\mathfrak{g})$ a commutative subalgebra, then $S_B^2(\mathfrak{g})$ is a $\mathfrak{g}$-module defined by the multiplication

$$x.(y \vee_B z) := [x, y] \vee_B z + y \vee_B [x, z] \tag{4.1}$$

for $x, y, z \in \mathfrak{g}$. In order to see that the multiplication is well-defined, we fix $x \in \mathfrak{g}$ and apply the universal property of $S_B^2(\mathfrak{g})$ to the symmetric B-bilinear map $\mathfrak{g} \times \mathfrak{g} \to S_B^2(\mathfrak{g})$, $(y, z) \mapsto [x, y] \vee_B z + y \vee_B [x, z]$. So there is a unique linear map $f_x : S_B^2(\mathfrak{g}) \to S_B^2(\mathfrak{g})$ such that $f_x(y \vee_B z) = [x, y] \vee_B z + y \vee_B [x, z]$ for all $y, z \in \mathfrak{g}$.

Definition 4.1.3. Let $\mathfrak{g}$ be a Lie algebra, $B \subseteq \mathrm{Cent}(\mathfrak{g})$ a commutative subalgebra and equip $S_B^2(\mathfrak{g})$ with the $\mathfrak{g}$-module structure defined in (4.1). Then we write $V_B(\mathfrak{g})$ for the quotient vector space $S_B^2(\mathfrak{g})/\mathfrak{g}.S_B^2(\mathfrak{g})$. Moreover, the subspace $\mathfrak{g}.S_B^2(\mathfrak{g})$ is a B-submodule of $S_B^2(\mathfrak{g})$:

$$b.(x.(y \vee_B z)) = b.([x,y] \vee_B z + y \vee_B [x,z]) = b\,[x,y] \vee_B z + y \vee_B b\,[x,z]$$
$$= [b(x),y] \vee_B z + y \vee_B [b(x),z] = (b(x)).(y \vee_B z),$$

thus $V_B(\mathfrak{g})$ inherits a natural B-module structure. We also define the B-bilinear map

$$\kappa_u^B : \mathfrak{g} \times \mathfrak{g} \longrightarrow V_B(\mathfrak{g})$$
$$(x,y) \longmapsto [x \vee_B y].$$

We can conclude that the group $\mathrm{Aut}(\mathfrak{g})$ acts on $V_B(\mathfrak{g})$ via $f.[x \vee_B y] := [f(x) \vee_B f(y)]$ and there is a derived $\mathrm{Der}(\mathfrak{g})$-module structure on $V_B(\mathfrak{g})$ via $\theta.[x \vee_B y] := [\theta(x) \vee_B y + x \vee_B \theta(y)]$ and $\mathrm{ad}(\mathfrak{g}) \subseteq \mathrm{Der}(\mathfrak{g})$ acts trivially on $V_B(\mathfrak{g})$.

The concept of universal invariant symmetric bilinear forms helps us to understand general invariant symmetric bilinear forms: The latters are the compositions of the formers with a linear map.

Definition 4.1.4.

1. Let $\mathrm{Sym}^2(\mathfrak{g}, V)$ denote the vector space of symmetric bilinear maps $\mathfrak{g} \times \mathfrak{g} \to V$. A map $\beta \in \mathrm{Sym}^2(\mathfrak{g}, V)$ is an *invariant symmetric bilinear form*, if $\beta([x,y], z) = \beta(x, [y,z])$ for all $x, y, z \in \mathfrak{g}$. The set of invariant symmetric bilinear forms to V is denoted by $\mathrm{Sym}^2(\mathfrak{g}, V)^{\mathfrak{g}}$.

2. An invariant symmetric bilinear form $\gamma : \mathfrak{g} \times \mathfrak{g} \to V$ is *universal*, if for each vector space W and each invariant symmetric bilinear form $\beta : \mathfrak{g} \times \mathfrak{g} \to W$ there is a unique linear map $\beta' : V \to W$ such that $\beta = \beta' \circ \gamma$.

Remark 4.1.5.

1. If two invariant symmetric bilinear forms $\gamma_i : \mathfrak{g} \times \mathfrak{g} \to V_i$, where $i = 1, 2$, are universal, then there are unique linear morphisms $f : V_1 \to V_2$ and $g : V_2 \to V_1$ such that $\gamma_1 = g \circ \gamma_2$ and $\gamma_2 = f \circ \gamma_1$. Applying this argument again, we see that $g \circ f : V_1 \to V_1$ is the unique linear map h such that $\gamma_2 = h \circ \gamma_1$, hence $g \circ f = \mathrm{id}_{V_1}$ and analogously $f \circ g = \mathrm{id}_{V_2}$, thus $V_1 \cong V_2$. In this sense, (γ_1, V_1) and (γ_2, V_2) are equivalent in the category of invariant symmetric bilinear forms.

2. The invariant symmetric bilinear form $\kappa_u^{\mathbb{K}} : \mathfrak{g} \times \mathfrak{g} \to V_{\mathbb{K}}(\mathfrak{g})$ for a $\mathbb{K}$-Lie algebra $\mathfrak{g}$ is universal:

   Let $\beta : \mathfrak{g} \times \mathfrak{g} \to W$ be an invariant symmetric bilinear form. Then there is a unique linear map $\check\beta : S_{\mathbb{K}}^2(\mathfrak{g}) \to W$ such that $\beta = \check\beta \circ \vee_{\mathbb{K}}$ and $\mathfrak{g}.S_{\mathbb{K}}^2(\mathfrak{g}) \subseteq \ker(\check\beta)$, since, for $x, y_1, y_2 \in \mathfrak{g}$, we can calculate:

   $$\check\beta(x.(y_1 \vee_{\mathbb{K}} y_2)) = \check\beta([x,y_1] \vee_{\mathbb{K}} y_2 + y_1 \vee_{\mathbb{K}} [x,y_2]) = \check\beta([x,y_1] \vee_{\mathbb{K}} y_2) + \check\beta(y_1 \vee_{\mathbb{K}} [x,y_2])$$
   $$= \beta([x,y_1], y_2) + \beta(y_1, [x,y_2]) = 0.$$

3. If $\mathfrak{g} = \mathfrak{a} \oplus \mathfrak{b}$ for Lie algebras $\mathfrak{a}$, $\mathfrak{b}$ with $\mathfrak{a}$ perfect, then $V_{\mathbb{K}}(\mathfrak{g}) \cong V_{\mathbb{K}}(\mathfrak{a}) \oplus V_{\mathbb{K}}(\mathfrak{b})$ because for any invariant symmetric bilinear form $\beta : \mathfrak{g} \times \mathfrak{g} \to W$ we have $\beta([x,y],z) = \beta(x,[y,z]) = 0$ for $x, y \in \mathfrak{a}$ and $z \in \mathfrak{b}$. Thus β takes the form $\beta(x+z, x'+z') = \beta_1(x,x') + \beta_2(z,z')$ for invariant symmetric bilinear forms $\beta_1 : \mathfrak{a} \times \mathfrak{a} \to W$ and $\beta_2 : \mathfrak{b} \times \mathfrak{b} \to W$.

4. The image of a universal invariant symmetric bilinear form $\kappa_{\mathfrak{g}} : \mathfrak{g} \times \mathfrak{g} \to V_{\mathbb{K}}(\mathfrak{g})$ spans $V_{\mathbb{K}}(\mathfrak{g})$: Otherwise, there would be a non-trivial linear map $f : V_{\mathbb{K}}(\mathfrak{g}) \to \mathbb{K}$ such that $0 = f \circ \kappa_{\mathfrak{g}}$, contradicting the universality of $\kappa_{\mathfrak{g}}$.

The following lemma is almost trivial, but important in its consequences, since it will help us to show that for perfect Lie algebras $\mathfrak{g}$ the form κ_u^A for $A = \mathrm{Cent}(\mathfrak{g})$ also is universal.

Lemma 4.1.6. *Let $\beta : \mathfrak{g} \times \mathfrak{g} \to V$ be an invariant symmetric bilinear form. Then*

$$\beta(ax, y) = \beta(x, ay)$$

for all $a \in \mathrm{Cent}(\mathfrak{g})$, $x \in [\mathfrak{g}, \mathfrak{g}]$ and $y \in \mathfrak{g}$. In particular, if $\mathfrak{g}$ is perfect, then for each invariant symmetric bilinear form β on $\mathfrak{g}$ every $a \in \mathrm{Cent}(\mathfrak{g})$ is β-symmetric.

Proof. There are elements $x_1^{(1)}, x_2^{(1)}, \ldots, x_1^{(n)}, x_2^{(n)} \in \mathfrak{g}$ such that $x = \sum_{i=1}^n \left[x_1^{(i)}, x_2^{(i)} \right]$. We calculate:

$$\beta\left(a\left[x_1^{(i)}, x_2^{(i)} \right], y \right) = \beta\left(\left[x_1^{(i)}, ax_2^{(i)} \right], y \right) = \beta\left(x_1^{(i)}, \left[ax_2^{(i)}, y \right] \right) = \beta\left(x_1^{(i)}, a\left[x_2^{(i)}, y \right] \right)$$
$$= \beta\left(x_1^{(i)}, \left[x_2^{(i)}, ay \right] \right) = \beta\left(\left[x_1^{(i)}, x_2^{(i)} \right], ay \right)$$

and

$$\beta(ax, y) = \sum_{i=1}^n \beta\left(a\left[x_1^{(i)}, x_2^{(i)} \right], y \right) = \sum_{i=1}^n \beta\left(\left[x_1^{(i)}, x_2^{(i)} \right], ay \right) = \beta(x, ay). \qquad \square$$

Proposition 4.1.7. *Let $A := \mathrm{Cent}(\mathfrak{g})$ be the commutative centroid of a perfect Lie algebra $\mathfrak{g}$. Then $\kappa_u^A : \mathfrak{g} \times \mathfrak{g} \to V_A(\mathfrak{g})$ is a universal invariant symmetric bilinear form. In particular, $V_{\mathbb{K}}(\mathfrak{g}) \cong V_A(\mathfrak{g})$ by Remark 4.1.5.1.*

Proof. We have to show the universality of κ_u^A. Let W be a vector space and $\beta : \mathfrak{g} \times \mathfrak{g} \to W$ an invariant symmetric bilinear form and $\check{\beta} : S_{\mathbb{K}}^2(\mathfrak{g}) \to W$ be the unique linear map such that $\beta = \check{\beta} \circ \vee_{\mathbb{K}}$. By Lemma 4.1.6, β is A-symmetric, thus $\check{\beta}(ax \vee_{\mathbb{K}} y) = \check{\beta}(x \vee_{\mathbb{K}} ay)$ for $a \in A$, $x, y \in \mathfrak{g}$. If $q : S_{\mathbb{K}}^2(\mathfrak{g}) \to S_A^2(\mathfrak{g}) = S_{\mathbb{K}}^2(\mathfrak{g})/I$ is the natural quotient map for the subspace I generated by the set $\{ax \vee_{\mathbb{K}} y - x \vee_{\mathbb{K}} ay : a \in A, x, y \in \mathfrak{g}\}$, then it immediately follows that there is a unique linear map $\tilde{\beta} : S_A^2(\mathfrak{g}) \to W$ such that $\check{\beta} = \tilde{\beta} \circ q$. The invariance of β yields that $\mathfrak{g}.S_A^2(\mathfrak{g}) \subseteq \ker(\tilde{\beta})$, thus there is a unique linear map $\beta'' : V_A(\mathfrak{g}) \to W$ such that $\tilde{\beta} = \beta'' \circ \left(S_A^2(\mathfrak{g}) \to V_A(\mathfrak{g}) \right)$, thus

$$\beta'' \circ \kappa_u^A = \tilde{\beta} \circ \vee_A = \tilde{\beta} \circ q \circ \vee_{\mathbb{K}} = \check{\beta} \circ \vee_{\mathbb{K}} = \beta. \qquad \square$$

Definition 4.1.8. Let $A := \mathrm{Cent}(\mathfrak{g})$ be the commutative centroid of a perfect Lie algebra $\mathfrak{g}$. We write $\kappa_u := \kappa_{\mathfrak{g}} := \kappa_u^A$, $V(\mathfrak{g}) := V_A(\mathfrak{g}) = V_{\mathbb{K}}(\mathfrak{g})$ and $x \vee y := x \vee_A y$.

Example 4.1.9. Let $\mathfrak{k}$ be a finite-dimensional semisimple (and hence perfect) Lie algebra with universal invariant symmetric bilinear form $\kappa_\mathfrak{k} : \mathfrak{k} \times \mathfrak{k} \to V(\mathfrak{k})$. By the universal property, there is a natural linear isomorphism

$$\zeta : V(\mathfrak{k})^* \longrightarrow \mathrm{Sym}^2(\mathfrak{k}, \mathbb{K})^\mathfrak{k}$$
$$\alpha \longmapsto \alpha \circ \kappa_\mathfrak{k}.$$

Let $\gamma : \mathfrak{k} \times \mathfrak{k} \to \mathbb{K}$, $(x,y) \mapsto \mathrm{tr}(\mathrm{ad}(x) \circ \mathrm{ad}(y))$ be the Cartan-Killing form of $\mathfrak{k}$. Then the map

$$\eta : \mathrm{Cent}(\mathfrak{k}) \longrightarrow \mathrm{Sym}^2(\mathfrak{k}, \mathbb{K})^\mathfrak{k}$$
$$\eta(\varphi)(x,y) := \gamma(\varphi(x), y)$$

is a linear isomorphism, too:

Note that it is well-defined because for $\varphi \in \mathrm{Cent}(\mathfrak{k})$ and $x, y, z \in \mathfrak{k}$ we have:

$$\eta(\varphi)([x,y],z) = \gamma(\varphi[x,y],z) = \gamma([\varphi(x),y],z) = \gamma(\varphi(x),[y,z]) = \eta(\varphi)(x,[y,z]).$$

The linearity of η is obvious. The kernel of η consists of all elements $\varphi \in \mathrm{Cent}(\mathfrak{k})$ such that $\mathrm{im}(\varphi) \subseteq \mathrm{rad}(\gamma) = \mathbf{0}$, thus η is injective. For the surjectivity, note that the map $\mathfrak{k} \to \mathfrak{k}^*$, $x \mapsto \gamma_x$, where $\gamma_x(y) := \gamma(x,y)$, is a linear isomorphism, so for every bilinear map $\beta : \mathfrak{k} \times \mathfrak{k} \to \mathbb{K}$ there is a unique linear map $L = L_\beta : \mathfrak{k} \to \mathfrak{k}$ such that $\beta(x,y) = \gamma(Lx, y)$ (cf. Section XIII.5 of [Lan02]). If β is invariant and $x, y, z \in \mathfrak{k}$, then:

$$\gamma(L[x,y],z) = \beta([x,y],z) = \beta(x,[y,z]) = \gamma(Lx,[y,z]) = \gamma([Lx,y],z),$$

hence $L \in \mathrm{Cent}(\mathfrak{k})$.

Thus $V(\mathfrak{k})^* \cong \mathrm{Cent}(\mathfrak{k})$. We will show that we may write the universal invariant symmetric bilinear form as

$$\widetilde{\kappa}_\mathfrak{k} : \mathfrak{k} \times \mathfrak{k} \to \mathrm{Cent}(\mathfrak{k})^*$$
$$\widetilde{\kappa}_\mathfrak{k}(x,y)(\varphi) := \gamma(\varphi(x),y). \tag{4.2}$$

To see that, let $\xi : (V(\mathfrak{k})^*)^* \to V(\mathfrak{k})$ be the inverse of the evaluation map $V(\mathfrak{k}) \to (V(\mathfrak{k})^*)^*$ and define the linear isomorphism $\psi : \mathrm{Cent}(\mathfrak{k})^* \to V(\mathfrak{k})$, $\delta \mapsto \xi(\delta \circ \eta^{-1} \circ \zeta)$. Since $\kappa_\mathfrak{k}$ is universal, there is a unique linear map $\theta : V(\mathfrak{k}) \to \mathrm{Cent}(\mathfrak{k})^*$ with $\theta \circ \kappa_\mathfrak{k} = \widetilde{\kappa}_\mathfrak{k}$. The following calculation for $[x \vee y] \in V(\mathfrak{k})$ shows that θ is injective, hence bijective, thus $\widetilde{\kappa}_\mathfrak{k}$ is universal:

$$(\psi \circ \theta)[x \vee y] = (\psi \circ \theta)(\kappa_\mathfrak{k}(x,y)) = \psi(\widetilde{\kappa}_\mathfrak{k}(x,y)) = \psi(\varphi \mapsto \gamma(\varphi(x),y)) = \psi(\varphi \mapsto \eta(\varphi)(x,y))$$
$$= \xi((\varphi \mapsto \eta(\varphi)(x,y)) \circ \eta^{-1} \circ \zeta) = \xi((\beta \mapsto \beta(x,y)) \circ \zeta) = \xi(\alpha \mapsto \alpha(\kappa_\mathfrak{k}(x,y)))$$
$$= \kappa_\mathfrak{k}(x,y) = [x \vee y].$$

Note that this also proves the explicit formula $\theta[x \vee y] = \mathrm{ev}_{(x,y)} \circ \eta$.

If $\mathfrak{k}$ is central simple, then $V(\mathfrak{k}) \cong \mathrm{Cent}(\mathfrak{k}) = \mathbb{K} \cdot \mathbf{1}$ and the Cartan-Killing form γ itself is a universal invariant symmetric bilinear form in the sense that for every invariant symmetric bilinear form $\beta : \mathfrak{k} \times \mathfrak{k} \to W$ there is a unique element $w \in W$ such that $\beta(x,y) = \gamma(x,y) \cdot w$. If

$\mathfrak{k}$ is real simple, but not central simple, then $\mathrm{Cent}(\mathfrak{k}) = \mathbb{R} \cdot 1 + \mathbb{R} \cdot J$ for a complex structure J on $\mathfrak{k}$ and the Cartan-Killing form

$$\gamma_{\mathbb{C}} : \mathfrak{k} \times \mathfrak{k} \longrightarrow \mathbb{C} \cong \mathbb{R}^2$$

$$(x, y) \longmapsto \frac{1}{2}(\gamma(x, y) - i\gamma(ix, y))$$

of $\mathfrak{k}$ considered as a complex Lie algebra is the universal invariant symmetric bilinear form of $\mathfrak{k}$ considered as a real Lie algebra.

Combining this and Remark 4.1.5.3, if $\mathfrak{k} = \bigoplus_{i=1}^{\ell} \mathfrak{k}_i \oplus \bigoplus_{j=1}^{m} \mathfrak{h}_j$ is the decomposition of a semisimple Lie algebra into the direct sum of central simple $\mathfrak{k}_i$'s and non-central $\mathfrak{h}_j$'s, then the Cartan-Killing forms on the simple ideals give rise to a universal invariant symmetric bilinear form $\kappa_{\mathfrak{k}} : \mathfrak{k} \times \mathfrak{k} \to \mathrm{V}(\mathfrak{k}) \cong \mathbb{K}^n$, where $n = \ell + 2m$. Furthermore, we know that in the semisimple case all derivations are inner, so by considering the left actions $\mathrm{Aut}(\mathfrak{k}) \times \mathrm{V}(\mathfrak{k}) \to \mathrm{V}(\mathfrak{k})$ and $\mathrm{Der}(\mathfrak{k}) \times \mathrm{V}(\mathfrak{k}) \to \mathrm{V}(\mathfrak{k})$ from Definiton 4.1.3, we conclude that the identity component $\mathrm{Aut}(\mathfrak{k})_0$ acts trivially on $\mathrm{V}(\mathfrak{k})$.

Definition 4.1.10. For a finite-dimensional Lie algebra bundle $\pi : \mathfrak{K} \to M$ with semisimple fiber $\mathfrak{k}$, we define the vector bundle $\nu : \mathrm{V}(\mathfrak{K}) \to M$ to be the bundle associated to the $\mathrm{Aut}(\mathfrak{k})$-principal bundle $\rho : \mathrm{P}(\mathfrak{K}) \to M$ with respect to $\lambda : \mathrm{Aut}(\mathfrak{k}) \times \mathrm{V}(\mathfrak{k}) \to \mathrm{V}(\mathfrak{k})$, $(f, [x \vee y]) \mapsto [f(x) \vee f(y)]$. For every $m \in M$, the fiber $\mathrm{V}(\mathfrak{K})_m$ is isomorphic to $\mathrm{V}(\mathfrak{K}_m)$ via $[p, [x \vee y]] \mapsto \kappa_{\mathfrak{K}_{\pi(p)}}([p, x], [p, y])$.

Definition 4.1.11. For a finite-dimensional Lie algebra bundle $\pi : \mathfrak{K} \to M$ with fiber $\mathfrak{k}$, we define the vector bundle $\widetilde{\nu} : \mathrm{Cent}^*(\mathfrak{K}) \to M$ to be the bundle associated to the $\mathrm{Aut}(\mathfrak{k})$-principal bundle $\rho : \mathrm{P}(\mathfrak{K}) \to M$ with respect to the left action $\widetilde{\lambda} : \mathrm{Aut}(\mathfrak{k}) \times \mathrm{Cent}(\mathfrak{k})^* \to \mathrm{Cent}(\mathfrak{k})^*$, $(f, \delta) \mapsto \delta \circ c_{f^{-1}}$ for the conjugation $c_g(\varphi) := g\varphi g^{-1}$, where $g \in \mathrm{Aut}(\mathfrak{k})$, $\varphi \in \mathrm{Cent}(\mathfrak{k})$. A typical element of $\mathrm{Cent}^*(\mathfrak{K})_m \subseteq \mathrm{Cent}^*(\mathfrak{K})$ takes the form

$$[(\psi, \delta)] := \mathrm{Aut}(\mathfrak{k}) \cdot (\psi, \delta) = \left\{ \left(\psi \circ f^{-1}, \delta \circ c_{f^{-1}} \right) : f \in \mathrm{Aut}(\mathfrak{k}) \right\}$$

for a Lie algebra isomorphism $\psi : \mathfrak{k} \to \mathfrak{K}_m$ and $\delta \in \mathrm{Cent}(\mathfrak{k})^*$. The map $F_m : \mathrm{Cent}(\mathfrak{K}_m) \to \mathbb{K}$, $\varphi \mapsto F_m(\psi, \delta)(\varphi) := \delta \left(\psi^{-1} \circ \varphi \circ \psi \right)$ is clearly well-defined and, by applying this construction to all classes $[(\psi, \delta)]$ for $\psi \in \mathrm{Iso}(\mathfrak{k}, \mathfrak{K}_m)$, where m runs through M and $\delta \in \mathrm{Cent}(\mathfrak{k})^*$ is fixed, we obtain an isomorphism $\mathrm{Cent}^*(\mathfrak{K})_m \cong \mathrm{Cent}(\mathfrak{K}_m)^*$ via $[(\psi, \delta)] \mapsto F_m(\psi, \delta)$.

Remark 4.1.12.

1. Let $\mathfrak{k}$ be a semisimple $\mathbb{K}$-Lie algebra. In this case the vector bundles $\mathrm{V}(\mathfrak{K})$ and $\mathrm{Cent}^*(\mathfrak{K})$ are equivalent, since, with the notation of Example 4.1.9, the vector bundles $\mathrm{V}(\mathfrak{K})$ and $\mathrm{Cent}^*(\mathfrak{K})$ are associated to the $\mathrm{Aut}(\mathfrak{k})$-principal bundle $\rho : \mathrm{P}(\mathfrak{K}) \to M$ with respect to the left actions λ and $\widetilde{\lambda}$, respectively, and the linear isomorphism $\theta : \mathrm{V}(\mathfrak{k}) \to \mathrm{Cent}(\mathfrak{k})^*$, $[x \vee y] \mapsto \mathrm{ev}_{(x,y)} \circ \eta$ is equivariant, as the following calculation for $f \in \mathrm{Aut}(\mathfrak{k})$, $x, y \in \mathfrak{k}$, $\varphi \in \mathrm{Cent}(\mathfrak{k})$ and the Cartan-Killing form $\gamma : \mathfrak{k} \times \mathfrak{k} \to \mathbb{K}$ shows:

$$\theta(f. [x \vee y])(\varphi) = \theta [fx \vee fy] (\varphi) = \mathrm{ev}_{(fx, fy)} \circ \eta(\varphi) = \gamma(\varphi(fx), fy) = \gamma \left(f^{-1} \varphi f(x), y \right)$$
$$= \eta \left(c_{f^{-1}} \varphi \right) (x, y) = \theta [x \vee y] (f.\varphi).$$

2. If $\mathfrak{k}$ is central simple, then the $\mathrm{Aut}(\mathfrak{k})$-invariant Cartan-Killing form is a universal invariant symmetric bilinear form, so the action $\mathrm{Aut}(\mathfrak{k}) \times \mathrm{V}(\mathfrak{k}) \to \mathrm{V}(\mathfrak{k})$ is trivial, thus the bundle $\mathrm{V}(\mathfrak{K}) \cong M \times \mathrm{V}(\mathfrak{k}) \cong M \times \mathbb{K}$ is so. This also follows from Theorem 2.3.39.3 and the fact that $\mathrm{Cent}^*(\mathfrak{K})$ and $\mathrm{V}(\mathfrak{K})$ are equivalent bundles.

From now on, for Lie algebra bundles $\pi : \mathfrak{K} \to M$ with semisimple fiber $\mathfrak{k}$, we will not distinguish between the universal invariant symmetric bilinear forms $\kappa_\mathfrak{k}$ and $\widetilde{\kappa}_\mathfrak{k}$ and the vector bundles $\mathrm{V}(\mathfrak{K})$ and $\mathrm{Cent}^*(\mathfrak{K})$. Even if the fiber $\mathfrak{k}$ is not semisimple, there is an obvious invariant symmetric bilinear form with values in $\mathrm{V}(\mathfrak{K})$ on $\Gamma(\mathfrak{K})$.

Definition 4.1.13. For $X, Y \in \Gamma(\mathfrak{K})$, $m \in M$ we define

$$\kappa(X,Y)(m) := \kappa_{\mathfrak{K}_m}(X_m, Y_m) \in \mathrm{V}(\mathfrak{K}_m) \cong \mathrm{V}(\mathfrak{K})_m.$$

The resulting smooth map $\kappa(X,Y) : M \to \mathrm{V}(\mathfrak{K})$, $m \mapsto \kappa(X,Y)(m)$ leads to an invariant symmetric bilinear form

$$\kappa : \Gamma(\mathfrak{K}) \times \Gamma(\mathfrak{K}) \longrightarrow \Gamma(\mathrm{V}(\mathfrak{K}))$$
$$(X,Y) \longmapsto \kappa(X,Y).$$

Note that, by definition, κ is $\mathrm{C}^\infty(M, \mathbb{K})$-bilinear.

Remark 4.1.14. If $\mathfrak{k}$ is central simple, then we know that $\Gamma(\mathrm{V}(\mathfrak{K})) \cong \mathrm{C}^\infty(M, \mathrm{V}(\mathfrak{k})) \cong \mathrm{C}^\infty(M, \mathbb{K})$ and $\kappa : \Gamma(\mathfrak{K}) \times \Gamma(\mathfrak{K}) \to \mathrm{C}^\infty(M, \mathbb{K})$ can be defined by $\kappa(X,Y)(m) := \gamma(\varphi(X_m), \varphi(Y_m))$ for $X, Y \in \Gamma(\mathfrak{K})$, any bundle chart (U, φ) in $m \in M$ and the Cartan-Killing form γ on $\mathfrak{k}$. The centroid of $\Gamma(\mathfrak{K})$ in this case is $\mathrm{Cent}(\Gamma(\mathfrak{K})) \cong \mathrm{C}^\infty(M, \mathbb{K})$.

In order to show that, for finite-dimensional perfect Lie algebras $\mathfrak{k}$, the invariant symmetric bilinear form $\kappa : \Gamma(\mathfrak{K}) \times \Gamma(\mathfrak{K}) \to \Gamma(\mathrm{V}(\mathfrak{K}))$ is universal, we will discuss certain commutative algebras in the next section, so we can generalize Lemma II.11 of [MN03] which states the universality of κ in the case $\Gamma(\mathfrak{K}) = \mathrm{C}^\infty(M, \mathfrak{k})$ for central simple $\mathfrak{k}$.

4.2. Commutative Pseudo-Unital Algebras

The motivation for the discussion of commutative pseudo-unital algebras is the following type of algebras.

Definition 4.2.1. Let $X \subseteq M$ be a subset of the manifold M and Y be a Fréchet space. We define $\mathrm{C}_X^\infty(M, Y) \subseteq \mathrm{C}^\infty(X, Y)$ to be the image of $\{f \in \mathrm{C}^\infty(M, Y) | \mathrm{supp}(f) \subseteq X\}$ under the restriction map $\mathrm{res}_X : \mathrm{C}^\infty(M, Y) \to \mathrm{C}^\infty(X, Y)$, $f \mapsto f_{|X}$.

Lemma 4.2.2. If $U \subseteq M$ is an open subset of a manifold M and Y is a Fréchet space, then the subspace $\mathrm{C}_U^\infty(M, Y) \subseteq \mathrm{C}^\infty(U, Y)$ is dense with respect to the smooth topology on $\mathrm{C}^\infty(U, Y)$ (cf. Definition B.1.4 and Definition B.1.7).

Proof. Since $\mathrm{C}^\infty(U, Y)$ is Fréchet, it is metrizable, so in order to show the density of $\mathrm{C}_U^\infty(M, Y)$ it suffices to show that for every $f \in \mathrm{C}^\infty(U, Y)$ there is a sequence $(f_k)_{k \in \mathbb{N}_0} \subseteq \mathrm{C}_U^\infty(M, Y)$ that converges to f. We now fix f.

Let $(V_{n,t}, D_{n,t})_{n,t \in \mathbb{N}_0}$ be a smooth cover of U and $\pi_{r,s}^M : T^r M \to T^s M$, where $r > s \in \mathbb{N}_0$, the iterated tangent projections (cf. Definition B.1.6). Note that $D_{n,t} = \overline{V_{n,t}}$ is, a priori, the closure in $T^n U$. But the inclusions $T^n U \to T^n M$ are continuous, so $D_{n,t}$ is also compact with respect to the subspace topology induced by $T^n M$, hence closed in the Hausdorff space $T^n M$. We now fix $n \in \mathbb{N}_0$.

Choose, for $s \in \mathbb{N}_0$, a smooth map $\gamma_{n,s} : U \to [0,1]$ with $\gamma_{n,s|D_{n,s}} = \mathbf{1}$ and $\operatorname{supp}(\gamma_{n,s}) \subseteq V_{n,s+1}$, and set $f'_{n,s} := \gamma_{n,s} \cdot f$. Then we can also choose a smooth map $\rho_{n,s} : M \to [0,1]$ such that $\rho_{n,s|D_{n,s}} = \mathbf{1}$ and $\operatorname{supp}(\rho_{n,s}) \subseteq T^n U$, since $T^n U$ is a neighborhood of $D_{n,s}$ with respect to the subspace topology induced by $T^n M$, and set $f_{n,s} := \rho_{n,s} \cdot f'_{n,s}$.

Let $(p_j)_{j \in \mathbb{N}_0}$ be a family of seminorms defining the topology of Y and the topology on the Fréchet space $\mathrm{C}^\infty(U, Y)$ defined by the seminorms

$$p_{j,k,t} : \mathrm{C}^\infty(U, Y) \to \mathbb{R}_+, \quad f \mapsto \sup_{x \in D_{k,t}} p_j(d^k f(x)),$$

where $j, k, t \in \mathbb{N}_0$. With these indices fixed, the sequence $(p_{j,k,t}(f - f_{n,s}))_{n,s \in \mathbb{N}_0}$ becomes 0 eventually (when $n > k$ and $s > t$), so $(f_{n,s})_{n,s \in \mathbb{N}_0}$ converges to f. $\qquad\square$

These algebras are examples of the following more general ones.

Definition 4.2.3. An associative algebra C is called *commutative pseudo-unital algebra*, if it is commutative and for each two elements $a, b \in C$ there is an element $c \in C$ such that $ac = a$ and $bc = b$.

Remark 4.2.4. Let C be a commutative pseudo-unital algebra. Then for every finite subset $F \subseteq C$, there is an element $u \in C$ such that $fu = f$ for all $f \in F$:

The statement for $\#F \leq 2$ follows from the definition of a commutative pseudo-unital algebra. If $F = \{f_1, f_2, \ldots, f_n\} \subseteq C$ and $f_i v = f_i$ for some $v \in C$ and all $i \leq n - 1$, then any element $u \in C$ given by the conditions $vu = v$ and $f_n u = f_n$ also satisfies the condition $fu = f$ for all $f \in F$.

Example 4.2.5. Let Y be a commutative unital Fréchet algebra and M a manifold.

1. If $U \subseteq M$ is an open subset, then the subalgebra $\mathrm{C}_U^\infty(M, Y) \subseteq \mathrm{C}^\infty(U, Y)$ is a commutative pseudo-unital algebra: For $f, g \in \mathrm{C}_U^\infty(M, Y)$ let Z be the union of the supports of f and g. Then U is an open neighborhood of the closed subset $Z \subseteq M$, thus there is a smooth map $h : M \to [0,1] \subseteq Y$ with $h_{|Z} = \mathbf{1}$, $\operatorname{supp}(h) \subseteq U$ and thus $fh = f$, $gh = g$.

2. The space $\mathrm{C}_c^\infty(M, Y)$ of compactly supported smooth maps is a commutative pseudo-unital algebra: For $f, g \in \mathrm{C}_c^\infty(M, Y)$ let $Z := \operatorname{supp}(f) \cup \operatorname{supp}(g)$. Since M is locally compact, it can be covered by relatively compact m-neighborhoods $U_m \subseteq M$ for $m \in M$. Since Z is compact, there is a finite set $F \subseteq M$ such that $Z \subseteq \bigcup_{m \in F} U_m =: U$. We then choose a smooth map $h : M \to [0,1] \subseteq Y$ with $h_{|Z} = \mathbf{1}$, $\operatorname{supp}(h) \subseteq U \subseteq \overline{U}$ and thus $fh = f$, $gh = g$.

The proof of the next lemma is basically the same as the proof of Lemma 4.2.2. Note that $\mathrm{C}_c^\infty(M, Y)$ is usually equipped with a different topology, turning it to a complete locally convex vector space (cf. Example B.2.7).

Lemma 4.2.6. *If M is a manifold and Y is a Fréchet space, then $\mathrm{C}_c^\infty(M,Y) \subseteq \mathrm{C}^\infty(M,Y)$ is a dense subspace.*

Remark 4.2.7. Here are some facts on commutative pseudo-unital algebras.

1. Unital commutative algebras are commutative pseudo-unital algebras and commutative pseudo-unital algebras are perfect in the sense that the image of the multiplication map generates the algebra. If C is a commutative pseudo-unital algebra and $\mathfrak{h}$ is a perfect Lie algebra, then $C \otimes \mathfrak{h}$ is a perfect Lie algebra with the bracket $[a \otimes x, b \otimes y] := ab \otimes [x,y]$ (cf. Definition 2.1.8):

   To see that, we take a generating element $a \otimes x \in C \otimes \mathfrak{h}$. By the perfectness of $\mathfrak{h}$, there is an $m \in \mathbb{N}$ and elements $y_1, z_1, \ldots, y_m, z_m \in \mathfrak{h}$ such that $x = \sum_{j=1}^m [y_j, z_j]$. Then, since C is a commutative pseudo-unital algebra, there is an element $c \in C$ such that $ac = a$. So we can write:

   $$a \otimes x = a \otimes \left(\sum_{j=1}^m [y_j, z_j] \right) = \sum_{j=1}^m ac \otimes [y_j, z_j] = \sum_{j=1}^m [a \otimes y_j, c \otimes z_j].$$

2. The multiplier algebra of $C := \mathrm{C}_U^\infty(M, \mathbb{K})$ is naturally isomorphic to $\mathrm{C}^\infty(U, \mathbb{K})$: For each $x \in U$ we choose a relatively compact open x-neighborhood $U_x \subseteq U$ and a smooth map $\rho_x \in C$ such that $\rho_{x|U_x} = 1$. For $L \in \mathrm{Mult}(C)$ and $x \in U$ we then define the number $e_L(x) := L(\rho_x)(x) \in \mathbb{K}$. We obtain a smooth function $e_L \in \mathrm{C}^\infty(U, \mathbb{K})$ because, if $x \in U$ then for all $y \in U_x$ we have:

   $$e_L(y) = L(\rho_y)(y) = L(\rho_y)(y) \cdot \rho_x(y) = L(\rho_y \rho_x)(y) = \rho_y(y) \cdot L(\rho_x)(y) = L(\rho_x)(y),$$

   thus e_L is, locally, a well-defined map of the form $L(\rho_x)$ for some $x \in U$. Clearly, the resulting map $\mathcal{E} : \mathrm{Mult}(C) \to \mathrm{C}^\infty(U, \mathbb{K})$, $L \mapsto e_L$ is an algebra morphism. It has the natural inverse $\mathcal{F} : \mathrm{C}^\infty(U, \mathbb{K}) \to \mathrm{Mult}(C)$, $g \mapsto (f \mapsto f \cdot g)$, as follows from the following calculation:

   $$(\mathcal{E} \circ \mathcal{F})(g)(x) = e_{\mathcal{F}(g)}(x) = (\mathcal{F}(g))(\rho_x)(x) = \rho_x(x) \cdot g(x) = g(x),$$
   $$(\mathcal{F} \circ \mathcal{E})(L)(f)(x) = (f \cdot \mathcal{E}(L))(x) = (f \cdot e_L)(x) = f(x) \cdot L(\rho_x)(x) = L(f)(x) \cdot \rho_x(x)$$
   $$= L(f)(x).$$

The centroid of the Lie algebra $C \otimes \mathfrak{h}$, where C is a commutative pseudo-unital algebra and $\mathfrak{h}$ is a finite-dimensional Lie algebra, is calculated in Proposition 4.2.9. As a first step, we need the following easy lemma.

Lemma 4.2.8. *Let V, W be vector spaces, where W is finite-dimensional. Then the linear map $\Phi : \mathrm{End}(V) \otimes \mathrm{End}(W) \to \mathrm{End}(V \otimes W)$ defined by $\Phi(L, \varphi)(v \otimes w) := L(v) \otimes \varphi(w)$ is an isomorphism of associative algebras.*

Proof. Let $(w_j)_{j=1}^n$ be a basis of W and define $W_j := \mathbb{K} \cdot w_j$ and the linear maps $w_j^* : W \to \mathbb{K}$ with $w_j^*(w_k) = \delta_{jk}$ for $j, k = 1, \ldots, n$. Obviously, Φ is a morphism of associative algebras. It is bijective, since the isomorphisms

$$\operatorname{End}(V) \otimes \operatorname{End}(W) \overset{\Phi_1}{\cong} \bigoplus_{j,k=1}^{n} \operatorname{End}(V) \overset{\Phi_2}{\cong} \bigoplus_{j,k=1}^{n} \operatorname{Hom}(V \otimes W_j, V \otimes W_k)$$

$$\overset{\Phi_3}{\cong} \operatorname{End}\left(\bigoplus_{j=1}^{n} V \otimes W_j\right) \cong \operatorname{End}\left(V \otimes \bigoplus_{j=1}^{n} W_j\right) = \operatorname{End}(V \otimes W)$$

are given by the following assignments:

$$\Phi_1(L \otimes \varphi) := \left(w_k^*(\varphi(w_j)) \cdot L\right)_{j,k=1}^{n},$$

$$\Phi_2\left(L_{st}\right)_{s,t=1}^{n} := \left(L_{jk} \otimes (w_j^* \otimes w_k)\right)_{j,k=1}^{n},$$

$$\Phi_3\left(T_{st}\right)_{s,t=1}^{n}\left(\sum_{j=1}^{n} v^{(j)} \otimes w_j\right) := \left(T_{st}\left(v^{(j)} \otimes w_j\right)\right)_{j=1}^{n},$$

and $(\Phi_3 \circ \Phi_2 \circ \Phi_1)(L \otimes \varphi)(v \otimes w) = L(v) \otimes \varphi(w) = \Phi(L \otimes \varphi)(v \otimes w)$ for $L \in \operatorname{End}(V)$, $\varphi \in \operatorname{End}(W)$, $v \in V$, $w \in W$. □

Proposition 4.2.9. *Let C be a commutative pseudo-unital algebra and $\mathfrak{h}$ a finite-dimensional perfect Lie algebra. Then $\operatorname{Mult}(C) \otimes \operatorname{Cent}(\mathfrak{h}) \cong \operatorname{Cent}(C \otimes \mathfrak{h})$ by the restriction of the isomorphism $\Phi : \operatorname{End}(C) \otimes \operatorname{End}(\mathfrak{h}) \to \operatorname{End}(C \otimes \mathfrak{h})$ (cf. Lemma 4.2.8). If, in particular, C is unital, then $C \cong \operatorname{Mult}(C)$ via $\mu : C \to \operatorname{Mult}(C)$, $\mu(c)(d) = cd$, and $C \otimes \operatorname{Cent}(\mathfrak{h})$ is then naturally isomorphic to $\operatorname{Cent}(C \otimes \mathfrak{h})$ via $c \otimes \varphi \longmapsto \Phi(\mu(c), \varphi)$.*

Proof. Let $\Psi : \operatorname{Mult}(C) \otimes \operatorname{Cent}(\mathfrak{h}) \to \operatorname{End}(C \otimes \mathfrak{h})$ be the restriction of Φ. To show that Ψ is an algebra isomorphism, it suffices to show that $\operatorname{im}(\operatorname{Mult}(C) \otimes \operatorname{Cent}(\mathfrak{h})) = \operatorname{Cent}(C \otimes \mathfrak{h})$:

- The image of Ψ is contained in $\operatorname{Cent}(C \otimes h)$. To see that, let $L \in \operatorname{Mult}(C)$, $\varphi \in \operatorname{Cent}(\mathfrak{h})$, $c, d \in C$, $x, y \in \mathfrak{h}$ and calculate:

$$\Psi(L \otimes \varphi)\left[c \otimes x, d \otimes y\right] = \Psi(L \otimes \varphi)\left(cd \otimes [x, y]\right) = L(cd) \otimes \varphi\left[x, y\right]$$
$$= cL(d) \otimes [x, \varphi(y)] = [c \otimes x, (L \otimes \varphi)(d \otimes y)].$$

- Conversely, let $\phi \in \operatorname{Cent}(C \otimes \mathfrak{h})$. By the surjectivity of Φ, there are morphisms $\phi_2^i \in \operatorname{End}(\mathfrak{h})$ and linearly independent morphisms $\phi_1^i \in \operatorname{End}(C)$ for $i = 1, \ldots, N_1$ such that, for all $c \in C$, $x \in \mathfrak{h}$, one has:

$$\phi(c \otimes x) = \sum_{i=1}^{N_1} \phi_1^i(c) \otimes \phi_2^i(x).$$

Let $x, y \in \mathfrak{h}$ and $c \in C$. Since C is a commutative pseudo-unital algebra, we can choose

$d \in C$ such that $cd = c$ and $\phi_1^i(c)d = \phi_1^i(c)$ for all $i = 1, \ldots, N_1$. We calculate:

$$\sum_{i=1}^{N_1} \phi_1^i(c) \otimes \phi_2^i[x, y] = \phi(c \otimes [x, y]) = \phi[c \otimes x, d \otimes y] = [\phi(c \otimes y), d \otimes y]$$

$$= \left[\sum_{i=1}^{N_1} \phi_1^i(c) \otimes \phi_2^i(x), d \otimes y \right] = \sum_{i=1}^{N_1} \phi_1^i(c)d \otimes \left[\phi_2^i(x), y \right]$$

$$= \sum_{i=1}^{N_1} \phi_1^i(c) \otimes \left[\phi_2^i(x), y \right].$$

Since the ϕ_1^i's were linearly independent, it follows $\phi_2^i \circ \mathrm{ad}_\mathfrak{h}(y) = \mathrm{ad}_\mathfrak{h}(y) \circ \phi_2^i$ for all $y \in \mathfrak{h}$, $i = 1, \ldots, N_1$. Thus $\phi_2^i \in \mathrm{Cent}(\mathfrak{h})$ for $i = 1, \ldots, N_1$. Let $\chi_2^1, \ldots, \chi_2^{N_2}$ be a basis of $\mathrm{Cent}(\mathfrak{h})$. Then there are morphisms $\chi_1^i \in \mathrm{End}(C)$ for $i = 1, \ldots, N_2$ such that, for all $c \in C$, $x \in \mathfrak{h}$, one has:

$$\phi(c \otimes x) = \sum_{i=1}^{N_2} \chi_1^i(c) \otimes \chi_2^i(x).$$

So for $x, y \in \mathfrak{h}$ and $c, d \in C$ we may also calculate:

$$\sum_{i=1}^{N} \chi_1^i(cd) \otimes \chi_2^i[x, y] = \phi(cd \otimes [x, y]) = \phi[c \otimes x, d \otimes y] = [\phi(c \otimes x), d \otimes y]$$

$$= \left[\sum_{i=1}^{n} \chi_1^i(c) \otimes \chi_2^i(x), d \otimes y \right] = \sum_{i=1}^{n} \left[\chi_1^i(c) \otimes \chi_2^i(x), d \otimes y \right]$$

$$= \sum_{i=1}^{n} \chi_1^i(c)d \otimes \chi_2^i[x, y],$$

leading, by the perfectness of $\mathfrak{h}$ and the linear independence of the χ_2^i's, to $\chi_1^i \in \mathrm{Mult}(C)$ for $i = 1, \ldots, N_2$. Thus $\mathrm{Cent}(C \otimes h) \subseteq \mathrm{Mult}(C) \otimes \mathrm{Cent}(\mathfrak{h})$.

$\square$

4.3. Universal Invariant Symmetric Bilinear Forms on Tensor Product Lie Algebras

In this section, we fix a commutative pseudo-unital algebra C, a manifold M and a finite-dimensional perfect $\mathbb{K}$-Lie algebra $\mathfrak{k}$ with $\kappa_\mathfrak{k} : \mathfrak{k} \times \mathfrak{k} \to \mathrm{V}(\mathfrak{k})$ being a universal invariant symmetric bilinear form. The next lemmas will help us to show that there is a universal invariant symmetric bilinear form $(C \otimes \mathfrak{k})^2 \to C \otimes \mathrm{V}(\mathfrak{k})$, hence $\mathrm{V}(C \otimes \mathfrak{k}) \cong C \otimes \mathrm{V}(\mathfrak{k})$.

Lemma 4.3.1. Let $\beta : (C \otimes \mathfrak{k})^2 \to W$ be an invariant symmetric bilinear form. Then there is a symmetric bilinear map $\alpha = \alpha_\beta : C \times C \to \mathrm{Hom}(\mathrm{V}(\mathfrak{k}), W)$ with

$$\beta(a \otimes x, b \otimes y) = \alpha(a, b)\left(\kappa_\mathfrak{k}(x, y)\right),$$

or, in other words, there is a linear map $\check{\alpha} = \check{\alpha}_\beta : S^2_{\mathbb{K}}(C) \to \operatorname{Hom}(V(\mathfrak{k}), W)$ with

$$\beta(a \otimes x, b \otimes y) = \check{\alpha}(a \vee_{\mathbb{K}} b)(\kappa_{\mathfrak{k}}(x,y))$$

for $a, b \in C$, $x, y \in \mathfrak{k}$.

Proof. We introduce the notation $\beta_{a,b}(x,y) := \beta(a \otimes x, b \otimes y)$ for fixed $a, b \in C$ and let $c \in C$ such that $ac = a$ and $bc = b$. Then $\beta_{a,b} : \mathfrak{k} \times \mathfrak{k} \to W$ is a bilinear form. Is also fulfills the following properties:

- Symmetry:
  To see that, let $x, y, y_1, y_2 \in \mathfrak{k}$. We calculate:

  $$\beta_{a,b}([y_1, y_2], x) = \beta(a \otimes [y_1, y_2], b \otimes x) = \beta([a \otimes y_1, c \otimes y_2], b \otimes x) = \beta(a \otimes y_1, [c \otimes y_2, b \otimes x])$$
  $$= -\beta(a \otimes y_1, [c \otimes x, b \otimes y_2]) = -\beta([a \otimes y_1, c \otimes x], b \otimes y_2)$$
  $$= \beta([a \otimes x, c \otimes y_1], b \otimes y_2) = \beta(a \otimes x, [c \otimes y_1, b \otimes y_2]) = \beta(a \otimes x, b \otimes [y_1, y_2])$$
  $$= \beta_{a,b}(x, [y_1, y_2]).$$

  So the symmetry follows from the perfectness of $\mathfrak{k}$.

- Invariance:
  To see that, let $x, y, z \in \mathfrak{k}$. We calculate:

  $$\beta_{a,b}([x, y], z) = \beta(a \otimes [x, y], b \otimes z) = \beta([a \otimes x, c \otimes y], b \otimes z) = \beta(a \otimes x, [c \otimes y, b \otimes z])$$
  $$= \beta(a \otimes x, b \otimes [y, z]) = \beta_{a,b}(x, [y, z]).$$

The universality of $\kappa_{\mathfrak{k}}$ then yields the existence of a map $\alpha : C \times C \to \operatorname{Hom}(V(\mathfrak{k}), W)$ such that $\beta_{a,b} = \alpha(a, b) \circ \kappa_{\mathfrak{k}}$. Since the image of $\kappa_{\mathfrak{k}}$ spans $V(\mathfrak{k})$ (cf. Remark 4.1.5.4) and the $\alpha(a, b)$'s are unique, the bilinearity and symmetry of α are shown by the following calculations:

$$\alpha(a, b)(\kappa_{\mathfrak{k}}(x, y)) = \beta(a \otimes x, b \otimes y) = \beta(b \otimes y, a \otimes x) = \alpha(b, a)(\kappa_{\mathfrak{k}}(y, x))$$
$$= \alpha(b, a)(\kappa_{\mathfrak{k}}(x, y)),$$
$$\alpha(\lambda a + a', b)(\kappa_{\mathfrak{k}}(x, y)) = \beta((\lambda a + a') \otimes x, b \otimes y)$$
$$= \left(\lambda \cdot \beta(a \otimes x, b \otimes y) + \beta(a' \otimes x, b \otimes y)\right)(\kappa_{\mathfrak{k}}(x, y))$$
$$= \left(\lambda \cdot \alpha(a, b) + \alpha(a', b)\right)(\kappa_{\mathfrak{k}}(x, y)). \qquad \square$$

Lemma 4.3.2. *Let $\beta : (C \otimes \mathfrak{k})^2 \to W$ be an invariant symmetric bilinear form with corresponding linear map $\check{\alpha} : S^2_{\mathbb{K}}(C) \to \operatorname{Hom}(V(\mathfrak{k}), W)$ (cf. Lemma 4.3.1). We then have $\check{\alpha}(ab \vee_{\mathbb{K}} c) = \check{\alpha}(ac \vee_{\mathbb{K}} b)$ for $a, b, c \in C$, and there is a linear map $\widetilde{\alpha} = \widetilde{\alpha}_\beta : C \to \operatorname{Hom}(V(\mathfrak{k}), W)$ with*

$$\beta(a \otimes x, b \otimes y) = \widetilde{\alpha}(ab)(\kappa_{\mathfrak{k}}(x, y))$$

for $a, b \in C$, $x, y \in \mathfrak{k}$.

Proof. Let $a, b, c \in C$ and $x, y, x_1, x_2 \in \mathfrak{k}$. We calculate:

$$\begin{aligned}
\breve{\alpha}(ab \vee_{\mathbb{K}} c)(\kappa_{\mathfrak{k}}([x_1, x_2], y)) &= \beta(ab \otimes [x_1, x_2], c \otimes y) = \beta([a \otimes x_1, b \otimes x_2], c \otimes y) \\
&= \beta(a \otimes x_1, cb \otimes [x_2, y]) = \beta([a \otimes x_1, c \otimes x_2], b \otimes y) \\
&= \alpha(ac \vee_{\mathbb{K}} b)(\kappa_{\mathfrak{k}}([x_1, x_2], y)).
\end{aligned}$$

So $\breve{\alpha}(ab \vee_{\mathbb{K}} c) = \breve{\alpha}(ac \vee_{\mathbb{K}} b)$ by the perfectness of $\mathfrak{k}$ and the fact that the image of $\kappa_{\mathfrak{k}}$ spans $V(\mathfrak{k})$.

We define $\widetilde{\alpha} : C \to \mathrm{Hom}(V(\mathfrak{k}), W)$, $c \mapsto \breve{\alpha}(c \vee_{\mathbb{K}} c')$ for any $c' \in C$ such that $cc' = c$. This is well-defined because the condition $cc' = c = cc''$ implies:

$$\breve{\alpha}(c \vee_{\mathbb{K}} c'') = \breve{\alpha}(cc' \vee_{\mathbb{K}} c'') = \breve{\alpha}(cc'' \vee_{\mathbb{K}} c') = \breve{\alpha}(c \vee_{\mathbb{K}} c'). \qquad \square$$

Proposition 4.3.3. *Let C be a commutative pseudo-unital algebra and $\mathfrak{k}$ be a finite-dimensional perfect Lie algebra. Then $\kappa_{C,\mathfrak{k}} : (C \otimes \mathfrak{k})^2 \to C \otimes V(\mathfrak{k})$ defined by $\kappa_{C,\mathfrak{k}}(a \otimes x, b \otimes y) := ab \otimes \kappa_{\mathfrak{k}}(x, y)$ is a universal invariant symmetric bilinear form.*

Proof. Let $\beta : (C \otimes \mathfrak{k})^2 \to W$ be an invariant symmetric bilinear form with corresponding linear map $\widetilde{\alpha} : C \to \mathrm{Hom}(V(\mathfrak{k}), W)$ (cf. Lemma 4.3.2). The latter map naturally corresponds to a linear map $\widehat{\alpha} : C \otimes V(\mathfrak{k}) \to W$ via $\widehat{\alpha}(c \otimes \delta) := \widetilde{\alpha}(c)(\delta)$. Hence, for $a, b \in C$, $x, y \in \mathfrak{k}$ we have

$$\widehat{\alpha} \circ \kappa_{C,\mathfrak{k}}(a \otimes x, b \otimes y) = \widehat{\alpha}(ab \otimes \kappa_{\mathfrak{k}}(x, y)) = \widetilde{\alpha}(ab)(\kappa_{\mathfrak{k}}(x, y)) = \beta(a \otimes x, b \otimes y),$$

yielding $\widehat{\alpha} \circ \kappa_{C,\mathfrak{k}} = \beta$. The universality of $\kappa_{C,\mathfrak{k}}$ now follows from that fact that its image spans $C \otimes V(\mathfrak{k})$, since the image of $\kappa_{\mathfrak{k}}$ spans $V(\mathfrak{k})$ and the multiplication of C is surjective. $\qquad \square$

Corollary 4.3.4. *For a commutative pseudo-unital algebra C and a finite-dimensional perfect Lie algebra $\mathfrak{k}$, we have $V(C \otimes \mathfrak{k}) \cong C \otimes V(\mathfrak{k})$. In particular, there are the universal invariant symmetric bilinear forms*

$$\kappa_U : C_U^\infty(M, \mathfrak{k}) \times C_U^\infty(M, \mathfrak{k}) \longrightarrow C_U^\infty(M, V(\mathfrak{k}))$$
$$\kappa_U(f, g)(m) := \kappa_{\mathfrak{k}}(f(m), g(m))$$

for open $U \subseteq M$ and

$$\kappa_c : C_c^\infty(M, \mathfrak{k}) \times C_c^\infty(M, \mathfrak{k}) \longrightarrow C_c^\infty(M, V(\mathfrak{k}))$$
$$\kappa_c(f, g)(m) := \kappa_{\mathfrak{k}}(f(m), g(m)).$$

4.4. Universal Invariant Symmetric Bilinear Forms on Lie Algebras of Smooth Sections

Now fix a manifold M and finite-dimensional perfect $\mathbb{K}$-Lie algebra $\mathfrak{k}$ with universal invariant symmetric bilinear form $\kappa_{\mathfrak{k}} : \mathfrak{k} \times \mathfrak{k} \to V(\mathfrak{k})$. Let $\pi : \mathfrak{K} \to M$ be a Lie algebra bundle with fiber $\mathfrak{k}$. We will show that the invariant symmetric bilinear form $\kappa : \Gamma(\mathfrak{K}) \times \Gamma(\mathfrak{K}) \to V(\Gamma(\mathfrak{K}))$ from Definition 4.1.13 is universal, i.e., $V(\Gamma(\mathfrak{K})) \cong \Gamma(V(\mathfrak{K}))$. First of all, we need the following lemma.

Lemma 4.4.1. *The image of κ spans $\Gamma(V(\mathfrak{K}))$.*

Proof. A section $\mathcal{F} \in \Gamma(V(\mathfrak{K}))$ is a smooth map

$$M \longrightarrow V(\mathfrak{K})$$
$$m \longmapsto [(\psi(m), \delta(m))] = \left\{ \left(\psi \circ f^{-1}, \delta(m) \circ c_{f^{-1}} \right) : f \in \mathrm{Aut}(\mathfrak{k}) \right\},$$

for smooth maps $\psi : M \to P(\mathfrak{K}) = \bigcup_{m \in M} \mathrm{Iso}(\mathfrak{k}, \mathfrak{K}_m)$ and $\delta : M \to V(\mathfrak{k})$. Since Remark 4.1.5.4 yields that the image of $\kappa_\mathfrak{k}$ spans $V(\mathfrak{k})$, after choosing a basis $(x_1, \ldots, x_N)$ of $\mathfrak{k}$, we can find smooth maps $f_i, g_j : M \to \mathbb{K}$ for $i, j \in \{1, \ldots, N\}$ such that $\delta(m) = \sum_{i,j=1}^{N} \kappa_\mathfrak{k}(f_i(m)x_i, g_j(m)x_j)$ for all $m \in M$. The following calculation proves the lemma:

$$\mathcal{F}(m) = [(\psi(m), \delta(m))] = \left[\left(\psi(m), \sum_{i,j=1}^{N} f_i(m)g_j(m)\kappa_\mathfrak{k}(x_i, x_j) \right) \right]$$

$$= \sum_{i,j=1}^{N} [(\psi(m), \kappa_\mathfrak{k}(f_i(m)x_i, g_j(m)x_j))] = \sum_{i,j=1}^{N} \kappa_{\mathfrak{K}_m}\left([\psi(m), f_i(m)x_i], [\psi(m), g_j(m)x_j] \right)$$

$$= \sum_{i,j=1}^{N} \kappa\left(m \mapsto [\psi(m), f_i(m)x_i], m \mapsto [\psi(m), g_j(m)x_j] \right). \qquad \square$$

Definition 4.4.2. Let (U, φ) and (U, ψ) be bundle charts of $\pi : \mathfrak{K} \to M$ and $\nu : V(\mathfrak{K}) \to M$, respectively, and let $X \in \Gamma(\mathfrak{K})$ and $\mathcal{G} \in \Gamma(V(\mathfrak{K}))$ be sections with support in $U \subseteq M$. We use the notation $X^\varphi := \varphi \circ X \in C_U^\infty(M, \mathfrak{k})$ and $\mathcal{G}^\psi := \psi \circ \mathcal{G} \in C_U^\infty(M, V(\mathfrak{k}))$ (cf. Definition 2.3.11). The inverse operation is given by

$$(\varphi^* f)(m) := \begin{cases} (\pi, \varphi)^{-1}(m, f(m)) & \text{if } m \in U, \\ 0 & \text{otherwise} \end{cases}$$

for $f \in C_U^\infty(M, \mathfrak{k})$ and

$$(\psi^* D)(m) := \begin{cases} (\nu, \psi)^{-1}(m, D(m)) & \text{if } m \in U, \\ 0 & \text{otherwise} \end{cases}$$

for $D \in C_U^\infty(M, V(\mathfrak{k}))$, respectively.

Remark 4.4.3. Let $\rho : P(\mathfrak{K}) \to M$ be the $\mathrm{Aut}(\mathfrak{k})$-principal bundle, to which the Lie algebra bundle $\pi : \mathfrak{K} \to M$ is associated. By Remark 2.2.3, we choose a finite bundle atlas of this principal bundle $(U_i, \theta_i)_{i \in F}$, so that there are bundle atlases $(U_i, \varphi_i)_{i \in F}$ and $(U_i, \psi_i)_{i \in F}$ of $\pi : \mathfrak{K} \to M$ and $\nu : V(\mathfrak{K}) \to M$, respectively, defined by $\varphi_i : \pi^{-1}(U_i) \to \mathfrak{k}$, $[p, x] \mapsto \theta_i(p).x = \theta(p)(x)$ and $\psi_i : \nu^{-1}(U_i) \to V(\mathfrak{k})$, $\kappa_{\mathfrak{K}_{\pi(p)}}([p, x], [p, y]) \mapsto \kappa_\mathfrak{k}(\theta_i(p)(x), \theta_i(p)(y))$, respectively.

With the notation of Definition 4.4.2, we want to compare the universal invariant symmetric bilinear form κ_U from Corollary 4.3.4 with the form $\kappa : \Gamma(\mathfrak{K}) \times \Gamma(\mathfrak{K}) \to \Gamma(V(\mathfrak{K}))$ for any $(U, \varphi, \psi) \in (U_i, \varphi_i, \psi_i)_{i \in F}$. We fix a basis $(x_1, \ldots, x_N)$ of $\mathfrak{k}$ and two sections $X, Y \in \Gamma(\mathfrak{K})$ with support in $U \subseteq M$. Then there are smooth maps $\xi_s, \eta_t \in C_U^\infty(M, \mathbb{K})$ such that $X^\varphi = \sum_{s=1}^{N} \xi_s \otimes x_s$ and $Y^\varphi = \sum_{t=1}^{N} \eta_t \otimes x_t$, yielding $\kappa_U(X^\varphi, Y^\varphi) = \sum_{s,t=1}^{N} \xi_s \eta_t \otimes \kappa_\mathfrak{k}(x_s, x_t)$.

4. Universal Invariant Symmetric Bilinear Forms

The section $\kappa(X,Y) \in \Gamma(V(\mathfrak{K}))$ is U-supported, too, and for $m \in U$ we calculate:

$$(\kappa(X,Y)^\psi)(m) = (\psi \circ \kappa(X,Y))(m) = \psi(\kappa_{\mathfrak{K}_m}(X_m, Y_m)) = \psi(\kappa_{\mathfrak{K}_m}(\varphi^*(X^\varphi)(m), \varphi^*(Y^\varphi)(m)))$$

$$= \psi\left(\kappa_{\mathfrak{K}_m}\left((\pi,\varphi)^{-1}\left(m, \sum_{s=1}^{N} \xi_s(m)x_s\right), (\pi,\varphi)^{-1}\left(m, \sum_{t=1}^{N} \eta_t(m)x_t\right)\right)\right)$$

$$= \sum_{s,t=1}^{N} \xi_s(m)\eta_t(m) \cdot \psi\left(\kappa_{\mathfrak{K}_m}\left(\left(\varphi_{|\mathfrak{K}_m}\right)^{-1}(x_s), \left(\varphi_{|\mathfrak{K}_m}\right)^{-1}(x_t)\right)\right)$$

$$= \sum_{s,t=1}^{N} \xi_s(m)\eta_t(m) \cdot \psi\left(\kappa_{\mathfrak{K}_m}\left(\left[p, \theta_i(p)^{-1}(x_s)\right], \left[p, \theta_i(p)^{-1}(x_t)\right]\right)\right)$$

$$= \sum_{s,t=1}^{N} \xi_s(m)\eta_t(m) \cdot \kappa_\mathfrak{k}\left(\theta_i(p)\left(\theta_i(p)^{-1}(x_s)\right), \theta_i(p)\left(\theta_i(p)^{-1}(x_t)\right)\right)$$

$$= \sum_{s,t=1}^{N} (\xi_s\eta_t)(m) \cdot \kappa_\mathfrak{k}(x_s, x_t) = \kappa_U(X^\varphi, Y^\varphi)(m).$$

Thus $\kappa_U(X^\varphi, Y^\varphi) = \kappa(X,Y)^\psi$.

Theorem 4.4.4. *Let $\pi : \mathfrak{K} \to M$ be a finite-dimensional Lie algebra bundle with perfect fiber $\mathfrak{k}$. Then the invariant symmetric bilinear form $\kappa : \Gamma(\mathfrak{K}) \times \Gamma(\mathfrak{K}) \to \Gamma(V(\mathfrak{K}))$ is universal. In particular $V(\Gamma(\mathfrak{K})) \cong \Gamma(V(\mathfrak{K}))$.*

Proof. We fix a finite bundle atlas of the $\mathrm{Aut}(\mathfrak{k})$-principal bundle corresponding to $\pi : \mathfrak{K} \to M$ and obtain the finite bundle atlases $(U_i, \varphi_i)_{i \in F}$ and $(U_i, \psi_i)_{i \in F}$ of $\pi : \mathfrak{K} \to M$ and $\nu : V(\mathfrak{K}) \to M$, respectively, and use the notation of Definition 4.4.2. Let $(\rho_i : M \to [0,1])_{i \in F}$ be a smooth partition of unity such that $\mathrm{supp}(\rho_i) \subseteq U_i$ for all $i \in F$. Note that the maps $L_\eta : \Gamma(\mathfrak{K}) \to \Gamma(\mathfrak{K})$, $X \mapsto \eta \cdot X$ for $\eta \in C^\infty(M, \mathbb{K})$ are in $\mathrm{Cent}(\mathfrak{K})$, by the local definition of the Lie bracket on $\Gamma(\mathfrak{K})$.

Let $\beta : \Gamma(\mathfrak{K}) \times \Gamma(\mathfrak{K}) \to W$ be an invariant symmetric bilinear form. For $i \in F$ define $\beta_i : C^\infty_{U_i}(M, \mathfrak{k}) \times C^\infty_{U_i}(M, \mathfrak{k}) \to W$, $(f_1, f_2) \mapsto \beta\left((\varphi_i)^* f_1, (\varphi_i)^* f_2\right)$, an invariant symmetric bilinear form. Then Corollary 4.3.4 gives a unique linear map $\widehat{\beta}_i : C^\infty_{U_i}(M, \mathbb{K}) \otimes V(\mathfrak{k}) \to W$ such that $\beta_i = \widehat{\beta}_i \circ \kappa_{U_i}$.

Define the linear map $\widehat{\beta} : \Gamma(V(\mathfrak{K})) \to W$, $\mathcal{G} \mapsto \sum_{i \in F} \widehat{\beta}_i\left((\rho_i \mathcal{G})^{\psi_i}\right)$ and fix $X, Y \in \Gamma(\mathfrak{K})$. Choose, for each $i \in F$, a map $\zeta_i \in C^\infty_{U_i}(M, [0,1])$ such that $\zeta_{i|\mathrm{supp}(\rho_i)} = 1$. The following calculation shows that $\beta = \widehat{\beta} \circ \kappa$:

$$\widehat{\beta}(\kappa(X,Y)) = \sum_{i \in F} \widehat{\beta}_i\left((\rho_i \zeta_i \cdot \kappa(X,Y))^{\psi_i}\right) = \sum_{i \in F} \widehat{\beta}_i\left((\kappa(\rho_i \cdot X, \zeta_i \cdot Y))^{\psi_i}\right)$$

$$\underset{\text{Remark } 4.4.3}{=} \sum_{i \in F}\left(\widehat{\beta}_i \circ \kappa_{U_i}\right)((\rho_i \cdot X)^{\varphi_i}, (\zeta_i \cdot Y)^{\varphi_i}) = \sum_{i \in F} \beta_i\left((\rho_i \cdot X)^{\varphi_i}, (\zeta_i \cdot Y)^{\varphi_i}\right)$$

$$= \sum_{i \in F} \beta\left(\rho_i \cdot X, L_{\zeta_i} Y\right) \underset{\text{Lemma } 4.1.6}{=} \sum_{i \in F} \beta\left(\rho_i \zeta_i \cdot X, Y\right) = \sum_{i \in F} \beta\left(\rho_i \cdot X, Y\right) = \beta(X,Y).$$

The uniqueness of $\widehat{\beta}$ now follows from Lemma 4.4.1, whence the universality of κ. $\square$

4.5. Continuous Universal Invariant Symmetric Bilinear Forms

In this section we will briefly present the continuous category of invariant symmetric bilinear forms by modifying the definitions of Section 4.1 and following [Mai02], where the missing proofs of the statements in this section can be found. Let $\mathcal{K}$ be a full subcategory of the category of locally convex $\mathbb{K}$-vector spaces: any morphism in the locally convex category (i.e. any continuous linear map) between objects in $\mathcal{K}$ is a morphism in $\mathcal{K}$.

Definition 4.5.1. For two $\mathcal{K}$-objects A, B the *($\mathcal{K}$-)tensor product* is a $\mathcal{K}$-object C with a continuous bilinear map $\otimes : A \times B \to C$ such that the following universal property is satisfied:

For any $\mathcal{K}$-object D and for any continuous bilinear map $\beta : A \times B \to D$ there exists a unique $\mathcal{K}$-morphism, i.e., a continuous linear map, $\ell_\beta : C \to D$ with $\ell_\beta \circ \otimes = \beta$.

The object C is unique up to a $\mathcal{K}$-isomorphism, i.e., a continuous linear isomorphism with continuous inverse, and one writes $A \otimes_{\mathcal{K}} B := C$.

Remark 4.5.2. For two $\mathcal{K}$-objects A, B, the vector space $A \otimes B := \operatorname{span}_{\mathbb{K}} \{ a \otimes b \, | \, a \in A, b \in B \}$ is dense in $A \otimes_{\mathcal{K}} B$ (cf. Chapter 43 of [Trè67]), but in general not the whole space.

We now restrict ourselves to the case $\mathcal{K} = \mathcal{F}$, the category of Fréchet spaces. Let $\mathfrak{g}$ be a perfect Fréchet-Lie algebra with commutative centroid $A := \operatorname{Cent}(\mathfrak{g})$. We assume A to be a Fréchet algebra such that the natural action $A \times \mathfrak{g} \to \mathfrak{g}$ is continuous.

Proposition 4.5.3. *Let $\sigma : \mathfrak{g} \otimes \mathfrak{g} \to \mathfrak{g} \otimes_{\mathcal{F}} \mathfrak{g}$ be the linear map defined by $\sigma(x, y) := (y, x)$ and let $\overline{\sigma} : \mathfrak{g} \otimes_{\mathcal{F}} \mathfrak{g} \to \mathfrak{g} \otimes_{\mathcal{F}} \mathfrak{g}$ be its continuous linear extension. For $\mathrm{S}^2_{\mathcal{F}}(\mathfrak{g}) := \ker(\mathbf{1} - \overline{\sigma})$ we have a continuous symmetric bilinear map $\vee_{\mathcal{F}} : \mathfrak{g} \times \mathfrak{g} \to \mathrm{S}^2_{\mathcal{F}}(\mathfrak{g})$ such that the linear span of its image is dense and the following universal property is satisfied:*

For each Fréchet space W and each continuous symmetric bilinear map $\beta : \mathfrak{g} \times \mathfrak{g} \to W$ there is a unique continuous linear map $\check{\beta} : \mathrm{S}^2_{\mathcal{F}}(\mathfrak{g}) \to W$ such that $\beta = \check{\beta} \circ \vee_{\mathcal{F}}$.

The Fréchet space $\mathrm{S}^2_{\mathcal{F}}(\mathfrak{g})$ is unique up to a continuous isomorphism. It is called the continuous symmetric square *of $\mathfrak{g}$.*

Remark 4.5.4. The Fréchet space $\mathrm{S}^2_{\mathcal{F}}(\mathfrak{g})$ is a $\mathfrak{g}$-Fréchet module defined[1] by

$$x.(y \vee_{\mathcal{F}} z) := [x, y] \vee_{\mathcal{F}} z + y \vee_{\mathcal{F}} [x, z]$$

for $x, y, z \in \mathfrak{g}$. The multiplication is well-defined by the universal property of $\mathrm{S}^2_{\mathcal{F}}(\mathfrak{g})$.

Definition 4.5.5. We write $\mathrm{V}_{\mathcal{F}}(\mathfrak{g})$ for the quotient Fréchet space $\mathrm{S}^2_{\mathcal{F}}(\mathfrak{g})/\overline{\mathfrak{g}. \mathrm{S}^2_{\mathcal{F}}(\mathfrak{g})}$. We also define the bilinear map

$$\kappa_{\mathcal{F}, \mathfrak{g}} : \mathfrak{g} \times \mathfrak{g} \longrightarrow \mathrm{V}_{\mathcal{F}}(\mathfrak{g})$$
$$(x, y) \longmapsto [x \vee_{\mathcal{F}} y].$$

Like in the algebraic category, the concept of universal continuous invariant symmetric bilinear forms helps us to understand general continuous invariant symmetric bilinear forms.

---

[1] The following defines the multiplication on a set of which the linear span is dense.

Definition 4.5.6. Let V be a locally convex vector space. A continuous invariant symmetric bilinear form $\gamma : \mathfrak{g} \times \mathfrak{g} \to V$ is *universal*, if for each topological vector space W and each continuous invariant symmetric bilinear form $\beta : \mathfrak{g} \times \mathfrak{g} \to W$ there is a unique continuous linear map $\beta' : V \to W$ such that $\beta = \beta' \circ \gamma$.

Remark 4.5.7. If $\gamma : \mathfrak{g} \times \mathfrak{g} \to V$ is a universal continuous invariant symmetric bilinear form, then $\mathrm{span}_{\mathbb{K}}(\mathrm{im}(\gamma))$ is dense.

Too see that, let $U := \overline{\mathrm{span}_{\mathbb{K}}(\mathrm{im}(\gamma))}$. If $U \neq V$, then, by the Hahn-Banach Theorem, there exists a $v \in V \backslash U$ and a continuous linear map $f : V \to \mathbb{K}$ such that $f(v) = 1$ and $f_{|U} = \mathbf{0}$. For the trivial continuous invariant symmetric bilinear form $\beta : \mathfrak{g} \times \mathfrak{g} \to W$, $(x, y) \mapsto 0$ and for any vector $w \in W$ we then have $\beta = \gamma \circ (f \cdot w)$. This is a contradiction for γ to be universal.

Remark 4.5.8.

1. If two continuous invariant symmetric bilinear forms $\gamma_i : \mathfrak{g} \times \mathfrak{g} \to V_i$, where $i = 1, 2$, are universal, then there are unique continuous linear morphisms $f : V_1 \to V_2$ and $g : V_2 \to V_1$ such that $\gamma_1 = g \circ \gamma_2$ and $\gamma_2 = f \circ \gamma_1$. Applying this argument again, we see that $g \circ f : V_1 \to V_1$ is the unique continuous linear map h such that $\gamma_2 = h \circ \gamma_1$, hence $g \circ f = \mathrm{id}_{V_1}$ and analogously $f \circ g = \mathrm{id}_{V_2}$, thus $V_1 \cong V_2$. In this sense, (γ_1, V_1) and (γ_2, V_2) are equivalent in the category of continuous invariant symmetric bilinear forms.

2. The continuous invariant symmetric bilinear form $\kappa_{\mathcal{F}, \mathfrak{g}} : \mathfrak{g} \times \mathfrak{g} \to V_{\mathcal{F}}(\mathfrak{g})$ is universal:

   Let $\beta : \mathfrak{g} \times \mathfrak{g} \to W$ be a continuous invariant symmetric bilinear form. Then there is a unique continuous linear map $\check{\beta} : S^2_{\mathcal{F}}(\mathfrak{g}) \to W$ such that $\beta = \check{\beta} \circ \vee_{\mathcal{F}}$ and $\mathfrak{g}.\,S^2_{\mathcal{F}}(\mathfrak{g}) \subseteq \ker(\check{\beta})$, since, for $x, y_1, y_2 \in \mathfrak{g}$, we can calculate:

   $$\check{\beta}(x.(y_1 \vee_{\mathcal{F}} y_2)) = \check{\beta}([x, y_1] \vee_{\mathcal{F}} y_2 + y_1 \vee_{\mathcal{F}} [x, y_2]) = \check{\beta}([x, y_1] \vee_{\mathcal{F}} y_2) + \check{\beta}(y_1 \vee_{\mathcal{F}} [x, y_2])$$
   $$= \beta([x, y_1], y_2) + \beta(y_1, [x, y_2]) = 0.$$

The next statement is the topological version of Lemma 4.1.6. Note that $\mathfrak{g}$ is supposed to be a perfect Fréchet-Lie algebra with commutative centroid $A = \mathrm{Cent}(\mathfrak{g})$.

Lemma 4.5.9. *Let* $\beta : \mathfrak{g} \times \mathfrak{g} \to V$ *be a continuous invariant symmetric bilinear form. Then* $\beta(ax, y) = \beta(x, ay)$ *for all* $a \in A$, $x, y \in \mathfrak{g}$.

Corollary 4.5.10. *The form* $\kappa_{\mathcal{F}, \mathfrak{g}} : \mathfrak{g} \times \mathfrak{g} \to V_{\mathcal{F}}(\mathfrak{g})$ *is A-bilinear with respect to the A-module structure defined by* $a.\,[x \vee_{\mathcal{F}} y] := [ax \vee_{\mathcal{F}} y]$ *for* $a \in A$, $x, y \in \mathfrak{g}$.

Proof. The A-module structure is well-defined and the A-bilinearity of $\kappa_{\mathcal{F}, \mathfrak{g}}$ is immediate. $\square$

4.6. Continuous Universal Invariant Symmetric Bilinear Forms on Lie Algebras of Smooth Sections

In this section, we want to show that the invariant symmetric bilinear form κ from Definition 4.1.13, which is universal without considering any topology (cf. Theorem 4.4.4), is a universal continuous invariant symmetric bilinear form for the Lie algebra of smooth sections $\Gamma(\mathfrak{K})$ of a Lie algebra bundle $\pi : \mathfrak{K} \to M$ with finite-dimensional perfect fiber $\mathfrak{k}$. In the following, the spaces $C^\infty(M, Y) \cong C^\infty(M, \mathbb{K}) \otimes Y$ for finite-dimensional Y's and $\Gamma(\mathfrak{K})$ are equipped with Fréchet topologies defined in the Definitions B.1.7, B.1.8 and B.1.12.

Proposition 4.6.1. *The invariant symmetric bilinear form* $\kappa : \Gamma(\mathfrak{K}) \times \Gamma(\mathfrak{K}) \to \Gamma(V(\mathfrak{K}))$ *is continuous.*

Proof. We use the notation of Definition 4.4.2 and fix finite bundle atlases $(U_i, \varphi_i)_{i \in F}$ and $(U_i, \psi_i)_{i \in F}$ of $\pi : \mathfrak{K} \to M$ and $\nu : V(\mathfrak{K}) \to M$, respectively, like in Remark 4.4.3. Observe that the maps $U_i \to \mathfrak{k} \times \mathfrak{k}$, $x \mapsto (X^{\varphi_i}(x), Y^{\varphi_i}(x))$ are continuous. Since $\mathfrak{k}$ is of finite dimension, there are norms $\|\cdot\|_1$ and $\|\cdot\|_2$ on $\mathfrak{k}$ and $V(\mathfrak{k})$, respectively, such that $\|\kappa_{\mathfrak{k}}(x, y)\|_2 \leq \|x\|_1 \cdot \|y\|_1$. So the same estimates as in Remark 2.3.4 show the continuity of $\kappa_{U_i}(X^{\varphi_i}, Y^{\varphi_i}) = \kappa(X, Y)^{\psi_i}$. $\square$

Theorem 4.6.2. *Let* $\pi : \mathfrak{K} \to M$ *be a finite-dimensional Lie algebra bundle with perfect fiber* $\mathfrak{k}$. *Then the continuous invariant symmetric bilinear form* $\kappa : \Gamma(\mathfrak{K}) \times \Gamma(\mathfrak{K}) \to \Gamma(V(\mathfrak{K}))$ *is universal. In particular* $V_{\mathcal{F}}(\Gamma(\mathfrak{K})) \cong \Gamma(V(\mathfrak{K}))$ *in the continuous category.*

Proof. Let $\kappa_{\mathcal{F},\mathfrak{g}} : \Gamma(\mathfrak{K}) \times \Gamma(\mathfrak{K}) \to V_{\mathcal{F}}(\Gamma(\mathfrak{K}))$ for $\mathfrak{g} = \Gamma(\mathfrak{K})$ be a universal continuous invariant symmetric bilinear form (cf. Definition 4.5.5). Note that both, κ and $\kappa_{\mathcal{F},\mathfrak{g}}$, are also A-bilinear for $A = \text{Cent}(\Gamma(\mathfrak{K}))$ (cf. Corollary 4.5.10), thus in particular we have $\kappa(\eta X, Y) = \eta \kappa(X, Y)$ and $\kappa_{\mathcal{F},\mathfrak{g}}(\eta X, Y) = \eta \kappa_{\mathcal{F},\mathfrak{g}}(X, Y)$ for all $\eta \in C^{\infty}(M, \mathbb{K})$ and $X, Y \in \Gamma(\mathfrak{K})$.

Since κ is continuous, there is a unique continuous linear map $f : V_{\mathcal{F}}(\Gamma(\mathfrak{K})) \to \Gamma(V(\mathfrak{K}))$ such that $f \circ \kappa_{\mathcal{F},\mathfrak{g}} = \kappa$. On the other hand, by Theorem 4.4.4, there is a unique linear map $g : \Gamma(V(\mathfrak{K})) \to V_{\mathcal{F}}(\Gamma(\mathfrak{K}))$ such that $g \circ \kappa = \kappa_{\mathcal{F},\mathfrak{g}}$.

By $f \circ g \circ \kappa = \kappa$, we deduce that $f \circ g$ is the only endomorphism h of $\Gamma(V(\mathfrak{K}))$ such that $h \circ \kappa = \kappa$, thus $f \circ g = \text{id}_{\Gamma(V(\mathfrak{K}))}$, showing the surjectivity of f and the injectivity of g. Since $V_{\mathcal{F}}(\Gamma(\mathfrak{K}))$ and $\Gamma(V(\mathfrak{K}))$ are Fréchet spaces, the Open Mapping Theorem B.2.6 ensures that f is open.

In order to show that the $\mathbb{K}$-linear map $g : \Gamma(V(\mathfrak{K})) \to V_{\mathcal{F}}(\Gamma(\mathfrak{K}))$ is also a morphism of $C^{\infty}(M, \mathbb{K})$-modules, we choose a section $\mathcal{G} \in \Gamma(V(\mathfrak{K}))$ and use the fact that $\text{im}(\kappa)$ algebraically spans $\Gamma(V(\mathfrak{K}))$ (cf. Remark 4.1.5.4), thus $\mathcal{G} = \sum_i \kappa(X_i, Y_i)$ for some $X_i, Y_i \in \Gamma(\mathfrak{K})$. So for $\rho \in C^{\infty}(M, \mathbb{K})$, , $X_i, Y_i \in \Gamma(\mathfrak{K})$, we calculate:

$$g(\rho\mathcal{G}) = g\left(\rho \sum_i \kappa(X_i, Y_i)\right) = \sum_i g(\kappa(\rho X_i, Y_i)) = \sum_i \kappa_{\mathcal{F},\mathfrak{g}}(\rho X_i, Y_i)$$
$$= \sum_i \rho \kappa_{\mathcal{F},\mathfrak{g}}(X_i, Y_i) = \rho \sum_i \kappa_{\mathcal{F},\mathfrak{g}}(X_i, Y_i) = \rho g(\mathcal{G}).$$

Since $V(\mathfrak{K})$ is a finite-dimensional vector bundle, the space of its sections $\Gamma(V(\mathfrak{K}))$ is a finitely generated $C^{\infty}(M, \mathbb{K})$-Fréchet module by the Theorem of Serre-Swan (cf., e.g., Theorem 3.2.14 of [Wag11]). Since $V_{\mathcal{F}}(\Gamma(\mathfrak{K}))$ is a $C^{\infty}(M, \mathbb{K})$-Fréchet module, too, the continuity of g follows (cf. Proposition 3.3.4.(a) of [Wag11]).

So $g \circ f \circ \kappa_{\mathcal{F},\mathfrak{g}} = \kappa_{\mathcal{F},\mathfrak{g}}$ implies that $e = g \circ f$ is the only continuous endomorphism e of $V_{\mathcal{F}}(\Gamma(\mathfrak{K}))$ such that $e \circ \kappa_{\mathcal{F},\mathfrak{g}} = \kappa_{\mathcal{F},\mathfrak{g}}$, thus $g \circ f = \text{id}_{\Gamma(V(\mathfrak{K}))}$, showing the injectivity of f. Thus f is an isomorphism of Fréchet spaces, whence the continuous universality of κ. $\square$

5. Universal Central Extensions

In this chapter we want to discuss central extensions of the Lie algebra of smooth sections $\Gamma(\mathfrak{K})$ of a bundle $\pi : \mathfrak{K} \to M$ with finite-dimensional semisimple[1] fiber $\mathfrak{k}$. The universal invariant symmetric bilinear form $\kappa : \Gamma(\mathfrak{K}) \times \Gamma(\mathfrak{K}) \to V(\mathfrak{K})$ that we have found in the previous chapter will help us to define a universal 2-cocycle.

5.1. Second Cohomology of Tensor Product Lie Algebras

In order to find a (weakly) universal 2-cocycle on a Lie algebra of smooth sections, we will examine the second cohomology classes of Lie algebras $C \otimes \mathfrak{k}$ for semisimple Lie algebras $\mathfrak{k}$ and commutative pseudo-unital algebras C with and without topology. The key concept is locality / diagonality (cf. Lemma 4.1 of [Tan10] and Definition II.1 of [JW10]) of 2-cocyles on $C \otimes \mathfrak{k}$. From now on, let X be a complete locally convex vector space, considered as a trivial module. For the definition and basic properties of Lie algebra extensions and of Lie algebra cohomology, we refer to Section A.2.

Definition 5.1.1. A bilinear map $\psi : (C \otimes \mathfrak{k})^2 \to X$ is called *diagonal*, if for any $\xi, \eta \in C \otimes \mathfrak{k}$ such that there exist $a, b \in C$ with $a.\xi = \xi$, $b.\eta = \eta$, $ab = 0$, we have:

$$\psi(\xi, \eta) = 0.$$

Here, C acts on $C \otimes \mathfrak{k}$ via multiplication by centroid elements.

Remark 5.1.2. We define diagonality to generalize the case of $C = C^\infty(M, \mathbb{K})$, where the condition $\xi, \eta \in C \otimes \mathfrak{k} \cong C^\infty(M, \mathfrak{k})$ with $a, b \in C^\infty(M, \mathbb{K})$ such that $a.\xi = \xi$, $b.\eta = \eta$, $ab = 0$ holds, if ξ, η have disjoint supports and a, b are smooth bump functions with constant value 1 on the support of ξ, η, respectively. For technical reasons, we need to be able to find them even constantly 1 on neighborhoods of the supports, thus we need neutral triples.

Definition 5.1.3. Let $(f_j)_{j \in F} \subseteq C \otimes \mathfrak{k}$ be a finite family, such that each f_j can be written as $f_j = \sum_{i=1}^n a_i^j \otimes x_i^j$ for some $a_i^j \in C$, $x_i^j \in \mathfrak{k}$. Then for $\mu \in C$ such that $\mu a_i^j = a_i^j$ for $i \in \{1, \dots, n\}$, $j \in F$, we have:

$$\mu.f_j = \mu. \sum_{i=1}^n a_i^j \otimes x_i^j = \sum_{i=1}^n \mu.(a_i \otimes x_i^j) = \sum_{i=1}^n \mu a_i \otimes x_i^j = f_j$$

for all $j \in F$. If, in addition, $\nu, \lambda \in C$ are chosen such that $\nu\mu = \mu$ and $\lambda\nu = \nu$, then (λ, ν, μ) is called a *neutral triple* for $(f_j)_{j \in F}$.

Lemma 5.1.4. *Every (not necessarily continuous) 2-cocycle $\psi \in \mathrm{Z}^2(C \otimes \mathfrak{k}, X)$ is diagonal.*

---

[1]The semisimplicity is needed to apply Whitehead's Second Lemma in Theorem 5.1.10.

Proof. Choose $\xi, \eta \in C \otimes \mathfrak{k}$ such that there exist $a, b \in C$ with $a.\xi = \xi$, $b.\eta = \eta$, $ab = 0$. Since $C \otimes \mathfrak{k}$ is perfect, there exist elements $\mu_j, \nu_j \in C \otimes \mathfrak{k}$ for $j \in \{1, \ldots, n\}$ such that $\eta = \sum_{j=1}^{n} [\mu_j, \nu_j]$. We then calculate:

$$\psi(\xi, \eta) = \psi\left(a.\xi, b^2 \cdot \sum_{j=1}^{n} [\mu_j, \nu_j]\right) = \sum_{j=1}^{n} \psi\left(a.\xi, [b.\mu_j, b.\nu_j]\right)$$

$$= -\sum_{j=1}^{n} \left(\psi\left(b.\mu_j, [b.\nu_j, a.\xi]\right) + \psi\left(b.\nu_j, [a.\xi, b.\mu_j]\right)\right)$$

$$= -\sum_{j=1}^{n} \left(\psi\left(b.\mu_j, ab.[\nu_j, \xi]\right) + \psi\left(b.\nu_j, ab.[\xi, \mu_j]\right)\right) = 0.$$

$\square$

From now on, let C be a commutative pseudo-unital algebra and a locally convex space with continuous multiplication and a well-behaved family $(p_i)_{i \in I}$ of seminorms (cf. Definition 2.1.7). Then there is a natural topology such that $C \otimes \mathfrak{k}$ is locally convex, too (cf. Remark 2.1.9).

Definition 5.1.5. We define $(C \otimes \mathfrak{k})^+ := (C \otimes \mathfrak{k}) \rtimes \mathfrak{k}$ as the semidirect sum of the locally convex Lie algebras $C \otimes \mathfrak{k}$ and $\mathfrak{k}$ with $[(f_1, y_1), (f_2, y_2)] := ([f_1, f_2] + [y_1, f_2] + [f_1, y_2], [y_1, y_2])$ for $f_1, f_2 \in C \otimes \mathfrak{k}$ and $y_1, y_2 \in \mathfrak{k}$. We also define the natural inclusion $i : C \otimes \mathfrak{k} \to (C \otimes \mathfrak{k})^+$, $f \mapsto (f, 0)$. Its image is closed because it is the kernel of the continuous projection $(C \otimes \mathfrak{k})^+ \to \mathfrak{k}$.

The next few statements can be easily shown by estimates like in Remark 2.1.9.

Lemma 5.1.6. *The Lie bracket from Definition 5.1.5 is continuous, and so is the inclusion i.*

Remark 5.1.7. The Lie algebra $(C \otimes \mathfrak{k})^+ = (C \otimes \mathfrak{k}) \rtimes \mathfrak{k}$ can also be realized as the tensor product $C^+ \otimes \mathfrak{k}$ for the unital associative algebra $C^+ := C \oplus \mathbb{K}$ with multiplication

$$(a_1, s_1) \cdot (a_2, s_2) := (a_1 a_2 + s_1 a_2 + s_2 a_1, s_1 s_2).$$

Indeed, for a basis $(v_1, \ldots, v_d)$ of $\mathfrak{k}$, the linear map

$$\Phi : C^+ \otimes \mathfrak{k} \longrightarrow (C \otimes \mathfrak{k}) \rtimes \mathfrak{k}$$

$$\sum_{i=1}^{d} (a_i, s_i) \otimes v_i \longmapsto \left(\sum_{i=1}^{d} a_i \otimes v_i, \sum_{j=1}^{d} s_j v_j\right)$$

is bijective and preserves the Lie bracket:

$$[\Phi((a_1, s_1) \otimes v_1), \Phi((a_2, s_2) \otimes v_2)] = [(a_1 \otimes v_1, s_1 v_1), (a_2 \otimes v_2, s_2 v_2)]$$

$$= ([a_1 \otimes v_1, a_2 \otimes v_2] + [a_1 \otimes v_1, s_2 v_2]$$

$$+ [s_1 v_1, a_2 \otimes v_2], [s_1 v_1, s_2 v_2])$$

$$= (a_1 a_2 \otimes [v_1, v_2] + s_2 a_1 \otimes [v_1, v_2]$$

$$+ s_1 a_2 \otimes [v_1, v_2], s_1 s_2 [v_1, v_2])$$

$$= ((a_1 a_2 + s_2 a_1 + s_1 a_2) \otimes [v_1, v_2], s_1 s_2 [v_1, v_2])$$

$$= (((a_1, s_1) \cdot (a_2, s_2)) \otimes [v_1, v_2], s_1 s_2 [v_1, v_2])$$

$$= \Phi\left(((a_1, s_1) \cdot (a_2, s_2)) \otimes [v_1, v_2]\right)$$

$$= \Phi\left([((a_1, s_1) \otimes v_1), ((a_2, s_2) \otimes v_2)]\right).$$

The construction in Remark 5.1.7 also gives topologically equivalent Lie algebras.

Lemma 5.1.8. *If C is a locally convex commutative pseudo-unital algebra, then the morphism $\Phi : C^+ \otimes \mathfrak{k} \to (C \otimes \mathfrak{k}) \rtimes \mathfrak{k}$ is a homeomorphism, hence an isomorphism of topological Lie algebras.*

There is a natural isomorphism of the algebraic cohomology groups of $(C \otimes \mathfrak{k})^+$ and $C \otimes \mathfrak{k}$.

Definition 5.1.9. The cofunctor $\mathrm{H}^2(\cdot, X)$ maps i from Definition 5.1.5 to a morphism

$$\mathrm{H}^2(i) : \mathrm{H}^2((C \otimes \mathfrak{k})^+, X) \longrightarrow \mathrm{H}^2(C \otimes \mathfrak{k}, X)$$
$$[\omega] \longmapsto [\omega \circ i].$$

Theorem 5.1.10. $\mathrm{H}^2(i) : \mathrm{H}^2((C \otimes \mathfrak{k})^+, X) \to \mathrm{H}^2(C \otimes \mathfrak{k}, X)$ *is bijective.*

Proof.

1. For the surjectivity of $\mathrm{H}^2(i)$, let $\omega_0 \in \mathrm{Z}^2(C \otimes \mathfrak{k}, X)$. We will define a corresponding $\omega \in \mathrm{Z}^2((C \otimes \mathfrak{k})^+, X)$ as follows:

   - Let $\omega(f_1, f_2) := \omega_0(f_1, f_2)$ and $\omega(x_1, x_2) := 0$ for $f_1, f_2 \in C \otimes \mathfrak{k}$ and $x_1, x_2 \in \mathfrak{k}$. The bilinear map $\omega : (C \otimes \mathfrak{k})^+ \times (C \otimes \mathfrak{k})^+ \to X$ is then determined by

     $$\omega(f, x) := \omega_0(f, \lambda \otimes x) \tag{5.1}$$

     for $(f, x) \in (C \otimes \mathfrak{k})^+$ and a neutral triple (λ, ν, μ) for f.

   - The right hand side of (5.1) does not depend on the choice of the neutral triple:

     Let $(\lambda_1, \nu_1, \mu_1)$, $(\lambda_2, \nu_2, \mu_2)$ be neutral triples for f. Then choose $\theta \in C$ such that $\theta \lambda_i = \lambda_i$ for $i = 1, 2$. Then we have the following identities:

     $$\mu_1 \mu_2 . f = \mu_1 . f = f,$$
     $$(\theta - \nu_1 \nu_2).((\lambda_1 - \lambda_2) \otimes x) = (\lambda_1 - \lambda_2 - \nu_1 \nu_2 \lambda_1 + \nu_1 \nu_2 \lambda_2) \otimes x$$
     $$= (\lambda_1 - \lambda_2 - \nu_1 \nu_2 + \nu_1 \nu_2) \otimes x = (\lambda_1 - \lambda_2) \otimes x,$$
     $$(\theta - \nu_1 \nu_2) \cdot \mu_1 \mu_2 = \theta \mu_1 \mu_2 - \nu_1 \nu_2 \mu_1 \mu_2 = \theta \lambda_1 \nu_1 \mu_1 \mu_2 - \mu_1 \mu_2$$
     $$= \mu_1 \mu_2 - \mu_1 \mu_2 = 0.$$

     Since ω_0 is diagonal by Lemma 5.1.4, we have: $\omega_0(f, (\lambda_1 - \lambda_2) \otimes x) = 0$.

   - ω is a 2-cocyle on $(C \otimes \mathfrak{k})^+$: For $f_1, f_2 \in C \otimes \mathfrak{k}$ and $x \in \mathfrak{k}$ choose a neutral triple (λ, ν, μ) for $\{f_1, f_2\}$ and calculate:

     $$\omega([f_1, f_2], x) + \omega([f_2, x], f_1) + \omega([x, f_2], f_1)$$
     $$= \omega_0([f_1, f_2], \lambda \otimes x) + \omega_0([f_2, \lambda \otimes x], f_1) + \omega_0([\lambda \otimes x, f_1], f_2) = 0.$$

   For $f \in C \otimes \mathfrak{k}$ and $x_1, x_2 \in \mathfrak{k}$ choose a neutral triple (λ, ν, μ) for f and $\lambda' \in C$ such that $\lambda \lambda' = \lambda$, thus (λ', ν, μ) is also a neutral triple for f, and calculate:

   $$\omega(f, [x_1, x_2]) + \omega(x_1, [x_2, f]) + \omega(x_2, [f, x_1])$$
   $$= \omega_0(f, \lambda \otimes [x_1, x_2]) + \omega(x_1, [\lambda' \otimes x_2, f]) + \omega(x_2, [f, \lambda \otimes x_1])$$
   $$= \omega_0(f, [\lambda \otimes x_1, \lambda' \otimes x_2]) + \omega_0(\lambda \otimes x_1, [\lambda' \otimes x_2, f]) + \omega_0(\lambda' \otimes x_2, [f, \lambda \otimes x_1]) = 0.$$

The second last equation follows from the fact that C acts on $C \otimes \mathfrak{k}$ via centroid elements, hence we have the identities $\mu. [\lambda' \otimes x_2, f] = [\lambda' \otimes x_2, \mu.f] = [\lambda' \otimes x_2, f]$ and $\mu. [f, \lambda \otimes x_1] = [\mu.f, \lambda \otimes x_1] = [f, \lambda \otimes x_1]$, so (λ, ν, μ) is a neutral triple for $[\lambda' \otimes x_2, f]$ and (λ, ν, μ) is a neutral triple for $[f, \lambda \otimes x_1]$.

Let $\Xi : \mathrm{Z}^2(C \otimes \mathfrak{k}, X) \to \mathrm{Z}^2((C \otimes \mathfrak{k})^+, X)$, $\omega_0 \mapsto \omega$ denote the above described map and $q^+ : \mathrm{Z}^2((C \otimes \mathfrak{k})^+, X) \to \mathrm{H}^2((C \otimes \mathfrak{k})^+, X)$ and $q : \mathrm{Z}^2(C \otimes \mathfrak{k}, X) \to \mathrm{H}^2(C \otimes \mathfrak{k}, X)$ the natural quotient maps. The next two facts show that $\mathrm{H}^2(i)$ is surjective.

- $q^+ \circ \Xi$ factors through a map $\varphi : \mathrm{H}^2(C \otimes \mathfrak{k}, X) \to \mathrm{H}^2((C \otimes \mathfrak{k})^+, X)$, i.e., $\varphi \circ q = q^+ \circ \Xi$: Let $\alpha_0 \in \mathrm{Hom}(C \otimes \mathfrak{k}, X)$, i.e., $\beta_0 := \mathrm{d}\,\alpha_0 \in \mathrm{B}^2(C \otimes \mathfrak{k}, X)$ and define a linear map $\alpha \in \mathrm{Hom}((C \otimes \mathfrak{k})^+, X)$ by $\alpha(f, x) := \alpha_0(f)$ for $f \in C \otimes \mathfrak{k}$, $x \in \mathfrak{k}$. For $f_1, f_2 \in C \otimes \mathfrak{k}$, $x_1, x_2 \in \mathfrak{k}$, neutral triples $(\lambda_i, \nu_i, \mu_i)$ for f_i and $\lambda \in C$ such that $\lambda \lambda_i = \lambda_i$, where $i = 1, 2$, we evaluate $\beta = \Xi(\beta_0)$:

$$\beta((f_1, x_1), (f_2, x_2)) = \beta_0(f_1, f_2) + \beta_0(\lambda_2 \otimes x_1, f_2) + \beta_0(f_1, \lambda_1 \otimes x_2)$$
$$= -\alpha_0 \left([f_1, f_2] + [\lambda \otimes x_1, f_2] + [f_1, \lambda \otimes x_2] \right)$$
$$= -\alpha \left([f_1, f_2] + [\lambda \otimes x_1, f_2] + [f_1, \lambda \otimes x_2], [x_1, x_2] \right)$$
$$= -\alpha \left[(f_1, x_1), (f_2, x_2) \right] = \mathrm{d}\,\alpha((f_1, x_1), (f_2, x_2)).$$

So $\varphi \circ q = q^+ \circ \Xi$.

- The map φ is a right inverse of $\mathrm{H}^2(i)$ because the assignment $\omega_0 \mapsto \Xi(\omega_0) = \omega \mapsto \omega \circ i$ is the identity on $\mathrm{Z}^2(C \otimes \mathfrak{k}, X)$.

2. For the injectivity, let $\omega \in \mathrm{Z}^2((C \otimes \mathfrak{k})^+, X)$ such that $q^+(\omega) \in \ker\left(\mathrm{H}^2(i)\right)$. Then there exists $\alpha_0 \in \mathrm{Hom}(C \otimes \mathfrak{k}, X)$ such that

$$(\omega \circ i)(f_1, f_2) = (\mathrm{d}\,\alpha_0)(f_1, f_2) = \alpha_0\,[f_1, f_2] \tag{5.2}$$

for $f_1, f_2 \in C \otimes \mathfrak{k}$. We define $\alpha \in \mathrm{Hom}((C \otimes \mathfrak{k})^+, X)$ by $\alpha(f, x) := \alpha_0(f)$ for $f \in C \otimes \mathfrak{k}$, $x \in \mathfrak{k}$ and another 2-cocycle $\omega' := \omega - \alpha \circ [\cdot, \cdot]$ on $(C \otimes \mathfrak{k})^+$. Thus on the left hand side of the cocycle identity

$$\omega'([f, g], x) + \omega'([g, x], f) + \omega'([x, f], g) = 0$$

the second and the third term vanish by (5.2) and so does the first. Since $C \otimes \mathfrak{k}$ is perfect, this means that ω' vanishs on $(C \otimes \mathfrak{k}) \times \mathfrak{k}$ and thus, by skew-symmetry of ω', also on $\mathfrak{k} \times (C \otimes \mathfrak{k})$, so it factors through a 2-cocycle ω'' on $\mathfrak{k} \times \mathfrak{k}$. But since $\mathfrak{k}$ is semisimple, Whitehead's Second Lemma (cf. Corollary A.2.9) yields that $\omega'' = \mathrm{d}\,\gamma$ for some $\gamma \in \mathrm{Hom}(\mathfrak{k}, X)$, thus ω' is a 2-coboundary on $(C \otimes \mathfrak{k})^+$ and so is ω, whence the injectivity of $\mathrm{H}^2(i)$.

Thus $\mathrm{H}^2(i)$ is bijective. $\qquad\qquad\qquad\qquad\qquad\qquad\qquad\qquad\qquad\qquad\qquad\qquad\qquad\square$

Special cases of commutative pseudo-unital algebras are $\mathrm{C}^\infty_c(M, \mathbb{K})$ and $\mathrm{C}^\infty_U(M, \mathbb{K})$ for an open subset $U \subseteq M$ of a manifold M.

Corollary 5.1.11. *There are group isomorphisms* $\mathrm{H}^2(\mathrm{C}^\infty_U(M, \mathfrak{k}) \rtimes \mathfrak{k}, X) \to \mathrm{H}^2(\mathrm{C}^\infty_U(M, \mathfrak{k}), X)$ *and* $\mathrm{H}^2(\mathrm{C}^\infty_c(M, \mathfrak{k}) \rtimes \mathfrak{k}, X) \to \mathrm{H}^2(\mathrm{C}^\infty_c(M, \mathfrak{k}), X)$, *induced by the inclusions* $\mathrm{C}^\infty_U(M, \mathfrak{k}) \to \mathrm{C}^\infty_U(M, \mathfrak{k}) \rtimes \mathfrak{k}$ *and* $\mathrm{C}^\infty_c(M, \mathfrak{k}) \to \mathrm{C}^\infty_c(M, \mathfrak{k}) \rtimes \mathfrak{k}$, *respectively.*

We want to prove a theorem similar to Theorem 5.1.10 on the continuous second cohomology groups of $(C \otimes \mathfrak{k})^+$ and $C \otimes \mathfrak{k}$ for a topological algebra C, but for that technical restrictions are necessary.

Definition 5.1.12. Let C be a commutative topological algebra which is a strict inductive limit of an increasing family of Fréchet subalgebras $\{A_n : n \in \mathbb{N}\}$ with respective units $1_n \in A_n$ and inclusion maps $h_{ji} : A_i \to A_j$ and $g_i : A_i \to C$ for $i \leq j \in \mathbb{N}$ (cf. Definition B.2.5). Then C is called a *CPUSLF-algebra*, standing for "commutative pseudo-unital strict LF-algebra". Indeed, a CPUSLF-algebra C is pseudo-unital, because for $a, b \in C$ there is some $n \in \mathbb{N}$ such that $a, b \in A_n$, yielding, for $c := 1_n$, the identities $ac = a$ and $bc = b$.

Definition 5.1.13. Let C be a CPUSLF-algebra. Then the covariant functor $\mathrm{H}^2_{\mathrm{ct}}(\cdot, X)$ maps i from Definition 5.1.5 to a morphism

$$\mathrm{H}^2_{\mathrm{ct}}(i) : \mathrm{H}^2_{\mathrm{ct}}((C \otimes \mathfrak{k})^+, X) \longrightarrow \mathrm{H}^2_{\mathrm{ct}}(C \otimes \mathfrak{k}, X)$$
$$[\omega] \longmapsto [\omega \circ i].$$

Theorem 5.1.14. *If C is a CPUSLF-algebra, then $\mathrm{H}^2_{\mathrm{ct}}(i) : \mathrm{H}^2_{\mathrm{ct}}((C \otimes \mathfrak{k})^+, X) \to \mathrm{H}^2_{\mathrm{ct}}(C \otimes \mathfrak{k}, X)$ is bijective.*

Proof. Let C be a strict inductive limit of unital Fréchet subalgebras $(A_n, 1_n)$ for $n \in \mathbb{N}$. Using the notation of the proof of Theorem 5.1.10, there is a linear map

$$\Xi : \mathrm{Z}^2(C \otimes \mathfrak{k}, X) \to \mathrm{Z}^2((C \otimes \mathfrak{k})^+, X), \quad \omega_0 \mapsto \omega,$$

where $\omega(f_1, f_2) = \omega_0(f_1, f_2)$, $\omega(x_1, x_2) = 0$ and $\omega(f_1, x_1) = \omega_0(f_1, \lambda_1 \otimes x_1)$ for $f_1, f_2 \in C \otimes \mathfrak{k}$, $x_1, x_2 \in \mathfrak{k}$ and any neutral triple $(\lambda_1, \nu_1, \mu_1)$ for f_1 (cf. Definition 5.1.3).
We will show that Ξ is inverse to $\mathrm{Z}^2_{\mathrm{ct}}(i)$ and induces an isomorphism on the cohomology level.

1. $\Xi\left(\mathrm{Z}^2_{\mathrm{ct}}(C \otimes \mathfrak{k}, X)\right) \subseteq \mathrm{Z}^2_{\mathrm{ct}}((C \otimes \mathfrak{k})^+, X)$: Let ω_0 be continuous. The continuity of $\Xi(\omega_0) = \omega$ would follow from the continuity of ω_0 and the continuity of ω on $((C \otimes \mathfrak{k}) \oplus \mathbf{0}) \times (\mathbf{0} \oplus \mathfrak{k})$. Let $(v_1, \ldots, v_d)$ be a basis of $\mathfrak{k}$. For fixed $i, j \in \{1, \ldots, d\}$ we consider the linear map

    $$L_{\omega,i,j} : C \longrightarrow X$$
    $$c \longmapsto \omega(c \otimes v_i, v_j).$$

    Since C is a strict LF-space, Proposition 13.1 of [Trè67] yields that the map $L_{\omega,i,j}$ would be continuous if and only if $L_{\omega,i,j|g_k(A_k)}$ was continuous for every $k \in \mathbb{N}$ (cf. Definition 5.1.12). But $L_{\omega,i,j}(c) = \omega_0(c \otimes v_i, g_k(1_k) \otimes v_j)$ defines a continuous map $g_k(A_k) \to X$ for every $k \in \mathbb{N}$, hence $L_{\omega,i,j}$ is continuous, thus ω is continuous on $((C \otimes \mathfrak{k}) \oplus \mathbf{0}) \times (\mathbf{0} \oplus \mathfrak{k})$.

2. Let $q^+ : \mathrm{Z}^2((C \otimes \mathfrak{k})^+, X) \to \mathrm{H}^2((C \otimes \mathfrak{k})^+, X)$ and $q : \mathrm{Z}^2(C \otimes \mathfrak{k}, X) \to \mathrm{H}^2(C \otimes \mathfrak{k}, X)$ be the natural quotient maps. The next two facts show that $\mathrm{H}^2_{\mathrm{ct}}(i)$ is surjective.

    - $q^+ \circ \Xi$ factors through a map $\varphi : \mathrm{H}^2_{\mathrm{ct}}(C \otimes \mathfrak{k}, X) \to \mathrm{H}^2_{\mathrm{ct}}((C \otimes \mathfrak{k})^+, X)$, i.e., $\varphi \circ q = q^+ \circ \Xi$: Let $\alpha_0 \in \mathrm{Hom}_{\mathrm{ct}}(C \otimes \mathfrak{k}, X)$, i.e., $\beta_0 := \mathrm{d}\, \alpha_0 \in \mathrm{B}^2_{\mathrm{ct}}(C \otimes \mathfrak{k}, X)$ and define a continuous linear map $\alpha' \in \mathrm{Hom}((C \otimes \mathfrak{k})^+, X)$ by $\alpha'(f, x) := \alpha_0(f)$ for $f \in C \otimes \mathfrak{k} \subseteq (C \otimes \mathfrak{k})^+$, $x \in \mathfrak{k} \subseteq (C \otimes \mathfrak{k})^+$. But then α' is also continuous since $(C \otimes \mathfrak{k}) \times \mathfrak{k}, (f, x) \mapsto f$ is so. Now the same calculations as in the proof of Theorem 5.1.10 show $\varphi \circ q = q^+ \circ \Xi$.

- The map φ is a right inverse of $\mathrm{H}^2_{ct}(i)$ because the assignment $\omega_0 \mapsto \Xi(\omega_0) = \omega \mapsto \omega \circ i$ is the identity on $\mathrm{Z}^2_{ct}(C \otimes \mathfrak{k}, X)$.

3. For the injectivity of $\mathrm{H}^2_{ct}(i)$, let $\omega \in \mathrm{Z}^2_{ct}((C \otimes \mathfrak{k})^+, X)$ such that $q^+(\omega) \in \ker\big(\mathrm{H}^2_{ct}(i)\big)$. Then there exists $\alpha_0 \in \mathrm{Hom}_{ct}(C \otimes \mathfrak{k}, X)$ such that

$$(\omega \circ i)(f_1, f_2) = (\mathrm{d}\,\alpha_0)(f_1, f_2) = \alpha_0\,[f_1, f_2]$$

for $f_1, f_2 \in C \otimes \mathfrak{k}$. We define $\alpha \in \mathrm{Hom}((C \otimes \mathfrak{k})^+, X)$ by $\alpha(f, x) := \alpha_0(f)$ for $f \in C \otimes \mathfrak{k}$, $x \in \mathfrak{k}$, which is again continuous, as is the 2-cocycle $\theta := \omega - \alpha \circ [\cdot, \cdot]$ on $(C \otimes \mathfrak{k})^+$. Now the same arguments that show the injectivity of $\mathrm{H}^2(i)$ in the proof of Theorem 5.1.10 also show the injectivity of $\mathrm{H}^2_{ct}(i)$.

Thus $\mathrm{H}^2_{ct}(i)$ is bijective. $\qquad\square$

Corollary 5.1.15. *There is a group isomorphism* $\mathrm{H}^2_{ct}(C^\infty_c(M, \mathfrak{k}) \rtimes \mathfrak{k}, X) \to \mathrm{H}^2_{ct}(C^\infty_c(M, \mathfrak{k}), X)$, *induced by* $C^\infty_c(M, \mathfrak{k}) \to C^\infty_c(M, \mathfrak{k}) \rtimes \mathfrak{k}$.

There is also a way to describe the second continuous Lie algebra cohomology space of $C^\infty_U(M, \mathbb{K})$, equipped with the subspace topology induced by $C^\infty(M, \mathbb{K})$, since it is dense. We will make use of the following statement, Theorem III.6.1 of [Bou89].

Theorem 5.1.16. *Let E, F, G be three complete Hausdorff commutative groups; let A be a dense subgroup of E and let B a dense subgroup of F. If f is a continuous biadditive mapping $A \times B \to G$, then f can be extended by continuity to a continuous biadditive mapping $E \times F \to G$.*

Corollary 5.1.17. *Let $\mathfrak{h}$ be a locally convex Lie algebra, X a complete locally convex vector space, $\mathfrak{f} \leq \mathfrak{h}$ a dense subalgebra, $p \in \mathbb{N}$ and $\mathrm{C}^p_{ct}(\mathfrak{h}, X)$, $\mathrm{C}^p_{ct}(\mathfrak{f}, X)$ the spaces of p-cochains (cf. Definition A.2.5). Then $\mathrm{C}^p_{ct}(\mathfrak{h}, X) \cong \mathrm{C}^p_{ct}(\mathfrak{f}, X)$ as vector spaces via restriction to $\mathfrak{f}^p$. Furthermore, these isomorphisms also induce isomorphisms $\mathrm{Z}^p_{ct}(\mathfrak{h}, X) \cong \mathrm{Z}^p_{ct}(\mathfrak{f}, X)$ and $\mathrm{B}^p_{ct}(\mathfrak{h}, X) \cong \mathrm{B}^p_{ct}(\mathfrak{f}, X)$ and $\mathrm{H}^p_{ct}(\mathfrak{h}, X) \cong \mathrm{H}^p_{ct}(\mathfrak{f}, X)$.*

Proof. We apply Theorem 5.1.16 p-times and obtain a continuation $\mathrm{Lin}^p_{ct}(\mathfrak{f}, X) \cong \mathrm{Lin}^p_{ct}(\mathfrak{h}, X)$, which maps alternating maps to alternating maps by the principle of extensions of identities, so $\mathrm{C}^p_{ct}(\mathfrak{f}, X) \cong \mathrm{C}^p_{ct}(\mathfrak{h}, X)$. The restriction to $\mathfrak{f}^p$ is an inverse to this continuation map and commutes with the differential d^p, proving the remaining isomorphies. $\qquad\square$

Proposition 5.1.18. *Let $\mathfrak{k}$ be a finite-dimensional Lie algebra, A a complete locally convex algebra, that is commutative and unital, and $C \leq A$ a dense pseudo-unital subalgebra. Then the inclusion $C \to A$ induces an isomorphism $\mathrm{H}^2_{ct}(A \otimes \mathfrak{k}, X) \to \mathrm{H}^2_{ct}(C \otimes \mathfrak{k}, X)$. In particular, the inclusion $C^\infty_U(M, \mathfrak{k}) \to C^\infty(U, \mathfrak{k})$ induces an isomorphism $\mathrm{H}^2_{ct}(C^\infty(U, \mathfrak{k}), X) \to \mathrm{H}^2_{ct}(C^\infty_U(M, \mathfrak{k}), X)$.*

Proof. This immediately follows from Corollary 5.1.17 and the fact that $C \otimes \mathfrak{k} \subseteq A \otimes \mathfrak{k}$ is dense by the density of $C \subseteq A$. $\qquad\square$

Universal central extensions of Lie algebras $A \otimes \mathfrak{k}$ for semisimple $\mathbb{K}$-Lie algebras $\mathfrak{k}$ and complete locally convex algebras A, that are unital and commutative, can be described with the help of universal differential modules.

5.2. Universal Differential Modules and 1-Forms

We will discuss universal differential modules of complete locally convex algebras A, that are unital and commutative, following [Mai02], where one can find the missing proofs in this section. Throughout this section, $\mathcal{K}$ will be a full subcategory of the category of locally convex $\mathbb{K}$-vector spaces.

Definition 5.2.1. A $\mathcal{K}$-*algebra* is $\mathcal{K}$-object A with a $\mathcal{K}$-morphism $\mu : A \otimes_{\mathcal{K}} A \to A$, $a \otimes b \mapsto ab$. It is called *unital* if there is an element $1 \in A$ such that $1a = a1 = a$ for all $a \in A$, *commutative* if $ab = ba$ for all $a, b \in A$ and *associative* if $a(bc) = (ab)c$ for all $a, b, c \in A$.

Throughout this section, $(A, \mu, 1)$, shortly A, will be a commutative and associative unital $\mathcal{K}$-algebra.

Definition 5.2.2.

1. An A-module (in the category $\mathcal{K}$) is an object M of $\mathcal{K}$ with a morphism $\nu : A \otimes_{\mathcal{K}} M \to M$ such that $\nu \circ (\mathrm{id}_A \otimes \nu) = \nu \circ (\mu \otimes \mathrm{id}_M)$ and $\nu(1 \otimes m) = m$ for all $m \in M$. We briefly write $am := \nu(a \otimes m)$ for $a \in A$, $m \in M$.

2. A *derivation* to an A-module M is a morphism $D : A \to M$ such that, for $a, b \in A$, we have $D(ab) = aD(b) + bD(a)$.

Remark 5.2.3. The $\mathcal{K}$-algebra A can itself be regarded as an A-module by $\nu := \mu$. Furthermore, $A \otimes_{\mathcal{K}} A$ is an A-module with $\widetilde{\nu} : A \otimes_{\mathcal{K}} (A \otimes_{\mathcal{K}} A) \to A \otimes_{\mathcal{K}} A$ defined by $\widetilde{\nu}(a \otimes (b \otimes c)) := ab \otimes c$. The multiplication $\mu : A \otimes_{\mathcal{K}} A \to A$ is a morphism of A-modules with respect to these discussed module structures, thus its kernel $I_A := \mu^{-1}(\mathbf{0})$ is an A-submodule of $A \otimes_{\mathcal{K}} A$.

Definition 5.2.4. Let $\Omega^1_{\mathcal{K}}(A)$ be the quotient $I_A / \overline{I_A^2}$, except for the case where $\mathcal{K}$ is a subcategory of the category of complete locally convex vector spaces, where $\Omega^1_{\mathcal{K}}(A)$ is defined as the closure of the quotient $I_A / \overline{I_A^2}$. Note that, if $\mathcal{K}$ is a subcategory of the category of Fréchet spaces, there is a natural isomorphism $\Omega^1_{\mathcal{K}}(A) \cong I_A / \overline{I_A^2}$. Moreover, define the derivation $d_A : A \to \Omega^1_{\mathcal{K}}(A)$ by $d_A(a) := [1 \otimes a - a \otimes 1]$. The pair $\left(\Omega^1_{\mathcal{K}}(A), d_A \right)$ is called $\mathcal{K}$-*universal differential module* of A.

The above definition is justified by the following theorem.

Theorem 5.2.5. *For any A-module E and for any derivation $D : A \to E$ there exists a unique continuous A-linear map $\overline{D} : \Omega^1_{\mathcal{K}}(A) \to E$ with $\overline{D} \circ d_A = D$.*

The notation $\Omega^1_{\mathcal{K}}$ reminds of 1-forms and, indeed, the Theorem 5.2.7, which is Theorem 9 of [Mai02], says that this is no coincidence. We recall some of the basic definitions on forms on M, a usual smooth finite-dimensional manifold.

Definition 5.2.6. Let $p \in \mathbb{N}_0$ and $\mathbb{V} \to M$ a finite-dimensional vector bundle with fiber V.

1. Let $\Omega^p(M, V)$ denote the space of smooth V-valued p-*forms* on M, i.e.

$$\Omega^p(M, V) := \left\{ \alpha \in \mathrm{C}^{\infty}((TM)^p, V) \,|\, \alpha_{|(T_m M)^p} \in \mathrm{Alt}^p(T_m M, V) \text{ for all } m \in M \right\},$$

which can be also described as the sections of a vector bundle $\Lambda^p(T^*M) \to M$, so it is a Fréchet space by Remark B.1.12. One can show that is space is naturally isomorphic as vector space to $\mathrm{Alt}^p_{\mathrm{C}^{\infty}(M, \mathbb{K})}(\Gamma(TM), \mathrm{C}^{\infty}(M, V))$, the space of alternating p-$\mathrm{C}^{\infty}(M, \mathbb{K})$-linear maps.

2. Let $\Omega^p(M, \mathbb{V})$ denote the space of smooth $\mathbb{V}$-valued *p-forms* on M, i.e.

$$\Omega^p(M, \mathbb{V}) := \left\{ \alpha \in C^\infty((TM)^p, \mathbb{V}) \mid \alpha_{|(T_m M)^p} \in \mathrm{Alt}^p(T_m M, \mathbb{V}_m) \text{ for all } m \in M \right\},$$

which can also be described as the sections of a vector bundle $\Lambda^p(T^*M \otimes \mathbb{V}) \to M$, so it is a Fréchet space. This space can be identified with $\mathrm{Alt}^p_{C^\infty(M, \mathbb{K})}(\Gamma(TM), \Gamma(\mathbb{V}))$.

3. Let $\Omega^p_c(M, V)$ and $\Omega^p_c(M, \mathbb{V})$ denote the spaces of p-forms with compact support. Both spaces are strict LF-spaces, hence complete (cf. Example B.2.7).

4. Let $d : C^\infty(M, V) \to \Omega^1(M, V)$ be the usual *exterior differential* of smooth maps. The image of d is closed in $\Omega^1(M, V)$, being, by Stoke's Theorem, the intersection

$$\bigcap_{\alpha \in C^\infty(\mathbb{S}^1, M)} \left\{ h \in \Omega^1(M, V) \middle| \int_\alpha h = 0 \right\}.$$

So we can define the Hausdorff locally convex space $\overline{\Omega}^1(M, V) := \Omega^1(M, V)/\operatorname{im} d$. Similarly, the quotient $\overline{\Omega}^1_c(M, V) := \Omega^1_c(M, V)/d\, C^\infty_c(M, V)$ carries a natural Hausdorff locally convex stucture (cf. Lemma IV.11 of [Nee04a]).

We can now state the following.

Theorem 5.2.7. *If $\mathcal{K} = \mathcal{F}$ is the category of Fréchet spaces, then $\overline{d} : \Omega^1_{\mathcal{F}}(C^\infty(M, \mathbb{K})) \to \Omega^1(M)$, the map induced by the exterior differential $d : C^\infty(M, \mathbb{K}) \to \Omega^1(M, \mathbb{K})$, is an isomorphism of $C^\infty(M, \mathbb{K})$-modules.*

Maier also showed the following, describing how the compactly supported 1-forms on M can be identified with the universal differential module of $C^\infty_c(M, \mathbb{K})^+$ (cf. Remark 5.1.7) for the CPUSLF-algebra of compactly supported maps on M.

Theorem 5.2.8. *If $\mathcal{K} = \mathcal{C}$ is the category of complete locally convex spaces, then the map $\overline{d} : \Omega^1_{\mathcal{C}}(C^\infty_c(M, \mathbb{K})^+) \to \Omega^1_c(M, \mathbb{K})$ induced by the exterior differential $d : C^\infty_c(M, \mathbb{K}) \to \Omega^1_c(M, \mathbb{K})$ is an isomorphism of $C^\infty_c(M, \mathbb{K})$-modules.*

From now on, let $\mathfrak{k}$ be a finite-dimensional semisimple $\mathbb{K}$-Lie algebra with universal invariant symmetric bilinear form $\kappa_{\mathfrak{k}} : \mathfrak{k} \times \mathfrak{k} \to \mathrm{V}(\mathfrak{k})$ and A a commutative and associative unital $\mathcal{K}$-algebra. In particular, $\mathfrak{k}$ is perfect. With the notion of the universal differential module, it is possible to define a 2-cocyle on $A \otimes \mathfrak{k}$.

Definition 5.2.9. For $\mathfrak{k}_A := A \otimes \mathfrak{k}$ we put $\mathfrak{z}_A := \left(\Omega^1_{\mathcal{K}}(A)/\overline{d_A A} \right) \otimes \mathrm{V}(\mathfrak{k})$ and define

$$\omega_A : \mathfrak{k}_A \times \mathfrak{k}_A \longrightarrow \mathfrak{z}_A$$
$$\omega_A(f \otimes x, g \otimes y) := [f d_A(g)] \otimes \kappa_{\mathfrak{k}}(x, y).$$

Since $\kappa_{\mathfrak{k}}$ is invariant, we easily see that $\omega_A \in Z^2_{\mathrm{ct}}(\mathfrak{k}_A, \mathfrak{z}_A)$.

The next statement is Theorem 16 of [Mai02] and is a topological version of Kassel's Theorem (cf. [Kas84]).

Theorem 5.2.10. *The 2-cocycle $\omega_A : \mathfrak{k}_A \times \mathfrak{k}_A \to \mathfrak{z}_A$ is weakly universal, i.e., for every locally convex space W, the linear map $\delta_W : \mathrm{Hom}_{\mathrm{ct}}(\mathfrak{z}_A, W) \to \mathrm{H}^2_{\mathrm{ct}}(\mathfrak{k}_A, W)$, $\theta \mapsto [\theta \circ \omega_A]$ is bijective.*

Corollary 5.2.11. *If $\mathcal{K} = \mathcal{F}$ is the category of Fréchet spaces, then 2-cocycle $\omega_A : \mathfrak{k}_A \times \mathfrak{k}_A \to \mathfrak{z}_A$ is universal.*

Proof. The perfect Lie algebra $\mathfrak{k}_A$ is Fréchet and so are $\Omega^1_{\mathcal{F}}(A)$ and $\mathfrak{z}_A$. Corollary A.2.14 and Theorem 5.2.10 then yield the universality of ω_A. $\qquad\square$

This result can be combined with Theorem 5.2.7 to the following, which generalizes the classical analogous result by Pressley and Segal (cf. Proposition 4.2.1.8 of [PS86]) for $M = \mathbb{S}^1$ and $\mathfrak{k}$ complex simple.

Corollary 5.2.12. *The cocycle $\omega_M := \omega_{\mathrm{C}^\infty(M,\mathbb{K})} \in \mathrm{Z}^2_{\mathrm{ct}}\left(\mathrm{C}^\infty(M,\mathfrak{k}), \overline{\Omega}^1(M,\mathbb{K}) \otimes \mathrm{V}(\mathfrak{k})\right)$, given by*

$$\omega_M(f \otimes x, g \otimes y) := [\alpha_M(f,g)] \otimes \kappa_{\mathfrak{k}}(x,y),$$
$$\alpha_M(f,g)(m)(v) := dg(m)(v) \cdot f(m),$$

is universal.

Analogous versions of Theorem 5.2.10 and Corollary 5.2.12 for compactly supported maps are shown by the construction in Theorem 5.1.14 and applying Theorem 5.2.8. This was done the first time in the proof of Theorem II.7 of [JW10].

Theorem 5.2.13. *If C is a CPUSLF-algebra, then the continuous 2-cocyle $\omega_C^+ := \omega_{C^+} \circ (j \times j)$ for the natural injection $j : C \otimes \mathfrak{k} \to C^+ \otimes \mathfrak{k}$ is weakly universal.*

Proof. If C is a CPUSLF-algebra, then $A := C^+$ is a commutative and associative unital $\mathcal{F}$-algebra for the category $\mathcal{F}$ of Fréchet spaces. By Theorem 5.1.14 and Lemma 5.1.8 there is a group isomorphism $\mathrm{H}^2_{\mathrm{ct}}(A \otimes \mathfrak{k}, X) \to \mathrm{H}^2_{\mathrm{ct}}(C \otimes \mathfrak{k}, X)$, $[\omega'] \mapsto [\omega' \circ (\Phi^{-1} \times \Phi^{-1}) \circ i]$. By Theorem 5.2.10, for every locally convex space W, the linear map $\delta_W : \mathrm{Hom}_{\mathrm{ct}}(\mathfrak{z}_A, W) \to \mathrm{H}^2_{\mathrm{ct}}(A \otimes \mathfrak{k}, W)$, $\theta \mapsto [\theta \circ \omega_A]$ is bijective and so is

$$\mathrm{Hom}_{\mathrm{ct}}(\mathfrak{z}_A, W) \to \mathrm{H}^2_{\mathrm{ct}}(C \otimes \mathfrak{k}, W), \quad \theta \mapsto \left[\theta \circ \omega_A \circ \left(\Phi^{-1} \times \Phi^{-1}\right) \circ i\right]. \qquad\square$$

Corollary 5.2.14. *The cocycle $\omega_{M,c} := \omega^+_{\mathrm{C}^\infty_c(M,\mathbb{K})} \in \mathrm{Z}^2_{\mathrm{ct}}\left(\mathrm{C}^\infty_c(M,\mathfrak{k}), \overline{\Omega}^1_c(M,\mathbb{K}) \otimes \mathrm{V}(\mathfrak{k})\right)$, given by*

$$\omega_{M,c}(f \otimes x, g \otimes y) := [\alpha_M(f,g)] \otimes \kappa_{\mathfrak{k}}(x,y),$$
$$\alpha_{M,c}(f,g)(m)(v) := dg(m)(v) \cdot f(m),$$

is weakly universal.

5.3. Universal Central Extensions of Lie Algebras of Smooth Sections

We turn back to the global bundle picture: Let $\mathfrak{k}$ be a semisimple Lie algebra, the fiber of a Lie algebra bundle $\pi : \mathfrak{K} \to M$ with section algebra $\Gamma(\mathfrak{K})$ and the bundle $\nu : \mathrm{V}(\mathfrak{K}) \to M$ from Definition 4.1.10. In order to find a weakly universal central extension on $\Gamma(\mathfrak{K})$, we need a more explicit notion of diagonality (cf. Remark 5.1.2). We will follow the article [JW10].

Definition 5.3.1. A bilinear map $\psi : \Gamma(\mathfrak{K}) \times \Gamma(\mathfrak{K}) \to X$ is called *diagonal*, if for any $\xi, \eta \in \Gamma(\mathfrak{K})$ such that $\operatorname{supp}(\xi) \cap \operatorname{supp}(\eta) = \emptyset$, we have:

$$\psi(\xi, \eta) = 0.$$

Remark 5.3.2. For the Lie algebra $C^\infty(M, \mathfrak{k}) \cong C^\infty(M, \mathbb{K}) \otimes \mathfrak{k}$, the two notions of diagonality from Definitions 5.1.1 and 5.3.1 coincide:

Let $\psi : C^\infty(M, \mathfrak{k}) \times C^\infty(M, \mathfrak{k}) \to X$ be diagonal in the first sense. If $\xi, \eta \in C^\infty(M, \mathfrak{k})$ have disjoint supports, then there are two smooth maps $a, b \in C^\infty(M, [0, 1]) \subseteq C^\infty(M, \mathbb{K})$ with disjoint supports, in particular $ab = 0$, such that $a\xi = \xi$, $b\eta = \eta$. Thus $\psi(\xi, \eta) = 0$.

Let $\psi : C^\infty(M, \mathfrak{k}) \times C^\infty(M, \mathfrak{k}) \to X$ be diagonal in the second sense. If $a, b \in C^\infty(M, \mathbb{K})$ and $\xi, \eta \in C^\infty(M, \mathfrak{k})$ such that $a\xi = \xi$, $b\eta = \eta$ and $ab = 0$, then a is constantly 1 on $\xi^{-1}(\mathfrak{k} \setminus \{0\})$, thus on $\operatorname{supp}(\xi)$. By the same argument, b is constantly 1 on $\operatorname{supp}(\eta)$. Since $a(m) \cdot b(m) = 0$ for all $m \in M$, there cannot be a point $m \in \operatorname{supp}(\xi) \cap \operatorname{supp}(\eta)$. Thus the supports are disjoint and $\psi(\xi, \eta) = 0$.

It is no surprise that there is a version of Lemma 5.1.4 in the new sense of diagonality.

Lemma 5.3.3. *Every (not necessarily continuous) 2-cocycle $\psi \in Z^2(\Gamma(\mathfrak{K}), X)$ is diagonal.*

Proof. Choose $\xi, \eta \in \Gamma(\mathfrak{K})$ with disjoint support and let U be a neighborhood of $\operatorname{supp}(\eta)$ such that $U \cap \operatorname{supp}(\xi) = \emptyset$. Then there exist elements $\mu_j, \nu_j \in \Gamma(\mathfrak{K})$ with $\operatorname{supp}(\mu_j) \cap \operatorname{supp}(\nu_j) \subseteq U$ for $j \in \{1, \ldots, n\}$ such that $\eta = \sum_{j=1}^n [\mu_j, \nu_j]$. We then calculate:

$$\psi(\xi, \eta) = \psi\left(\xi, \sum_{j=1}^n [\mu_j, \nu_j]\right) = \sum_{j=1}^n \psi\left(\xi, [\mu_j, \nu_j]\right) = -\sum_{j=1}^n \psi\left(\nu_j, [\xi, \mu_j]\right) - \sum_{j=1}^n \psi\left(\mu_j, [\nu_j, \xi]\right) = 0.$$

$\square$

Definition 5.3.4. Let $\nabla : \Gamma(TM) \to \operatorname{Der}(\Gamma(\mathfrak{K}))$, $X \mapsto \nabla_X$ be a Lie connection of $\pi : \mathfrak{K} \to M$ as in Definition 2.3.13.2, inducing a connection $d_\nabla : \Gamma(TM) \to \operatorname{End}(\Gamma(V(\mathfrak{K})))$, $X \mapsto d_\nabla X$ by $d_\nabla X [\xi \vee \eta] := [\nabla_X \xi \vee \eta] + [\xi \vee \nabla_X \eta]$ of $\nu : V(\mathfrak{K}) \to M$.

Remark 5.3.5. Note that d_∇ is independent from the choice of the Lie connection: Let d_∇ and $d_{\nabla'}$ be defined by the Lie connections ∇ and ∇', respectively. Then, by Proposition 2.3.33, there exists $\zeta \in \Gamma(\mathfrak{K})$ such that $(\nabla - \nabla')_X \xi = [\zeta, \xi]$ for all $X \in \Gamma(TM)$, $\xi \in \Gamma(\mathfrak{K})$. So

$$(d_\nabla - d_{\nabla'}) X [\xi \vee \eta] = [\nabla_X \xi \vee \eta] + [\xi \vee \nabla_X \eta] - [\nabla'_X \xi \vee \eta] - [\xi \vee \nabla'_X \eta]$$
$$= [[\zeta, \xi] \vee \eta + \xi \vee [\zeta, \eta]] = 0 \in \Gamma(V(\mathfrak{K})).$$

So we may write $d = d_\nabla$.

We have the following formula for the universal invariant symmetric bilinear form κ from Definition 4.1.13, helping us to define a 2-cocyle ω on $\Gamma(\mathfrak{K})$, similar to the 2-cocycles ω_M and $\omega_{M,c}$, weakly universal for $C^\infty(M, \mathfrak{k})$ and $C_c^\infty(M, \mathfrak{k})$, respectively, (cf. Corollaries 5.2.14 and 5.2.12).

Lemma 5.3.6. *We have the identity $d\kappa(\xi, \eta) = \kappa(\nabla\xi, \eta) + \kappa(\xi, \nabla\eta)$ for all $\xi, \eta \in \Gamma(\mathfrak{K})$. In other words,*

$$(d\, X)\kappa(\xi, \eta) = \kappa(\nabla_X \xi, \eta) + \kappa(\xi, \nabla_X \eta)$$

for all $X \in \Gamma(TM)$, $\xi, \eta \in \Gamma(\mathfrak{K})$.

Definition 5.3.7.

1. Since d is a map $\Gamma(V(\mathfrak{K})) \cong \Omega^0(M, V(\mathfrak{K})) \to \Omega^1(M, V(\mathfrak{K}))$, $[\xi \vee \eta] \mapsto [\nabla \xi \vee \eta] + [\xi \vee \nabla \eta]$, we can write $\overline{\Omega}^1(M, V(\mathfrak{K})) := \Omega^1(M, V(\mathfrak{K}))/\,\mathrm{d}\,\Omega^0(M, V(\mathfrak{K}))$, which is Hausdorff since $\mathrm{d}\,\Omega^0(M, V(\mathfrak{K}))$ is closed, as seen in Lemma 4.8 of [NW09], and hence even Fréchet.

2. We define the map $\omega_\nabla : \Gamma(\mathfrak{K}) \times \Gamma(\mathfrak{K}) \to \overline{\Omega}^1(M, V(\mathfrak{K}))$, $(\xi, \eta) \mapsto [\kappa(\xi, \nabla \eta)]$, which is continuous, bilinear and is alternating by Lemma 5.3.6. Furthermore, ω_∇ fulfills the cocycle identity by the invariance of κ.

3. We write $\overline{\Omega}_c^1(M, V(\mathfrak{K})) := \Omega_c^1(M, V(\mathfrak{K}))/\,\mathrm{d}\,\Omega_c^0(M, V(\mathfrak{K}))$, obtaining a Hausdorff locally convex space, and define the continuous 2-cocycle $\omega_{\nabla,c} : \Gamma_c(\mathfrak{K}) \times \Gamma_c(\mathfrak{K}) \to \overline{\Omega}_c^1(M, V(\mathfrak{K}))$, $(\xi, \eta) \mapsto [\kappa(\xi, \nabla \eta)]$.

The following is a recent result on weakly universal central extensions, Proposition I.1 of [JW10].

Theorem 5.3.8. *If $\pi : \mathfrak{K} \to M$ is a finite-dimensional Lie algebra bundle with semisimple fiber $\mathfrak{k}$, then $\omega_{\nabla,c} \in Z_{ct}^2\left(\Gamma_c(\mathfrak{K}), \overline{\Omega}_c^1(M, V(\mathfrak{K}))\right)$ is weakly universal, i.e., for every locally convex space W, the following linear map is bijective:*

$$\mathrm{Hom}_{ct}\left(\overline{\Omega}_c^1(M, V(\mathfrak{K})), W\right) \longrightarrow H_{ct}^2(\Gamma_c(\mathfrak{K}), W)$$

$$\theta \longmapsto [\theta \circ \omega_{\nabla,c}].$$

Corollary 5.3.9. *If $\pi : \mathfrak{K} \to M$ is a finite-dimensional Lie algebra bundle with semisimple fiber $\mathfrak{k}$ and compact basis manifold M, then $\omega_\nabla \in Z_{ct}^2\left(\Gamma(\mathfrak{K}), \overline{\Omega}^1(M, V(\mathfrak{K}))\right)$ is weakly universal, i.e., for every locally convex space W, the following linear map is bijective:*

$$\mathrm{Hom}_{ct}\left(\overline{\Omega}^1(M, V(\mathfrak{K})), W\right) \longrightarrow H_{ct}^2(\Gamma(\mathfrak{K}), W)$$

$$\theta \longmapsto [\theta \circ \omega_\nabla].$$

6. Section Algebras as Kac-Moody Algebras

In this chapter we want to apply the facts on section algebras to explore how the well-known results on Kac-Moody algebras can be extended to generalizations of Kac-Moody algebras.

6.1. The Loop Realization of Affine Kac-Moody Algebras

In the large field of infinite-dimensional complex Lie algebras, the family of affine Kac-Moody algebras plays an important role by providing Lie algebras with many of the important properties of finite-dimensional complex semisimple Lie algebras: They possess universal invariant symmetric bilinear forms and can be classified by root systems and generalized Dynkin diagrams. For details on the definition of complex Kac-Moody algebras $\mathfrak{g}(A)$ with generalized Cartan matrices A and their classification, we refer to Section A.1. In this section we recapitulate some facts from [Kac90] to show parallels to the section algebra $\Gamma(\mathfrak{K})$, for instance the following proposition is Exercise 2.5, and gives us the uniqueness of non-degenerate invariant symmetric bilinear forms on a Kac-Moody algebra.

Proposition 6.1.1. *If A is an indecomposable generalized Cartan matrix with non-degenerate invariant symmetric bilinear forms β and γ on $\mathfrak{g}(A) \times \mathfrak{g}(A)$, then there is a Lie algebra automorphism $\phi : \mathfrak{g}(A) \to \mathfrak{g}(A)$ pointwise fixing the commutator algebra $\mathfrak{g}'(A)$ such that $\beta \circ (\phi \times \phi)$ and γ are proportional. Furthermore, two invariant symmetric bilinear forms on $\mathfrak{g}'(A) \times \mathfrak{g}'(A)$ are always proportional.*

Not only can affine Kac-Moody algebras be abstractly defined similarly to complex simple Lie algebras, but there is also a "concrete" construction as a double extension of an (algebraical) loop algebra. This is the link between affine Kac-Moody algebras and section algebras, since Lie algebras of smooth sections are generalizations of smooth loop algebras (cf. Example 2.3.2).

Definition 6.1.2. Let $\mathfrak{g}$ be a complex simple Lie algebra and $\mathcal{L} := \mathbb{C}\left[t, t^{-1}\right]$ the commutative algebra of Laurent polynomials. Then define the *algebraic loop Lie algebra* or just *loop algebra* of $\mathfrak{g}$ to be $\mathcal{L}(\mathfrak{g}) := \mathcal{L} \otimes \mathfrak{g}$, i.e., for $x, y \in \mathfrak{g}$ and $m, n \in \mathbb{Z}$ the Lie bracket on $\mathcal{L}(\mathfrak{g})$ satisfies the relation $[t^m \otimes x, t^n \otimes y] = t^{m+n} \otimes [x, y]$. There is also a bilinear form on $\mathcal{L}(\mathfrak{g})$ defined by $\langle t^m \otimes x, t^n \otimes y \rangle_0 := \delta_{m+n,0} \cdot \kappa(x, y)$ and a linear map of $\mathcal{L}(\mathfrak{g})$ defined by $D\left(t^m \otimes x\right) := m\left(t^m \otimes x\right)$, where κ is the Cartan-Killing form on $\mathfrak{g}$ and $x, y \in \mathfrak{g}$ and $m, n \in \mathbb{Z}$.

Easy calculations show the following.

Lemma 6.1.3. *The map $\langle \cdot, \cdot \rangle_0 : \mathcal{L}(\mathfrak{g}) \times \mathcal{L}(\mathfrak{g}) \to \mathbb{C}$ is an invariant symmetric bilinear form and $D : \mathcal{L}(\mathfrak{g}) \to \mathcal{L}(\mathfrak{g})$ is a derivation.*

Definition 6.1.4. Let $\mathfrak{g}$ be a complex simple Lie algebra. We define $\widetilde{\mathcal{L}}(\mathfrak{g})$ to be the one-dimensional extension $\mathcal{L}(\mathfrak{g}) \oplus \mathbb{C}c$ with Lie bracket $[a + \lambda c, b + \mu c] := [a, b] + \langle D(a), b \rangle_0 c$ for $a, b \in \mathcal{L}(\mathfrak{g})$ and $\lambda, \mu \in \mathbb{C}$.

After extending D to $\widetilde{\mathcal{L}}(\mathfrak{g})$ by $D(c) := 0$, we equip $\widehat{\mathcal{L}}(\mathfrak{g}) := \widetilde{\mathcal{L}}(\mathfrak{g}) \oplus \mathbb{C}d = \mathcal{L}(\mathfrak{g}) \oplus \mathbb{C}c \oplus \mathbb{C}d$ with the following Lie bracket: $[a + \lambda d, b + \mu d] := [a, b] + \lambda D(b) - \mu D(a)$ for $a, b \in \widetilde{\mathcal{L}}(\mathfrak{g})$ and $\lambda, \mu \in \mathbb{C}$.

Again, the following can easily be calculated.

Lemma 6.1.5. *We have* $\mathfrak{z}\left(\widehat{\mathcal{L}}(\mathfrak{g})\right) = \mathbb{C}c$ *and* $\left[\widehat{\mathcal{L}}(\mathfrak{g}), \widehat{\mathcal{L}}(\mathfrak{g})\right] = \widetilde{\mathcal{L}}(\mathfrak{g})$.

The following was shown in Theorem 2.36 of [Gar80].

Theorem 6.1.6. *The central extension* $\mathbf{0} \to \mathbb{C}c \to \widetilde{\mathcal{L}}(\mathfrak{g}) \to \mathcal{L}(\mathfrak{g}) \to \mathbf{0}$ *is universal. A corresponding universal 2-cocycle is given by*

$$\omega : \mathcal{L}(\mathfrak{g}) \times \mathcal{L}(\mathfrak{g}) \longrightarrow \mathbb{C}c$$
$$\omega(P \otimes x, Q \otimes y) = \langle D(P \otimes x), Q \otimes y \rangle_0\, c.$$

In order to state Theorem 6.1.8 on concrete realizations of untwisted affine Kac-Moody algebras (cf. Theorem 7.4 and Section 7.5 of [Kac90]), we first define the notion of an extended Cartan matrix.

Definition 6.1.7. Let $\mathfrak{g}$ be a complex simple Lie algebra of type X_ℓ with Cartan matrix A. Then the generalized Cartan matrix of type $X_\ell^{(1)}$ (cf. Theorem A.1.12) is its *extended untwisted Cartan matrix* with respect to A, denoted by $\widehat{A}$.

Theorem 6.1.8. *Let $\mathfrak{g}$ be a complex simple Lie algebra with Cartan matrix A and Cartan-Killing form $\kappa : \mathfrak{g} \times \mathfrak{g} \to \mathbb{C}$. Then there is a Lie algebra isomorphism $\Psi : \widehat{\mathcal{L}}(\mathfrak{g}) \to \mathfrak{g}(\widehat{A})$, hence the commutator algebras $\widetilde{\mathcal{L}}(\mathfrak{g})$ and $\mathfrak{g}'(A)$ are isomorphic, too. Furthermore, there is a non-degenerate invariant symmetric bilinear form on $\widehat{\mathcal{L}}(\mathfrak{g})$ defined by*

$$\langle P \otimes x, Q \otimes y \rangle = \mathrm{Res}\left(t^{-1}PQ\right) \cdot \kappa(x, y),$$
$$\langle \mathbb{C}c + \mathbb{C}d, \mathcal{L}(\mathfrak{g}) \rangle = \mathbf{0}, \quad \langle c, c \rangle = \langle d, d \rangle = 0, \quad \langle c, d \rangle = 1$$

for the residue map $\mathrm{Res} : \mathcal{L} \to \mathbb{C}$, $\sum_{\ell \in \mathbb{Z}} a_\ell t^\ell \mapsto a_{-1}$ *and* $x, y \in \mathfrak{g}$ *and* $P, Q \in \mathcal{L}$.

As as consequence of this realization of affine Kac-Moody algebras, we have the following on the loop algebra $\mathcal{L}(\mathfrak{g})$.

Proposition 6.1.9. *Let $\mathfrak{g}$ be a complex simple Lie algebra. The form $\langle \cdot, \cdot \rangle_0 : \mathcal{L}(\mathfrak{g}) \times \mathcal{L}(\mathfrak{g}) \to \mathbb{C}$ from Definition 6.1.2 is a universal invariant symmetric bilinear form.*

Proof. Let $\delta : \mathcal{L}(\mathfrak{g}) \times \mathcal{L}(\mathfrak{g}) \to V$ be another invariant symmetric bilinear form. Then the extension $\widetilde{\delta} : \widetilde{\mathcal{L}}(\mathfrak{g}) \times \widetilde{\mathcal{L}}(\mathfrak{g}) \to V$, defined by

$$\widetilde{\delta}(P \otimes x, Q \otimes y) = \delta(P \otimes x, Q \otimes y),$$
$$\widetilde{\delta}(\mathbb{C}c, \mathcal{L}(\mathfrak{g})) = \mathbf{0}, \quad \widetilde{\delta}(c, c) = 0$$

for $x, y \in \mathfrak{g}$, $P, Q \in \mathcal{L}$, is also invariant:

$$\widetilde{\delta}([P \otimes x + \lambda c, Q \otimes y + \mu c], R \otimes z + \nu c) = \widetilde{\delta}([P \otimes x, Q \otimes y] + \langle D(P \otimes x), Q \otimes y \rangle c, R \otimes z)$$
$$= \widetilde{\delta}([P \otimes x, Q \otimes y], R \otimes z)$$
$$= \delta([P \otimes x, Q \otimes y], R \otimes z)$$
$$= \delta(P \otimes x, [Q \otimes y, R \otimes z])$$
$$= \widetilde{\delta}(P \otimes x + \lambda c, [Q \otimes y + \mu c, R \otimes z + \nu c]).$$

Proposition 6.1.1 gives us a $v \in V$ such that $\langle \cdot, \cdot \rangle \cdot v = \widetilde{\delta}$ on $\widetilde{\mathcal{L}}(\mathfrak{g}) \times \widetilde{\mathcal{L}}(\mathfrak{g})$ for the bilinear form $\langle \cdot, \cdot \rangle$ from Theorem 6.1.8. By factorization to $\mathcal{L}(\mathfrak{g})$ we obtain $\langle \cdot, \cdot \rangle_0 \cdot v = \delta$. $\qquad\square$

We now discuss the so-called twisted affine Kac-Moody algebras.

Definition 6.1.10. Let $\mathfrak{g}$ be a complex simple Lie algebra of type X_ℓ with Cartan matrix A and $r \in \{2, 3\}$. If the generalized Cartan matrix of type $X_\ell^{(r)}$ exists (cf. Theorem A.1.12), then it is called the *extended j-twisted Cartan matrix* with respect to A, denoted by $\widehat{A}_r$ and the corresponding affine Kac-Moody algebra $\mathfrak{g}(\widehat{A}_r)$ is called *r-twisted*. Note that there is an r-twisted affine Kac-Moody algebra $\mathfrak{g}(\widehat{A}_j)$, if and only if the complex simple Lie algebra $\mathfrak{g} = \mathfrak{g}(A)$ has outer automorphisms of order r (cf. Table 3.1).

Kac showed in Theorem 8.3 of [Kac90], that the realization in Theorem 6.1.8 can be extended to twisted affine Kac-Moody algebras. Before stating this fact in Theorem 6.1.12, we define twisted loop algebras.

Definition 6.1.11. Let $\mathfrak{g}$ be a complex simple Lie algebra and $\sigma \in \mathrm{Aut}(\mathfrak{g})$ an automorphism of positive order m. Let $\epsilon := e^{\frac{2\pi i}{m}}$. Every eigenvalue of σ is of the form ϵ^j for some $0 \le j \le m - 1$. Since σ is diagonalizable, we have the following decomposition of $\mathfrak{g}$ into σ-eigenspaces:

$$\mathfrak{g} = \bigoplus_{j=0}^{m-1} \mathfrak{g}_j.$$

For each $j \in \mathbb{Z}$ we define the space $\mathcal{L}(\mathfrak{g}, \sigma, m)_j := t^j \otimes \mathfrak{g}_\ell$, where $0 \le \ell \le m - 1$ with $j - \ell \in m\mathbb{Z}$. The (σ, m)-*loop algebra* of $\mathfrak{g}$ is the direct sum

$$\mathcal{L}(\mathfrak{g}, \sigma, m) := \bigoplus_{j \in \mathbb{Z}} \mathcal{L}(\mathfrak{g}, \sigma, m)_j$$

with the Lie bracket defined by

$$\left[t^\ell \otimes x, t^n \otimes y \right] := t^{\ell+n} \otimes [x, y].$$

Theorem 6.1.12. *Let $\mathfrak{g}$ be a complex simple Lie algebra with Cartan matrix A. If there is an outer automorphism $\sigma \in \mathrm{Aut}(\mathfrak{g}) \backslash \mathrm{Inn}(\mathfrak{g})$ of order $r \in \{2, 3\}$, then the twisted affine Kac-Moody algebra $\mathfrak{g}(\widehat{A}_r)$ is isomorphic to $\widehat{\mathcal{L}}(\mathfrak{g}, \sigma, r)$, a one-dimensional extension of the universal (one-dimensional) central extension $\widetilde{\mathcal{L}}(\mathfrak{g}, \sigma, r)$ of $\mathcal{L}(\mathfrak{g}, \sigma, r)$.*

6.2. Applying Results to Generalized Kac-Moody Algebras

We will reformulate results on Lie algebras of smooth sections from the previous chapters and see that they generalize many of the facts presented in the previous section. Finally, we will define a possible generalization of Kac-Moody algebras.

Let M be a closed manifold and $\mathfrak{k}$ a central simple Lie algebra which is the fiber of a bundle $\pi : \mathfrak{K} \to M$. In this case, the vector bundle $\nu : \mathrm{V}(\mathfrak{K}) \to M$ with fiber $\mathrm{V}(\mathfrak{k}) \cong \mathbb{K}$ is trivial and there is an invariant symmetric bilinear form $\kappa : \Gamma(\mathfrak{K}) \times \Gamma(\mathfrak{K}) \to \mathrm{C}^\infty(M, \mathbb{K})$ (cf. Remark 4.1.14). The topological Lie algebra $\Gamma(\mathfrak{K})$ of the sections of $\pi : \mathfrak{K} \to M$ is a generalization of the smooth mapping algebra $\mathrm{C}^\infty(M, \mathfrak{k})$.

The following is a generalization of Proposition 6.1.9.

6. Section Algebras as Kac-Moody Algebras

Proposition 6.2.1. *By Theorems 4.4.4 and 4.6.2, the smooth section algebra $\Gamma(\mathfrak{K})$ posseses a universal continuous invariant symmetric bilinear form $\kappa : \Gamma(\mathfrak{K}) \times \Gamma(\mathfrak{K}) \to \mathrm{C}^\infty(M, \mathbb{K})$, defined by $\kappa(X, Y)(m) := \gamma\left(\varphi(X_m), \varphi(Y_m)\right)$ for $X, Y \in \Gamma(\mathfrak{K})$, any bundle chart (U, φ) in $m \in M$ and the Cartan-Killing form γ on $\mathfrak{k}$. In particular, the image of κ spans a dense subspace of $\mathrm{C}^\infty(M, \mathbb{K})$.*

The following two statements generalize Theorem 6.1.6.

Proposition 6.2.2. *Let $\nabla : \Gamma(TM) \to \mathrm{Der}\left(\Gamma(\mathfrak{K})\right)$ be a Lie connection. There is a well-defined linear map $\mathrm{d} : \Gamma(TM) \to \mathrm{End}(\mathrm{C}^\infty(M, \mathbb{K}))$, $X \mapsto \mathrm{d}_\nabla X$, which can also be considered as a linear map $\mathrm{d} : \mathrm{C}^\infty(M, \mathbb{K}) \to \Omega^1(M, \mathbb{K})$, satisfying the condition $\mathrm{d}_X \kappa(\xi, \eta) := \kappa(\nabla_X \xi, \eta) + \kappa(\xi, \nabla_X \eta)$ for $X \in \Gamma(TM)$, $\xi, \eta \in \Gamma(\mathfrak{K})$. The map d does not depend on the choice of the Lie connection ∇ (cf. Definition 5.3.4, Remark 5.3.5 and Lemma 5.3.6).*

Theorem 6.2.3. *By Corollaries 5.3.9 and A.2.14, for any Lie connection ∇, the continuous 2-cocyle $\omega_\nabla : \Gamma(\mathfrak{K}) \times \Gamma(\mathfrak{K}) \to \Omega^1(M, \mathbb{K})/\mathrm{d}\,\mathrm{C}^\infty(M, \mathbb{K})$, $(\xi, \eta) \mapsto [\kappa(\nabla\xi, \eta)]$ is universal.*

Remark 6.2.4. Let ∇, ∇' be two Lie connections. By Proposition 2.3.33 and the fact that $\mathfrak{k}$ is simple, there exists a $\zeta \in \Gamma(\mathfrak{K})$ such that $\nabla_X \xi = \nabla'_X \xi + [\zeta, \xi]$ for all $X \in \Gamma(TM)$, $\xi \in \Gamma(\mathfrak{K})$. Thus, for all $\xi, \eta \in \Gamma(\mathfrak{K})$, the following calculation shows that the cohomology class of the 2-cocycle from Theorem 6.2.3 is canonical:

$$(\omega_\nabla - \omega_{\nabla'})(\xi, \eta) = [\kappa(\nabla\xi, \eta)] - [\kappa(\nabla'\xi, \eta)] = [\kappa(\nabla'\xi, \eta)] + [\kappa([\zeta, \xi], \eta)] - [\kappa(\nabla'\xi, \eta)]$$
$$= [\kappa([\zeta, \xi], \eta)] = [\kappa([\xi, \eta], \zeta)] = ([\kappa(\cdot, \zeta)] \circ [\cdot, \cdot])(\xi, \eta)$$

That is why we write $[\omega] = [\omega_\nabla]$.

We now give the definition of a possible generalization of Kac-Moody algebras, which is associated to the Lie algebra bundle $\pi : \mathfrak{K} \to M$.

Definition 6.2.5. Let $\widetilde{\Gamma}(\mathfrak{K})$ be the universal central extension $\left(\Omega^1(M, \mathbb{K})/\mathrm{d}\,\mathrm{C}^\infty(M, \mathbb{K})\right) \oplus_\omega \Gamma(\mathfrak{K})$. We define the Lie algebra $\widehat{\Gamma}(\mathfrak{K}) := \widetilde{\Gamma}(\mathfrak{K}) \oplus \mathrm{C}^\infty(M, \mathbb{K})$ to be the vector space sum with the Lie bracket defined by

$$[(([\alpha], \xi), f), (([\beta], \eta), g)] := [([\alpha], \xi), ([\beta], \eta)] + f \cdot \nabla\xi - g \cdot \nabla\eta$$
$$= (\omega(\xi, \eta), [\xi, \eta]) + f \cdot \nabla\xi - g \cdot \nabla\eta$$

for $[\alpha], [\beta] \in \Omega^1(M, \mathbb{K})/\mathrm{d}\,\mathrm{C}^\infty(M, \mathbb{K})$, $\xi, \eta \in \Gamma(\mathfrak{K})$ and $f, g \in \mathrm{C}^\infty(M, \mathbb{K})$.

A. More on Lie Algebras

A.1. The Classification of Semisimple Lie Algebras and Affine Kac-Moody Algebras

A full classification of finite-dimensional complex semisimple Lie algebras has been given by Wilhelm Killing and Élie Cartan (cf. [Car52], [Dyn52], [Ser66]). To this end, one shows that there are bijections between the

1. isomorphism classes $[\mathfrak{g}]$ of semisimple complex Lie algebras of rank $n \in \mathbb{N}$,

2. isomorphism classes of finite root systems Δ with bases $\Pi = \{\alpha_1, \ldots, \alpha_n\}$,

3. matrices[1] $A = (a_{ij})$ defined by $a_{ij} := \alpha_j(\check{\alpha}_j)$, the so-called *Cartan matrices*,

4. disjoint unions of *Dynkin diagrams*, i.e., oriented graphs $\Gamma(A)$ defined as follows:

    - $\Gamma(A)$ has the $n = \text{rank}(A)$ vertices $v_1, \ldots, v_n$.
    - If $a_{ij}a_{ji} \leq 4$ for some $i \neq j$, then v_i, v_j are connected by $\max(|a_{ij}|, |a_{ji}|)$ edges and there is an arrow from v_j to v_i if $|a_{ij}| > 1$.

The simple complex Lie algebras correspond to the root systems with indecomposable[2] root bases, Cartan matrices, Dynkin diagrams, respectively. During this classification the so-called "exceptional" simple complex Lie algebras $\mathfrak{e}_6, \mathfrak{e}_7, \mathfrak{e}_8, \mathfrak{f}_4, \mathfrak{g}_2$ are defined and the classification can then be expressed as in the following theorem.

Theorem A.1.1. *Every finite-dimensional complex simple Lie algebra is isomorphic to exactly one in the following list. For the type X_n, the integer n is the rank of the Lie algebra, the number of columns in the Cartan matrix and the number of vertices in the Dynkin diagram.*

1. Classical:

    - $\mathfrak{sl}(n+1, \mathbb{C})$ *for $n \geq 1$, type A_n, with Cartan matrix* $\begin{pmatrix} 2 & -1 & & & & \\ -1 & 2 & -1 & & & \\ & -1 & 2 & -1 & & \\ & & \ddots & \ddots & \ddots & \\ & & & -1 & 2 & -1 \\ & & & & -1 & 2 \end{pmatrix}$,

    and with Dynkin diagram $\begin{matrix} v_n & v_{n-1} & & v_3 & v_2 & v_1 \\ \circ\!\!\!-\!\!\!-\!\!\!-\!\!\!\circ\!\!\!-\!\!\!\cdots\!\!\!-\!\!\!\circ\!\!\!-\!\!\!-\!\!\!-\!\!\!\circ\!\!\!-\!\!\!-\!\!\!-\!\!\!\circ & & & & & \end{matrix}$,

---

[1]Rigorously said, one has the classes of matrices $\left\{ A_\sigma = \left(a_{ij}^\sigma\right) \text{ defined by } a_{ij}^\sigma := \alpha_{\sigma(j)}\left(\check{\alpha}_{\sigma(i)}\right) \middle| \sigma \in \mathcal{S}_n \right\}$.

[2]A root basis Π is *indecomposable*, if there is no decomposition $\Pi = \Pi_1 \dot{\cup} \Pi_2$ such that $\alpha(\check{\beta}) = 0$ for all $\alpha \in \Pi_1$, $\beta \in \Pi_2$. A matrix $A \in \text{M}(n, \mathbb{K})$ is *indecomposable*, if there is no permutation $\sigma \in \mathcal{S}_n$ such that the matrix A_σ is a direct sum of matrices. A Dynkin diagram is *indecomposable* if it is connected.

- $\mathfrak{so}(2n+1,\mathbb{C})$ for $n \geq 2$, type B_n, with Cartan matrix $\begin{pmatrix} 2 & -2 & & & & \\ -1 & 2 & -1 & & & \\ & -1 & 2 & -1 & & \\ & & \ddots & \ddots & \ddots & \\ & & & -1 & 2 & -1 \\ & & & & -1 & 2 \end{pmatrix}$,

and with Dynkin diagram

$$\overset{v_n}{\circ}\!\!-\!\!-\!\!\overset{v_{n-1}}{\circ}\cdots\overset{v_3}{\circ}\!\!-\!\!-\!\!\overset{v_2}{\circ}\!\!\Longrightarrow\!\!\overset{v_1}{\circ},$$

- $\mathfrak{sp}(2n,\mathbb{C})$ for $n \geq 3$, type C_n, with Cartan matrix $\begin{pmatrix} 2 & -1 & & & & \\ -2 & 2 & -1 & & & \\ & -1 & 2 & -1 & & \\ & & \ddots & \ddots & \ddots & \\ & & & -1 & 2 & -1 \\ & & & & -1 & 2 \end{pmatrix}$,

and with Dynkin diagram

$$\overset{v_n}{\circ}\!\!-\!\!-\!\!\overset{v_{n-1}}{\circ}\cdots\overset{v_3}{\circ}\!\!-\!\!-\!\!\overset{v_2}{\circ}\!\!\Longleftarrow\!\!\overset{v_1}{\circ},$$

- $\mathfrak{so}(2n,\mathbb{C})$ for $n \geq 4$, type D_n, with Cartan matrix $\begin{pmatrix} 2 & & -1 & & & \\ & 2 & -1 & & & \\ -1 & -1 & 2 & -1 & & \\ & & \ddots & \ddots & \ddots & \\ & & & -1 & 2 & -1 \\ & & & & -1 & 2 \end{pmatrix}$,

and with Dynkin diagram

$$\overset{v_n}{\circ}\!\!-\!\!-\!\!\overset{v_{n-1}}{\circ}\cdots\overset{v_4}{\circ}\!\!-\!\!-\!\!\overset{v_3}{\circ}\!\!-\!\!-\!\!\overset{v_1}{\circ}.$$
$$\underset{v_2}{\overset{|}{\circ}}$$

2. Exceptional:

- $\mathfrak{e}_6$, type E_6, with Cartan matrix $\begin{pmatrix} 2 & -1 & & & & \\ -1 & 2 & & -1 & & \\ & & 2 & -1 & & \\ & -1 & -1 & 2 & -1 & \\ & & & -1 & 2 & -1 \\ & & & & -1 & 2 \end{pmatrix}$,

and with Dynkin diagram $\circ\!\!-\!\!-\!\!-\!\!\circ\!\!-\!\!-\!\!-\!\!\circ\!\!-\!\!-\!\!-\!\!\circ\!\!-\!\!-\!\!-\!\!\circ,$
$$\underset{}{\overset{|}{\circ}}$$

- $\mathfrak{e}_7$, *type E_7, with Cartan matrix* $\begin{pmatrix} 2 & -1 & & & & & \\ -1 & 2 & & -1 & & & \\ & & 2 & -1 & & & \\ & -1 & -1 & 2 & -1 & & \\ & & & -1 & 2 & -1 & \\ & & & & -1 & 2 & -1 \\ & & & & & -1 & 2 \end{pmatrix}$,

and with Dynkin diagram ○———○———○———○———○———○,

- $\mathfrak{e}_8$, *type E_8, with Cartan matrix* $\begin{pmatrix} 2 & -1 & & & & & & \\ -1 & 2 & & -1 & & & & \\ & & 2 & -1 & & & & \\ & -1 & -1 & 2 & -1 & & & \\ & & & -1 & 2 & -1 & & \\ & & & & -1 & 2 & -1 & \\ & & & & & -1 & 2 & -1 \\ & & & & & & -1 & 2 \end{pmatrix}$,

and with Dynkin diagram ○———○———○———○———○———○———○,

- $\mathfrak{f}_4$, *type F_4, with Cartan matrix* $\begin{pmatrix} 2 & -1 & & \\ -1 & 2 & -2 & \\ & -1 & 2 & -1 \\ & & -1 & 2 \end{pmatrix}$,

and with Dynkin diagram ○———○⟹○———○,

- $\mathfrak{g}_2$, *type G_2, with Cartan matrix* $\begin{pmatrix} 2 & -3 \\ -1 & 2 \end{pmatrix}$, *and with Dynkin diagram* ○⟹○.

The first step towards Kac-Moody algebras are the generalized Cartan matrices and their realizations. We follow Kac' book [Kac90].

Definition A.1.2. A *generalized Cartan matrix* is a matrix $A = (a_{ij}) \in \mathrm{M}(n, \mathbb{Z})$ fulfilling the following conditions:

1. $a_{ii} = 2$ for $i = 1, \ldots, n$,

2. $a_{ij} \leq 0$ for $i \neq j = 1, \ldots, n$,

3. $a_{ij} = 0 \Leftrightarrow a_{ji} = 0$ for $i \neq j = 1, \ldots, n$.

A *realization* of a generalized Cartan matrix A is a triple $(\mathfrak{h}, \Pi, \check{\Pi})$, where $\mathfrak{h}$ is a finite-dimensional complex vector space, $\Pi = \{\alpha_1, \ldots, \alpha_n\} \subseteq \mathfrak{h}^*$ and $\check{\Pi} = \{\check{\alpha}_1, \ldots, \check{\alpha}_n\} \subseteq \mathfrak{h}$ are indexed subsets, satisfying the following conditions:

1. Π and $\check{\Pi}$ are linearly independent,

2. $a_{ij} = \alpha_j(\check{\alpha}_i)$ for $i, j = 1, \ldots, n$,

3. $n - \mathrm{rank}(A) = \dim(\mathfrak{h}) - n$.

Definition A.1.3. If $w = (w_1, \ldots, w_n) \in \mathbb{R}^n$ is a vector, then we write

- $w > 0$ if $w_i > 0$ for all $i = 1, \ldots, n$,

- $w \geq 0$ if $w_i \geq 0$ for all $i = 1, \ldots, n$,

- $w < 0$ if $w_i < 0$ for all $i = 1, \ldots, n$.

The following statement is basically Theorem 3.4 of [Kac90] and gives a classification of generalized Cartan matrices into three types: finite, affine and indefinite.

Theorem A.1.4. *A generalized Cartan matrix $A \in \mathrm{M}(n, \mathbb{Z})$ fulfills exactly one of the following conditions:*

1. $\mathrm{rank}(A) = n$, *there is a vector $u > 0$ such that $Au > 0$, and $Av \geq 0$ implies $v \geq 0$.*
   Then A is called of finite *type.*

2. $\mathrm{rank}(A) = n - 1$, *there is a vector $u > 0$ such that $Au = 0$, and $Av \geq 0$ implies $Av = 0$.*
   Then A is called of affine *type.*

3. *There is a vector $u > 0$ such that $Au < 0$, and $Av \geq 0$ with $v \geq 0$ implies $v = 0$.*
   Then A is called of indefinite *type.*

The next step on the way to the Kac-Moody algebra $\mathfrak{g}(A)$ associated to the generalized Cartan matrix A is the auxiliary Lie algebra $\widetilde{\mathfrak{g}}(A)$. For its construction we need the concept of Lie algebras defined by generators and relations.

Definition A.1.5. Let X be a set.

1. Define $X_1 := X$ and inductively for $n \geq 2$:

$$X_n := \bigcup_{p,q \in \mathbb{N} \text{ with } p+q=n} X_p \times X_q.$$

Let $M(X) := \bigcup_{n \in \mathbb{N}} X_n$ be the disjoint union, consisting of all parentheses expressions: (x_1), (x_2, x_3), $((x_3, x_4), x_5)$, $(x_6, (x_7, x_8))$, ... for all $x_i \in X$. Let $FM(X)$ be the vector space with basis $M(X)$, i.e., $FM(X) := \{f : M(X) \to \mathbb{K}\} = \mathbb{K}^{M(X)}$ with pointwise addition and multiplication by a scalar in $\mathbb{K}$. We define a $\mathbb{K}$-bilinear map on $FM(X)$, turning it into an algebra, as follows: We extend the map

$$X_n \times X_m \longrightarrow X_{n+m}$$
$$((x_1, \ldots, x_n), (y_1, \ldots, y_m)) \longmapsto (x_1, \ldots, x_n) \cdot (y_1, \ldots, y_m) := (x_1, \ldots, x_n, y_1, \ldots, y_m)$$

for $n, m \in \mathbb{N}$ "lexicographically" to a map $M(X) \times M(X) \to M(X)$ and extend it bilinearly to a map $\mu_X : FM(X) \times FM(X) \to FM(X)$, $(u, v) \mapsto u \cdot v$.

2. Define $J(X) = \{J(u, v, w) := u \cdot (v \cdot w) + v \cdot (w \cdot u) + w \cdot (u \cdot v) \mid u, v, w \in FM(X)\}$ and $Q(X) := \{Q(u) := u \cdot u \mid u \in FM(X)\}$. We the define the two-sided ideal

$$I(X) := \bigcap \{I \trianglelefteq FM(X) \mid I \supseteq J(X) \cup Q(X)\}$$

and the quotient algebra $\mathrm{FL}(X) := FM(X)/I(X)$.

Lemma A.1.6. $\mathrm{FL}(X)$ *is a Lie algebra and* $\eta : X \to \mathrm{FL}(X), x \mapsto \overline{x} := x + I(X)$ *fulfills the following universal property: For any Lie algebra* $\mathfrak{g}$ *and for any map* $\varphi : X \to \mathfrak{g}$ *there exists a unique morphism of Lie algebras* $\mathrm{FL}(\varphi) : \mathrm{FL}(X) \to \mathfrak{g}$ *with* $\mathrm{FL}(\varphi) \circ \eta = \varphi$. *Such a pair* $(\mathrm{FL}(X), \eta)$ *is called* free Lie algebra with generators X.

Definition A.1.7. If $R \subseteq \mathrm{FL}(X)$ is a subset[3] and $\mathfrak{a}_R := \bigcap \{ \mathfrak{a} \trianglelefteq \mathrm{FL}(X) | \, \mathfrak{a} \supseteq R \}$ the ideal generated by R, then the quotient Lie algebra $\mathrm{FL}(X, R) := \mathrm{FL}(X)/\mathfrak{a}_R$ is called *Lie algebra with generators A and relations R.*

Definition A.1.8. Let A be a generalized Cartan matrix with realization[4] $(\mathfrak{h}, \Pi, \check{\Pi})$. We denote the set $\{ h_i, e_i, f_i | \, h \in \mathfrak{h}, i = 1, \ldots, n \}$ of abstract symbols by X. Let $\mathrm{FL}(X)$ be the free Lie algebra generated by X and $\mathfrak{a} \trianglelefteq \mathrm{FL}(X)$ the ideal defined by the relations

$$\big[h, h' \big] = 0 \text{ for } h, h' \in \mathfrak{h},$$
$$[h, e_i] = \alpha(h) e_i \text{ for } h \in \mathfrak{h}, i = 1, \ldots, n,$$
$$[h, f_i] = -\alpha(h) f_i \text{ for } h \in \mathfrak{h}, i = 1, \ldots, n,$$
$$[e_i, f_j] = \delta_{ij} \check{\alpha} \text{ for } i, j = 1, \ldots, n.$$

We define: $\widetilde{\mathfrak{g}}(A) := \mathrm{FL}(X)/\mathfrak{a}$.

The following proposition, which is essentially Theorem 1.2 of [Kac90], gives us the possibility to define the Kac-Moody algebra $\mathfrak{g}(A)$.

Proposition A.1.9. *Let* A *be a generalized Cartan matrix with realization* $(\mathfrak{h}, \Pi, \check{\Pi})$. *Then* $\mathfrak{h}$ *can be naturally embedded into* $\widetilde{\mathfrak{g}}(A)$ *and there is a unique maximal ideal* $\mathfrak{r}(A) \trianglelefteq \widetilde{\mathfrak{g}}(A)$ *intersecting* $\mathfrak{h}$ *trivially.*

Definition A.1.10. We define $\mathfrak{g}(A) := \widetilde{\mathfrak{g}}(A)/\mathfrak{r}(A)$ to be the *Kac-Moody algebra* associated to the generalized Cartan matrix A. If A is of affine type, then $\mathfrak{g}(A)$ is called an *affine Kac-Moody algebra*. We also write $\mathfrak{g}'(A) := [\mathfrak{g}(A), \mathfrak{g}(A)]$ for its commutator algebra.

Remark A.1.11.

1. If A is of finite type, then $\mathfrak{g}(A)$ is a finite-dimensional complex semisimple Lie algebra and A is its ordinary Cartan matrix (cf. Chapitre V of [Ser66]). In this case, $\mathfrak{g}(A)$ is perfect, i.e., $\mathfrak{g}(A) = \mathfrak{g}'(A)$. The indecomposable components of A correspond to the simple ideals of $\mathfrak{g}(A)$.

2. If A is of affine or indefinite type, then $\mathfrak{g}(A)$ is infinite-dimensional.

3. Like for Cartan matrices, i.e., generalized Cartan matrices of finite type, there are Dynkin diagrams for generalized Cartan matrices of affine type. The corresponding construction rules are similar to those in the finite case:
   - $\Gamma(A)$ has $n = \ell + 1 = \mathrm{rank}(A) + 1$ vertices $v_0, \ldots, v_\ell$.

---

[3]The elements in R are often described by elements $x \in X$ and not by $\overline{x} \in \eta(X) \subseteq \mathrm{FL}(X)$ etc.
[4]It is easy to verify that the following definitions are independent of the choice of a realization of A up to isomorphism.

- If $a_{ij}a_{ji} \leq 4$ for some $i \neq j$, then v, v_j are connected by $\max(|a_{ij}|, |a_{ji}|)$ edges and there is an arrow from v_j to v_i if $|a_{ij}| > 1$.

The classification of generalized Cartan matrices of affine type and thus of affine Kac-Moody algebras then goes as follows (cf. Tables Aff1, Aff2, Aff3 in Chapter 4 of [Kac90]).

Theorem A.1.12. *The following list provides every indecomposable generalized Cartan matrix of affine type up to concurrent permutation of the columns and rows and the corresponding generalized Dynkin diagram. The integer $\ell + 1 = \mathrm{rank}(A) + 1 = n$ is the number of columns in the generalized Cartan matrix (GCM) and the number of vertices in the Dynkin diagram.*

1. Untwisted:

- *type* $A_1^{(1)}$, *with GCM* $\begin{pmatrix} 2 & -2 \\ -2 & 2 \end{pmatrix}$, *and with Dynkin diagram* ∘⟸⟹∘,

- *for $\ell \geq 2$, type* $A_\ell^{(1)}$, *with GCM*
$$\begin{pmatrix} 2 & -1 & & & & & -1 \\ -1 & 2 & -1 & & & & \\ & -1 & 2 & -1 & & & \\ & & -1 & 2 & -1 & & \\ & & & \ddots & \ddots & \ddots & \\ & & & & -1 & 2 & -1 \\ -1 & & & & & -1 & 2 \end{pmatrix},$$

and with Dynkin diagram

- *for $\ell \geq 3$, type* $B_\ell^{(1)}$, *with GCM*
$$\begin{pmatrix} 2 & & & & & & -1 \\ & 2 & -2 & & & & \\ & -1 & 2 & -1 & & & \\ & & -1 & 2 & -1 & & \\ & & & \ddots & \ddots & \ddots & \\ -1 & & & & -1 & 2 & -1 \\ & & & & & -1 & 2 \end{pmatrix},$$

and with Dynkin diagram

- *for $\ell \geq 2$, type $C_\ell^{(1)}$, with GCM*

$$\begin{pmatrix} 2 & & & & & & -1 \\ & 2 & -1 & & & & \\ & -2 & 2 & -1 & & & \\ & & -1 & 2 & -1 & & \\ & & & \ddots & \ddots & \ddots & \\ & & & & -1 & 2 & -1 \\ -2 & & & & & -1 & 2 \end{pmatrix},$$

and with Dynkin diagram

$$\underset{v_0}{\circ}\!\Longrightarrow\!\underset{v_\ell}{\circ}\!-\!-\!\underset{v_{\ell-1}}{\circ}\!-\cdots\!-\underset{v_3}{\circ}\!-\!-\!\underset{v_2}{\circ}\!\Longleftarrow\!\underset{v_1}{\circ},$$

- *for $\ell \geq 4$, type $D_\ell^{(1)}$, with GCM*

$$\begin{pmatrix} 2 & & & & & & -1 \\ & 2 & & -1 & & & \\ & & 2 & -1 & & & \\ & -1 & -1 & 2 & -1 & & \\ & & & \ddots & \ddots & \ddots & \\ -1 & & & & -1 & 2 & -1 \\ & & & & & -1 & 2 \end{pmatrix},$$

and with Dynkin diagram

$$\underset{v_\ell}{\circ}\!-\!-\!\underset{v_{\ell-1}}{\circ}\!-\cdots\!-\underset{v_4}{\circ}\!-\!-\!\underset{v_3}{\circ}\!-\!-\!\underset{v_1}{\circ},$$
with v_0 below $v_{\ell-1}$ and v_2 below v_3.

- *type $E_6^{(1)}$, with GCM*

$$\begin{pmatrix} 2 & & -1 & & & & \\ & 2 & -1 & & & & \\ & -1 & 2 & & -1 & & \\ -1 & & & 2 & -1 & & \\ & & -1 & -1 & 2 & -1 & \\ & & & & -1 & 2 & -1 \\ & & & & & -1 & 2 \end{pmatrix},$$

and with Dynkin diagram

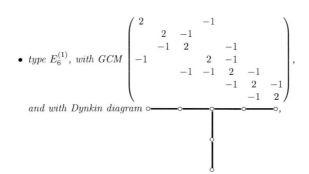

- type $E_7^{(1)}$, with GCM
$$\begin{pmatrix} 2 & & & & -1 & & & \\ & 2 & -1 & & & & & \\ & -1 & 2 & & -1 & & & \\ -1 & & & 2 & -1 & & & \\ & & -1 & -1 & 2 & -1 & & \\ & & & & -1 & 2 & -1 & \\ & & & & & -1 & 2 & -1 \\ & & & & & & -1 & 2 \end{pmatrix},$$
and with Dynkin diagram

- type $E_8^{(1)}$, with GCM
$$\begin{pmatrix} 2 & & & & -1 & & & \\ & 2 & -1 & & & & & \\ & -1 & 2 & & -1 & & & \\ -1 & & & 2 & -1 & & & \\ & & -1 & -1 & 2 & -1 & & \\ & & & & -1 & 2 & -1 & \\ & & & & & -1 & 2 & -1 \\ & & & & & & -1 & 2 \end{pmatrix},$$
and with Dynkin diagram

- type $F_4^{(1)}$, with GCM
$$\begin{pmatrix} 2 & & & & -1 \\ & 2 & -1 & & \\ & -1 & 2 & -2 & \\ & & -1 & 2 & -1 \\ -1 & & & -1 & 2 \end{pmatrix},$$
and with Dynkin diagram

- type $G_2^{(1)}$, with GCM
$$\begin{pmatrix} 2 & & -1 \\ & 2 & -3 \\ -1 & -1 & 2 \end{pmatrix},$$
and with Dynkin diagram

2. Twisted:

- type $A_2^{(2)}$, with GCM $\begin{pmatrix} 2 & -4 \\ -1 & 2 \end{pmatrix}$, and with Dynkin diagram

- *for $\ell \geq 2$, type $A_{2\ell}^{(2)}$, with GCM*
$$\begin{pmatrix} 2 & & & & & & -2 \\ & 2 & -1 & & & & \\ & -2 & 2 & -1 & & & \\ & & -1 & 2 & -1 & & \\ & & & \ddots & \ddots & \ddots & \\ & & & & -1 & 2 & -1 \\ -1 & & & & & -1 & 2 \end{pmatrix},$$

and with Dynkin diagram

$$
\begin{array}{ccccccc}
v_0 & v_\ell & v_{\ell-1} & & v_3 & v_2 & v_1
\end{array}
$$

- *for $\ell \geq 3$, type $A_{2\ell-1}^{(2)}$, with GCM*
$$\begin{pmatrix} 2 & & & & & -1 \\ & 2 & -1 & & & \\ & -2 & 2 & -1 & & \\ & & -1 & 2 & -1 & \\ & & & \ddots & \ddots & \ddots \\ -1 & & & & -1 & 2 & -1 \\ & & & & & -1 & 2 \end{pmatrix},$$

and with Dynkin diagram

$$
\begin{array}{cccccc}
v_\ell & v_{\ell-1} & & v_3 & v_2 & v_1
\end{array}
$$

$$v_0$$

- *for $\ell \geq 2$, type $D_{\ell+1}^{(2)}$, with GCM*
$$\begin{pmatrix} 2 & & & & & & -2 \\ & 2 & -2 & & & & \\ & -1 & 2 & -1 & & & \\ & & -1 & 2 & -1 & & \\ & & & \ddots & \ddots & \ddots & \\ & & & & -1 & 2 & -1 \\ -1 & & & & & -1 & 2 \end{pmatrix},$$

and with Dynkin diagram

$$
\begin{array}{cccccc}
v_0 & v_\ell & v_{\ell-1} & & v_3 & v_2 & v_1
\end{array}
$$

- *type $E_6^{(2)}$, with GCM*
$$\begin{pmatrix} 2 & & & & -1 \\ & 2 & -1 & & \\ & -1 & 2 & -1 & \\ & & -2 & 2 & -1 \\ -1 & & & -1 & 2 \end{pmatrix},$$
and with Dynkin diagram

137

- type $D_4^{(3)}$, with GCM $\begin{pmatrix} 2 & & -1 \\ & 2 & -1 \\ -1 & -3 & 2 \end{pmatrix}$,

  and with Dynkin diagram $\circ\!\!\Longrightarrow\!\!\circ\!\!\rule{1.5cm}{0.4pt}\!\!\circ$.

A.2. Universal Central Extensions and Lie Algebra Cohomology

A central extension $\widehat{\mathfrak{g}}$ of a Lie algebra $\mathfrak{g}$ by a vector space V is a Lie algebra which contains V in its center such that $\widehat{\mathfrak{g}}/V \cong \mathfrak{g}$. The extensions can be described by Lie algebra cohomology sets and the universal objects in each category correspond. In this section, $\mathfrak{g}$ is a topological Lie algebra and V a topological vector space regarded as an abelian Lie algebra and a trivial $\mathfrak{g}$-module. All statements in this section are, if not stated otherwise, also true for finite-dimensional $\mathfrak{g}$'s and V's without explicit topology by equipping $\mathfrak{g}$ with a submultiplicative norm and V with the topology given by any norm.

Definition A.2.1.

1. A *(continuous Lie algebra) extension* $\widehat{\mathfrak{g}}$ of $\mathfrak{g}$ by V is a short exact sequence of topological Lie algebras as follows:
$$0 \longrightarrow V \overset{e}{\longrightarrow} \widehat{\mathfrak{g}} \overset{f}{\longrightarrow} \mathfrak{g} \longrightarrow 0,$$
   where $\widehat{\mathfrak{g}} \cong V \oplus \mathfrak{g}$ as topological vector spaces. [5] The extension is called *central*, if $e(V) \subseteq \mathfrak{z}(\widehat{\mathfrak{g}})$

2. Two continuous Lie algebra extensions $0 \to V \overset{e}{\to} \widehat{\mathfrak{g}} \overset{f}{\to} \mathfrak{g} \to 0$ and $0 \to V \overset{\iota}{\to} \mathfrak{h} \overset{\varphi}{\to} \mathfrak{g} \to 0$ are *equivalent*, if there exists a continuous Lie algebra isomorphism $\Phi : \widehat{\mathfrak{g}} \to \mathfrak{h}$ such that $f = \varphi \circ \Phi$ and $e = \Phi \circ \epsilon$. Let $\mathrm{CentExt}_{\mathrm{ct}}(\mathfrak{g}, V)$ denote the corresponding set of equivalence classes.

3. A continuous Lie algebra extension equivalent to $0 \longrightarrow V \overset{(\mathrm{id}_V, 0)}{\longrightarrow} V \oplus \mathfrak{g} \overset{\mathrm{pr}_2}{\longrightarrow} \mathfrak{g} \longrightarrow 0$ is called *trivial*.

Example A.2.2. With $V = \mathbb{K}$, $\mathfrak{g} = \mathfrak{sl}(n, \mathbb{K})$, $\widehat{\mathfrak{g}} = \mathfrak{gl}(n, \mathbb{K})$ and the morphisms $e : \mathbb{K} \to \mathfrak{gl}(n, \mathbb{K})$, $r \mapsto r \cdot \mathbf{1}_n$ and $f : \mathfrak{gl}(n, \mathbb{K}) \to \mathfrak{sl}(n, \mathbb{K})$, $x \mapsto x - \frac{\mathrm{tr}(x)}{n} \cdot \mathbf{1}_n$, there is a central extension

$$0 \longrightarrow \mathbb{K} \overset{e}{\longrightarrow} \mathfrak{gl}(n, \mathbb{K}) \overset{f}{\longrightarrow} \mathfrak{sl}(n, \mathbb{K}) \longrightarrow 0.$$

There is a concept of a universal central extensions and the condition of weak universality, which turns out to be sufficient for universality for the most interesting Lie algebras.

Definition A.2.3. Let W be a locally convex space.

1. A central extension $0 \to V \overset{e}{\to} \widehat{\mathfrak{g}} \overset{f}{\to} \mathfrak{g} \to 0$ is called *universal for W*, if the following universal property is fulfilled:

   For each central extension $0 \to W \overset{\epsilon}{\to} \mathfrak{h} \overset{\varphi}{\to} \mathfrak{g} \to 0$ there exists a unique continuous morphism $\psi : \widehat{\mathfrak{g}} \to \mathfrak{h}$ such that $\psi(e(V)) \subseteq \epsilon(W)$ and $\varphi \circ \psi = f$.

---

[5]We are only interested in extensions where we need not worry about the existence of complements.

2. A central extension $0 \to V \xrightarrow{e} \widehat{\mathfrak{g}} \xrightarrow{f} \mathfrak{g} \to 0$ is called *weakly universal for* W, if the following property is fulfilled:

   For each central extension $0 \to W \xrightarrow{\epsilon} \mathfrak{h} \xrightarrow{\varphi} \mathfrak{g} \to 0$ there exists a continuous morphism $\psi : \widehat{\mathfrak{g}} \to \mathfrak{h}$ such that $\psi(e(V)) \subseteq \epsilon(W)$ and $\varphi \circ \psi = f$.

3. A central extension is called *weakly universal* if it is weakly universal for all locally convex spaces and *universal* if it is universal for all locally convex spaces.

Example A.2.4. If $\mathfrak{g}$ is semisimple, then the trivial central extension $0 \to 0 \to \mathfrak{g} \to \mathfrak{g} \to 0$ is universal for any vector space W:

Any central extension $0 \to W \xrightarrow{e} \widehat{\mathfrak{g}} \to \mathfrak{g} \to 0$ splits with a Lie algebra section $\mathfrak{g} \to \widehat{\mathfrak{g}}$ by Levi's Theorem 2.1.37. For the uniqueness of the section, let σ, τ be two sections and $\nu := \tau - \sigma$, a Lie algebra morphism $\mathfrak{g} \to e(W) \subseteq \mathfrak{z}\,(\widehat{\mathfrak{g}})$. So $\nu\,[x,y] = [\nu(x), \nu(y)] = 0$ for all $x, y \in \mathfrak{g}$, hence the perfectness of $\mathfrak{g}$ implies $\tau = \sigma$.

To understand when universal central extensions exist, one uses the *Lie algebra cohomology by Chevalley and Eilenberg* (cf. [CE48]). The corresponding spaces are denoted by $\mathrm{C}^*_{\mathrm{ct}}$, $\mathrm{Z}^*_{\mathrm{ct}}$, $\mathrm{B}^*_{\mathrm{ct}}$, $\mathrm{H}^*_{\mathrm{ct}}$; leaving out the "ct", the statements remain true when we do not care about continuity.

Definition A.2.5. The following are basic definitions of Lie algebra cohomology. In this context V is a (not necessarily trivial) continuous $\mathfrak{g}$-module.

1. For $p \in \mathbb{N}_0$ define the space of *p-cochains* by $\mathrm{C}^p_{\mathrm{ct}}(\mathfrak{g}, V) := \mathrm{Alt}^p_{\mathrm{ct}}(\mathfrak{g}, V)$ for

   $$\mathrm{Alt}^p_{\mathrm{ct}}(\mathfrak{g}, V) := \{\omega : \mathfrak{g}^p \to V \,|\, \omega \text{ continuous, } p\text{-linear and } \omega \circ \sigma = \mathrm{sgn}(\sigma) \cdot \omega \text{ for } \sigma \in \mathcal{S}_p\}\,.$$

   In particular, $\mathrm{C}^0_{\mathrm{ct}}(\mathfrak{g}, V) = V$ and $\mathrm{C}^1_{\mathrm{ct}}(\mathfrak{g}, V) = \mathrm{Hom}_{\mathrm{ct}}(\mathfrak{g}, V)$.

2. Define a family of linear maps $\mathrm{d} = \left(\mathrm{d}^p : \mathrm{C}^p_{\mathrm{ct}}(\mathfrak{g}, V) \to \mathrm{C}^{p+1}_{\mathrm{ct}}(\mathfrak{g}, V)\right)_{p \in \mathbb{N}_0}$ by

   $$\mathrm{d}\,\omega(x_0, \ldots, x_p) := \mathrm{d}^p\,\omega(x_0, \ldots, x_p) := \sum_{j=0}^{p} (-1)^j x_j.\omega(x_0, \ldots, x_{j-1}, x_{j+1}, \ldots, x_p)$$
   $$+ \sum_{i<j} (-1)^{i+j} \omega([x_i, x_j]\,, x_0, \ldots, x_{i-1}, x_{i+1}, \ldots, x_{j-1}, x_{j+1}, \ldots, x_p)$$
   $$\stackrel{\text{if } \mathfrak{g}.V=0}{=} \sum_{i<j} (-1)^{i+j} \omega([x_i, x_j]\,, x_0, \ldots, x_{i-1}, x_{i+1}, \ldots, x_{j-1}, x_{j+1}, \ldots, x_p)$$

   for $p \in \mathbb{N}_0$, $\omega \in \mathrm{C}^p_{\mathrm{ct}}(\mathfrak{g}, V)$, $x_0, \ldots, x_p \in \mathfrak{g}$.

3. The image of d^p is called the space of $(p+1)$-*coboundaries*, denoted by $\mathrm{B}^{p+1}_{\mathrm{ct}}(\mathfrak{g}, V)$ and its kernel is the space of *p-cocycles*, denoted by $\mathrm{Z}^p_{\mathrm{ct}}(\mathfrak{g}, V)$.

4. Since $\mathrm{d}^{p+1} \circ \mathrm{d}^p = 0$ for all $p \in \mathbb{N}_0$, we can define the quotient vector spaces

   $$\mathrm{H}^p_{\mathrm{ct}}(\mathfrak{g}, V) := \mathrm{Z}^p_{\mathrm{ct}}(\mathfrak{g}, V) / \mathrm{B}^p_{\mathrm{ct}}(\mathfrak{g}, V),$$

   the spaces of *p-cohomology classes*. The topological Lie algebra $\mathfrak{g}$ is called *centrally closed*, if $\mathrm{H}^2_{\mathrm{ct}}(\mathfrak{g}, \mathbb{K}) = \mathbf{0}$.

Remark A.2.6. We discuss the relation between central extensions and 2-cohomology classes.

1. There is a map $\Theta : \mathrm{H}^2_{\mathrm{ct}}(\mathfrak{g}, V) \to \mathrm{CentExt}(\mathfrak{g}, V)$, $[\omega] \mapsto V \oplus_\omega \mathfrak{g}$:

   For a 2-cocycle $\omega \in \mathrm{Z}^2_{\mathrm{ct}}(\mathfrak{g}, V)$ we define as a topological vector space $\widehat{\mathfrak{g}} := V \oplus \mathfrak{g}$. We set $[(r, x), (s, y)]_\omega := [(r, x), (s, y)] := (\omega(x, y), [x, y])$ for $x, y \in \mathfrak{g}$, $r, s \in V$ and obtain a topological Lie algebra $\widehat{\mathfrak{g}}$. By using the obvious continuous morphisms $V \to \widehat{\mathfrak{g}}$, $r \mapsto (r, 0)$ and $\mathfrak{g} \to \widehat{\mathfrak{g}}$, $(s, x) \mapsto x$, we obtain the extension $0 \to V \to \widehat{\mathfrak{g}} \to \mathfrak{g} \to 0$, where V is mapped to the center of $\widehat{\mathfrak{g}}$.

   Note that if $\omega' = \omega + \varepsilon$ for a 2-coboundary $\varepsilon \in \mathrm{B}^2_{\mathrm{ct}}(\mathfrak{g}, V)$, i.e., there is an $\eta \in \mathrm{Hom}(\mathfrak{g}, V)$ such that $\varepsilon(x, y) = -\eta([x, y])$, then there is a map $\Phi : V \oplus_\omega \mathfrak{g} \to V \oplus_{\omega'} \mathfrak{g}$, $(r, x) \mapsto (r - \eta(x), x)$ which is an equivalence of central extensions, since

   $$\Phi\left[(r, x), (s, y)\right]_\omega = \Phi\left(\omega(x, y), [x, y]\right) = (\omega(x, y) - \eta([x, y]), [x, y]) = \left(\omega'(x, y), [x, y]\right)$$
   $$= [(r - \eta(x), x), (s - \eta(y), y)]_{\omega'} = [\Phi(r, x), \Phi(s, y)]_{\omega'}$$

   for $x, y \in \mathfrak{g}$, $r, s \in V$. Obviously, Φ is a homomorphism and fulfills the equivalence conditions from Definition A.2.1.2.

2. There is also a map $\Omega : \mathrm{CentExt}(\mathfrak{g}, V) \to \mathrm{H}^2_{\mathrm{ct}}(\mathfrak{g}, V)$:

   Let $0 \to V \xrightarrow{e} \widehat{\mathfrak{g}} \xrightarrow{f} \mathfrak{g} \to 0$ be a central extension. An appropriate 2-cocycle $\omega \in \mathrm{Z}^2_{\mathrm{ct}}(\mathfrak{g}, V)$ can be found as follows: Let $h : \mathfrak{g} \to \widehat{\mathfrak{g}}$ be a continuous linear map with $f \circ h = \mathrm{id}_\mathfrak{g}$. Then the continuous form $\omega : \mathfrak{g} \times \mathfrak{g} \to \widehat{\mathfrak{g}}$, $(x, y) \mapsto -h[x, y] + [h(x), h(y)]$ maps to $e(V)$ because $f(\omega(x, y)) = -(f \circ h)[x, y] + [(f \circ h)(x), (f \circ h)(y)] = 0$, i.e., $\omega(x, y) \in \ker(f) = \mathrm{im}(e)$ for all $x, y \in \mathfrak{g}$. So we have a corresponding map $\omega_h \in \mathrm{Alt}^2_{\mathrm{ct}}(\mathfrak{g}, V)$. It is even a 2-cocycle because of the following calculation, where we make multiple use of the the Jacobi identity:

   $$\sum_{\sigma \in \mathcal{A}_3} \omega\left(\left[x_{\sigma(1)}, x_{\sigma(2)}\right], x_{\sigma(3)}\right)$$

   $$= \sum_{\sigma \in \mathcal{A}_3} -h\left[\left[x_{\sigma(1)}, x_{\sigma(2)}\right], x_{\sigma(3)}\right] + \left[h\left[x_{\sigma(1)}, x_{\sigma(2)}\right], h(x_{\sigma(3)})\right]$$

   $$= -h\left(\sum_{\sigma \in \mathcal{A}_3} \left[\left[x_{\sigma(1)}, x_{\sigma(2)}\right], x_{\sigma(3)}\right]\right) + \sum_{\sigma \in \mathcal{A}_3} \left[h\left[x_{\sigma(1)}, x_{\sigma(2)}\right], h(x_{\sigma(3)})\right]$$

   $$= \left[h\left[x_1, x_2\right], h(x_3)\right] + \left[h\left[x_2, x_3\right], h(x_1)\right] + \left[h\left[x_3, x_1\right], h(x_2)\right]$$

   $$= \underbrace{\left[h\left[x_1, x_2\right], h(x_3)\right] - \left[[h(x_1), h(x_2)], h(x_3)\right]}_{=0 \text{ because } \mathrm{im}(\omega) \subseteq \mathfrak{z}(\mathfrak{g})} + \left[[h(x_1), h(x_2)], h(x_3)\right]$$

   $$+ \underbrace{\left[h\left[x_2, x_3\right], h(x_1)\right] - \left[[h(x_2), h(x_3)], h(x_1)\right]}_{=0 \text{ because } \mathrm{im}(\omega) \subseteq \mathfrak{z}(\mathfrak{g})} + \left[[h(x_2), h(x_3)], h(x_1)\right]$$

   $$+ \underbrace{\left[h\left[x_3, x_1\right], h(x_2)\right] - \left[[h(x_3), h(x_1)], h(x_2)\right]}_{=0 \text{ because } \mathrm{im}(\omega) \subseteq \mathfrak{z}(\mathfrak{g})} + \left[[h(x_3), h(x_1)], h(x_2)\right]$$

   $$= \sum_{\sigma \in \mathcal{A}_3} \left[[h(x_{\sigma(1)}), h(x_{\sigma(2)})], h(x_{\sigma(3)})\right] = 0.$$

   Note that the 2-cohomology class $[\omega_h] \in \mathrm{H}^2_{\mathrm{ct}}(\mathfrak{g}, V)$ does not depend on the choice of the vector space section h because if h_1, h_2 are two different continuous linear maps with

$f \circ h_1 = f \circ h_2 = \mathrm{id}_{\mathfrak{g}}$ and $h_1 - h_2 : \mathfrak{g} \to e(V) \subseteq \mathfrak{z}(\widehat{\mathfrak{g}})$ is their difference, then

$$
\begin{aligned}
(\omega_{h_2} - \omega_{h_1})(x, y) &= h_1 [x, y] - [h_1(x), h_1(y)] - h_2 [x, y] + [h_2(x), h_2(y)] \\
&= (h_1 - h_2)[x, y] - [h_1(x), h_1(y)] + [h_1(x), h_2(y)] \\
&\quad - [h_1(x), h_2(y)] + [h_2(x), h_2(y)] \\
&= (h_1 - h_2)[x, y] - [h_1(x), (h_1 - h_2)(y)] - [(h_1 - h_2)(x), h_2(y)] \\
&= (h_1 - h_2)[x, y]
\end{aligned}
$$

for $x, y \in \mathfrak{g}$.

3. The maps Θ and Ω are inverse to each other, hence $\mathrm{H}^2_{\mathrm{ct}}(\mathfrak{g}, V) \cong \mathrm{CentExt}(\mathfrak{g}, V)$:

   Firstly, choose an $\omega \in \mathrm{Z}^2_{\mathrm{ct}}(\mathfrak{g}, V)$ and let $e : V \to V \oplus_{\omega} \mathfrak{g}$, $r \mapsto (r, 0)$ and $h : \mathfrak{g} \to V \oplus_{\omega} \mathfrak{g}$, $x \mapsto (0, x)$, inducing the 2-cocycle ω_h. Let $x, y \in \mathfrak{g}$. Then:

   $$
   \begin{aligned}
   e(\omega(x, y) - \omega_h(x, y)) &= (\omega(x, y), 0) + h [x, y] - [h(x), h(y)] \\
   &= (\omega(x, y) - \omega(x, y), [x, y] - [x, y]) = (0, 0).
   \end{aligned}
   $$

   Thus $\omega - \omega_h \in \mathrm{B}^2_{\mathrm{ct}}(\mathfrak{g}, V)$. So $\Omega(\Theta[\omega]) = [\omega]$.

   Secondly, let $0 \to V \xrightarrow{e} \widehat{\mathfrak{g}} \xrightarrow{f} \mathfrak{g} \to 0$ be a central extension with continuous vector space section $h : \mathfrak{g} \to \widehat{\mathfrak{g}}$. Then the extensions $V \oplus_{\omega_h} \mathfrak{g}$ and $\widehat{\mathfrak{g}}$ are equivalent by $(r, x) \mapsto e(r) - h(x)$ since $[e(r) - h(x), e(s) - h(y)] = [h(x), h(y)] = h [x, y] + e(\omega(x, y))$. So $\Theta(\Omega[\widehat{\mathfrak{g}}]) = [\widehat{\mathfrak{g}}]$.

In the light of this alternative description of central extensions, the following statement, which is Proposition 2.10 of [Nee02], explains Definition A.2.5.7.

Proposition A.2.7. *If $\mathfrak{g}$ is a perfect Fréchet-Lie algebra, then it is centrally closed if and only if $\mathrm{H}^2_{\mathrm{ct}}(\mathfrak{g}, V) = 0$ for all complete locally convex spaces V.*

The following statement is Whitehead's Second Lemma, a classsical result (cf. Lemma 6.4.25 of [HN11]) telling us that semisimple Lie algebras cannot be extended non-trivially by finite-dimensional trivial modules.

Theorem A.2.8 (Whitehead). *If $\mathfrak{g}$ is a semisimple Lie algebra and V is a finite-dimensional $\mathfrak{g}$-module, then $\mathrm{H}^2_{\mathrm{ct}}(\mathfrak{g}, V) = 0$. In particular, semisimple Lie algebras are centrally closed.*

Corollary A.2.9. *If $\mathfrak{g}$ is a semisimple Lie algebra and V is a complete locally convex trivial $\mathfrak{g}$-module, then $\mathrm{H}^2_{\mathrm{ct}}(\mathfrak{g}, V) = 0$.*

We now discuss the concept of universality for 2-cocycles.

Definition A.2.10. Let $\omega \in \mathrm{Z}^2_{\mathrm{ct}}(\mathfrak{g}, V)$.

1. The 2-cocycle ω is called *universal for* a locally convex space W, if the central extension $0 \to V \to V \oplus_{\omega} \mathfrak{g} \to \mathfrak{g} \to 0$ is universal for W.

2. The 2-cocycle ω is called *weakly universal for* a locally convex space W, if the group morphism $\delta_W : \mathrm{Hom}_{\mathrm{ct}}(V, W) \to \mathrm{H}^2_{\mathrm{ct}}(\mathfrak{g}, W)$, $\theta \mapsto [\theta \circ \omega]$ is bijective.

3. A 2-cocycle is called is called *weakly universal* if it is weakly universal for all locally convex spaces and *universal* if it is universal for all locally convex spaces.

Remark A.2.11. We now discuss the relation between the concepts of universality for central extensions and 2-cocycles.

1. Let $0 \to V \xrightarrow{e} \widehat{\mathfrak{g}} \xrightarrow{f} \mathfrak{g} \to 0$ be weakly universal with continuous vector space section $h : \mathfrak{g} \to \widehat{\mathfrak{g}}$ and 2-cocyle ω_h. Any extension of $\mathfrak{g}$ by a locally convex space W is, without loss of generality, of the type $0 \to W \xrightarrow{\iota} W \oplus_\eta \mathfrak{g} \xrightarrow{\varphi} \mathfrak{g} \to 0$ for some $\eta \in Z^2_{ct}(\mathfrak{g}, W)$. By weakly universality of the former sequence, there exists a continuous morphism $\psi : \widehat{\mathfrak{g}} \to \mathfrak{h}$ such that $\varphi \circ \psi = f$ and $\psi(e(V)) \subseteq \epsilon(W)$, thus $\psi \circ h$ is a continuous vector space section of the latter sequence. If $\theta : V \to W$ denotes the linear map induced by ψ, then the cohomology class $[\theta \circ \omega_h] = [\eta] \in H^2_{ct}(\mathfrak{g}, W)$. Note that θ is unique with this last condition because $[\theta \circ \omega_h] = [\eta]$ implies $\epsilon \circ \theta = \psi \circ e$. Thus ω_h is weakly universal.

2. Conversely, let $\omega \in Z^2_{ct}(\mathfrak{g}, V)$ be weakly universal and $0 \to W \xrightarrow{\epsilon} \mathfrak{h} \xrightarrow{\varphi} \mathfrak{g} \to 0$ a central extension with continuous linear section $j : \mathfrak{g} \to \mathfrak{h}$ and 2-cocycle $\omega_j \in Z^2_{ct}(\mathfrak{g}, W)$. Let $\theta : V \to W$ be the unique continuous linear map such that $[\omega_j] = [\theta \circ \omega] \in H^2_{ct}(\mathfrak{g}, W)$, i.e., there is an $\eta \in \mathrm{Hom}_{ct}(W, \mathfrak{g})$ such that $\omega_j(x, y) = \theta(\omega(x, y)) - \eta([x, y])$ for all $x, y \in \mathfrak{g}$. Now consider the extension $0 \to V \to V \oplus_\omega \mathfrak{g} \to \mathfrak{g} \to 0$. The continuous linear map $\psi : V \oplus_\omega \mathfrak{g} \to W \oplus_{\omega_j} \mathfrak{g}$, $(r, x) \mapsto (\theta(r) - \eta(x), x)$, is a morphism of Lie algebras as the following calculation shows:

$$\psi\left[(r, x), (s, y)\right]_\omega = \psi(\omega(x, y), [x, y]) = (\theta(\omega(x, y)) - \eta([x, y]), [x, y]) = (\omega_j(x, y), [x, y])$$
$$= \left[(\theta(r) - \eta(x), x), (\theta(s) - \eta(y), y)\right]_{\omega_j} = \left[\psi(r, x), \psi(s, y)\right]_{\omega_j}.$$

By using the equivalence between $\mathfrak{h}$ and $W \oplus_{\omega_j} \mathfrak{g}$, we obtain an appropriate continuous morphism $\psi' : V \oplus_\omega \mathfrak{g} \to \mathfrak{h}$. Thus the extension $V \oplus_\omega \mathfrak{g}$ of $\mathfrak{g}$ by V is weakly universal.

3. Since we already know that 2-cohomology classes and central extensions are in bijective correspondence, we conclude that 2-cohomology classes with weakly universal representatives are in bijective correspondence with weakly universal central extensions. The bijective correspondence between universal central extensions and 2-cohomology classes with universal representatives is clear by definition.

The next lemma determines conditions for weakly universal 2-cocycles to be universal.

Lemma A.2.12. *Let $\omega \in Z^2_{ct}(\mathfrak{g}, V)$ be a 2-cocycle.*

1. *If $V \oplus_\omega \mathfrak{g}$ is topologically perfect and ω is weakly universal, then ω is already universal.*

2. *If $\mathfrak{g}$ is topologically perfect and $V \subseteq \overline{[V \oplus_\omega \mathfrak{g}, V \oplus_\omega \mathfrak{g}]}$, then $V \oplus_\omega \mathfrak{g}$ is topologically perfect.*

3. *If $\mathfrak{g}$ and V are locally convex and $\omega \in Z^2_{ct}(\mathfrak{g}, V)$ is universal for $W \neq 0$, then $V \oplus_\omega \mathfrak{g}$ and $\mathfrak{g}$ are topologically perfect.*

Proof.

1. Let $\mathbf{0} \to W \xrightarrow{\epsilon} \mathfrak{h} \xrightarrow{\varphi} \mathfrak{g} \to \mathbf{0}$ be a central extension. Then there are continuous Lie algebra morphisms $\psi_i : V \oplus_\omega \mathfrak{g} \to \mathfrak{h}$ such that $\psi(V \oplus \mathbf{0}) \subseteq \epsilon(W)$ and $\varphi(\psi_i(r,x)) = x$ for all $r \in V$, $x \in \mathfrak{g}$ and $i = 1, 2$. The morphism $\nu := \psi_2 - \psi_1$ maps $V \oplus_\omega \mathfrak{g}$ to $\epsilon(W) \subseteq \mathfrak{z}(\mathfrak{h})$. So $\nu\,[(r,x),(s,y)] = [\nu(r,x), \nu(s,y)] = 0$ for all $r, s \in V$ and $x, y \in \mathfrak{g}$, hence the closed subset $\nu^{-1}(\mathbf{0}) \subseteq V \oplus_\omega \mathfrak{g}$ includes the commutator $[V \oplus_\omega \mathfrak{g}, V \oplus_\omega \mathfrak{g}]$. Since $V \oplus_\omega \mathfrak{g}$ is topologically perfect, we obtain $\nu^{-1}(\mathbf{0}) = V \oplus_\omega \mathfrak{g}$ and $\psi_1 = \psi_2$ and $\omega \in Z^2(\mathfrak{g}, V)$ is universal.

2. + 3. These are the statements of Lemma 1.7 and Lemma 1.11.(iii) of [Nee02]. $\square$

The next statement, which is the result of Remark 1.6 and Corollary 2.9 of [Nee02], helps us to argue that we do not have to distinguish between weakly universal and universal central extensions or 2-cocycles, as long as the extended Lie algebra $\mathfrak{g}$ is a perfect Fréchet-Lie algebra and the space V we extend by is complete locally convex.

Proposition A.2.13. *If $\mathfrak{g}$ is a perfect Fréchet-Lie algebra and V is a complete locally convex space, then for any $\omega \in Z^2_{\mathrm{ct}}(\mathfrak{g}, V)$, the space $V \oplus_\omega \mathfrak{g}$ is topologically perfect.*

Corollary A.2.14. *For a perfect Fréchet-Lie algebra $\mathfrak{g}$ and a complete locally convex space V, a weakly universal cocycle $\omega \in Z^2_{\mathrm{ct}}(\mathfrak{g}, V)$ is already universal.*

B. Notions of Differential Topology

B.1. The Smooth Topology on Spaces of Smooth Mappings

We want to discuss a natural topology on the space of smooth mappings $C^\infty(M, N)$ for manifolds M, N and on $\Gamma(\mathbb{V})$, the space of smooth sections of a vector bundle $\pi : \mathbb{V} \to M$. A discussion of these and the proofs of the facts in this section can be found in Chapter 2 of Christoph Wockel's dissertation [Woc06] and in Chapter 4 of the lecturs notes [Glö05]. We start with the definitions of two topologies on mapping spaces.

Definition B.1.1. Let X and Z be Hausdorff topological spaces. On $C(X, Z)$, the space of continuous maps $X \to Z$, there is a topology defined by the subbasis

$$\{\langle C, W \rangle := \{ f \in C(X, Z) |\, f(C) \subseteq W \} : C \subseteq X \text{ compact}, W \subseteq Z \text{ open} \},$$

called *compact-open topology*. Then $C(X, Z)_{co}$ denotes the corresponding topological space.

Definition B.1.2. Let X be a Hausdorff topological space and Y a locally convex space with continuous seminorms $(p_i)_{i \in I}$. On $C(X, Y)$ we then define seminorms

$$p_{C,i} : C(X, Y) \to \mathbb{R}_+, \quad f \mapsto \sup_{x \in C} p_i(f(x)),$$

for $i \in I$ and $C \subseteq X$ compact, defining the *topology of compact convergence*. Then $C(X, Y)_{cc}$ denotes the corresponding topological space.

If $Z = Y$ is a locally convex space, then the compact-open topology and the topology of compact convergence on $C(X, Z)$ coincide and we obtain the following (cf. Theorem X.3.4.2 of [Bou89] and Lemma 4.1.4 of [Glö05]).

Proposition B.1.3. *Let X be an Hausdorff topological spaces and Y a locally convex space.*

1. *The identity map $C(X, Y)_{cc} \to C(X, Y)_{co}$, $f \mapsto f$ is a homeomorphism.*

2. *The space $C(X, Y)_{co}$ is a locally convex vector space with respect to the pointwise operations $\mathbb{K} \times C(X, Y)_{co} \to C(X, Y)_{co}$, $(\lambda, f) \mapsto \lambda \cdot f$ and $C(X, Y)_{co} \times C(X, Y)_{co} \to C(X, Y)_{co}$, $(f, g) \mapsto f + g$.*

Definition B.1.4. Let M, N be manifolds. The tangent functor T induces, for each smooth map $f \in C^\infty(M, N)$, a sequence of continuous maps $(T^n f : T^n M \to T^n N)_{n \in \mathbb{N}_0}$ on the iterated tangent bundles. The initial topology with respect to the inclusion

$$C^\infty(M, N) \hookrightarrow \prod_{n=0}^{\infty} C\left(T^n M, T^n N\right)_{co}, \quad f \mapsto (T^n f)_{n \in \mathbb{N}_0}$$

is called C^∞- or *smooth topology* on $C^\infty(M, N)$. Let $C^\infty(M, N)_\infty$ denote the corresponding topological space.

Remark B.1.5. Let M be a manifold and Y a locally convex space. Then the tangent functor T maps Y to $Y \times Y$ and $Tf = \mathrm{id}_Y \times df$ for some smooth map $df \in C^\infty(TM, Y)$. Iterating this, there is, for all $n \in \mathbb{N}_0$, a map $d^n : C^\infty(M, Y) \to C(T^n M, Y)$ defined by $d^n f := \mathrm{pr}_{2^n} \circ T^n f$ and the C^∞-topology may then also be defined as the initial topology with respect to the inclusion

$$C^\infty(M, Y) \hookrightarrow \prod_{n=0}^{\infty} C(T^n M, Y)_{co}, \quad f \mapsto (d^n f)_{n \in \mathbb{N}_0}. \tag{B.1}$$

The $C(T^n M, Y)_{co}$'s are locally convex with some seminorms $p_{j,n}$ for $j \in J_n$, thus $C^\infty(M, Y)_\infty$ is locally convex with seminorms $p_{j,n}$ for $j \in J_n$, $n \in \mathbb{N}_0$ and the inclusion (B.1) is a topological embedding with closed image by Proposition 4.2.11 of [Glö05].

In order to describe, for locally convex Y, the smooth topology on $C^\infty(M, Y)_\infty$ with convenient seminorms, we define a useful type of cover of the iterated tangent spaces of M.

Definition B.1.6. Let M be a smooth manifold. For all $n \in \mathbb{N}_0$, there is a natural manifold structure on $T^n M$ and the tangent projection $TM \to M$ is smooth.

1. We denote the iterated projections $T^j M \to T^i M$ for $i \le j \in \mathbb{N}_0$ by $\pi_{j,i}^M$.

2. By the paracompactness and connectivity of M, we can define, for each $n \in \mathbb{N}_0$, a cover of relatively compact open sets $(U_{n,\ell})_{\ell \in \mathbb{N}_0}$ of $T^n M$ such that $U_{n,\ell} \subseteq C_{n,\ell} \subseteq U_{n,\ell+1}$ for the compact sets $C_{n,\ell} := \overline{U_{n,\ell}}$ and we can choose the $U_{n,\ell}$'s such that $\pi_{j,i}^M(U_{j,\ell}) \subseteq U_{i,\ell}$ and hence $\pi_{j,i}^M(C_{j,\ell}) \subseteq C_{i,\ell}$. We then call $(U_{n,\ell}, C_{n,\ell})_{n,\ell \in \mathbb{N}_0}$ a *smooth cover* of M.

Definition B.1.7. Let M be a manifold with a smooth cover $(U_{n,\ell}, C_{n,\ell})_{n,\ell \in \mathbb{N}_0}$ and Y be a Fréchet space with seminorms $(p_j)_{j \in \mathbb{N}_0}$. By Proposition 4.2.15 of [Glö05], the space $C^\infty(M, Y)_\infty$ is a Fréchet space with seminorms

$$p_{j,n,\ell} : C^\infty(M, Y) \to \mathbb{R}_+, \quad f \mapsto \sup_{x \in C_{n,\ell}} p_j(d^n f(x)).$$

Definition B.1.8. Let M be a manifold with a smooth cover $(U_{n,\ell}, C_{n,\ell})_{n,\ell \in \mathbb{N}_0}$ and $(V, \|\cdot\|)$ be a finite-dimensional normed vector space. Let $|\cdot|$ be the usual absolute value on $\mathbb{K}$. By Definition B.1.7, there are Fréchet topologies on $C^\infty(M, \mathbb{K})$ and $C^\infty(M, V)$ defined by the seminorms

$$q_{n,\ell} : C^\infty(M, \mathbb{K}) \to \mathbb{R}_+, \quad f \mapsto \sup_{x \in C_{n,\ell}} |d^n f(x)|,$$

and

$$p_{m,t} : C^\infty(M, V) \to \mathbb{R}_+, \quad g \mapsto \sup_{x \in C_{m,t}} \|d^m g(x)\|,$$

respectively. On the linear space $C^\infty(M, \mathbb{K}) \otimes V$ we then define a Fréchet topology with the seminorms

$$r_{k,s} : C^\infty(M, \mathbb{K}) \otimes V \to \mathbb{R}_+, \quad f \otimes y \mapsto q_{k,s}(f) \cdot \|y\|.$$

By classical results like Proposition III.44.1 of [Trè67] we know that the linear isomorphism $\Psi : C^\infty(M, \mathbb{K}) \otimes V \to C^\infty(M, V)$ given by $\Psi(f \otimes y)(m) := f(m)y$ is an isomorphism of Fréchet spaces. This construction is compatible with the topological tensor product $C \otimes \mathfrak{h}$ for a commutative and associative locally convex algebra C and a finite-dimensional Lie algebra $\mathfrak{h}$ as seen in Remark 2.1.9.

We now consider a smooth vector bundle $\pi : \mathbb{V} \to M$ with fiber Y, where Y is locally convex.

Definition B.1.9. The inclusion $\Gamma(\mathbb{V}) \hookrightarrow C^\infty(M, \mathbb{V})_\infty$ induces the C^∞- or smooth topology on $\Gamma(\mathbb{V})$. Let $\Gamma(\mathbb{V})_\infty$ denote the corresponding topological space.

Remark B.1.10. If $\pi : \mathbb{V} = M \times Y \to M$ is trivial, then there is a natural vector space isomorphism $\Gamma(\mathbb{V}) \to C^\infty(M, Y)$, $\sigma \mapsto \text{pr}_2 \circ \sigma$, which induces, by $T^n \mathbb{V} \cong T^n M \times T^n Y$, the isomorphism $\Gamma(\mathbb{V})_\infty \cong C^\infty(M, Y)_\infty$ as topological vector spaces.

The next statement, which combines Proposition 2.2.7 and Corollary 2.2.10 of [Woc06], simplifies the treatment of $\Gamma(\mathbb{V})_\infty$.

Proposition B.1.11. $\Gamma(\mathbb{V})_\infty$ *is a locally convex space with pointwise operations and its topology coincides, for any open cover* $\mathfrak{U} = (U_i)_{i \in I}$, *with the initial topology with respect to the restriction map*

$$\text{res} : \Gamma(\mathbb{V}) \hookrightarrow \prod_{i \in I} \Gamma(\mathbb{V}_i)_\infty, \quad \sigma \mapsto \left(\sigma_{|U_i} \right)_{i \in I},$$

where $\mathbb{V}_i$ *is the bundle* $\pi^{-1}(U_i) \to U_i$. *The set*

$$\Gamma_{\mathfrak{U}}(\mathbb{V}) := \left\{ (\sigma_i)_{i \in I} \in \bigoplus_{i \in I} \Gamma(\mathbb{V}_i) \,\middle|\, \sigma_i(x) = \sigma_j(x) \text{ for all } x \in U_i \cap U_j \right\}$$

is closed in $\bigoplus_{i \in I} \Gamma(\mathbb{V}_i)_\infty$ *and the continuous gluing map* $\text{glue} : \Gamma_{\mathfrak{U}}(\mathbb{V}) \to \Gamma(\mathbb{V})$ *defined by*

$$\text{glue} \left((\sigma_i)_{i \in I} \right)(x) = \sigma_j(x) \text{ if } x \in U_j$$

is an inverse to $\text{res}^{|\Gamma_{\mathfrak{U}}(\mathbb{V})}$, *so* $\Gamma(\mathbb{V})_\infty \cong \Gamma_{\mathfrak{U}}(\mathbb{V})$ *as topological vector spaces.*

In particular, if $(U_i, \varphi_i)_{i \in I}$ *is a bundle atlas, then* $\sigma \mapsto \varphi_i \circ \sigma_{|U_i}$ *induces an isomorphism*

$$\Gamma(\mathbb{V})_\infty \cong \left\{ (f_i)_{i \in I} \in \bigoplus_{i \in I} C^\infty(U_i, Y)_\infty \,\middle|\, f_i(x) = f_j(x) \text{ for all } x \in U_i \cap U_j \right\}.$$

Remark B.1.12. Let M be a manifold and $(V, \|\cdot\|)$ be a finite-dimensional normed vector space. Let $\pi : \mathbb{V} \to M$ be a vector bundle with fiber V. By Remark 2.2.3, there is a finite bundle atlas $(U_i, \varphi_i)_{i=1}^n$, so by Proposition B.1.11, the space $\Gamma(\mathbb{V})_\infty$ can then be considered as a closed subspace of $\bigoplus_{i=1}^n C^\infty(U_i, V)_\infty$, becoming a Fréchet space with seminorms

$$p_{i,n,\ell} : C^\infty(U_i, V) \to \mathbb{R}_+, \quad f \mapsto \sup_{x \in C_{i,n,\ell}} \|d^n f(x)\|,$$

where $(U_{i,n,\ell}, C_{i,n,\ell})_{n,\ell \in \mathbb{N}_0}$ is a smooth cover of U_i for $i \in \{1, \dots n\}$.

B.2. The Open Mapping Theorem for strict LF-spaces

We will present the concept of inductive limits of locally convex spaces and, in particular, the so-called strict LF-spaces.[1] These are generalizations of Fréchet spaces with some of the properties (complete, barrelled, etc.) of Fréchet spaces, but in general not being Baire, so in particular not metrizable.[2] For $C_c^\infty(M, \mathbb{K})$, the space of compactly supported smooth maps on a manifold M, the strict LF-topology is the natural complete topology, which is used in Distribution Theory. Our main interest in strict LF-spaces comes from the fact that there is an Open Mapping Theorem for them. In the following, we will use the notations, concepts and results of Chapter 13 of [Trè67] and Section II.6 of [SW99].

Definition B.2.1. Let E be a vector space which is the union of a family of vector subspaces $\{E_\alpha : \alpha \in A\}$ such that $E_\alpha \neq E_\beta$ unless $\alpha = \beta$. Then there are natural inclusions $h_{\beta\alpha} : E_\alpha \to E_\beta$ for $E_\alpha \subseteq E_\beta$ and $g_\alpha : E_\alpha \to E$ for $\alpha, \beta \in A$. Also suppose that the E_α's are topological vector spaces such that the $h_{\beta\alpha}$'s are continuous. Then the *inductive limit topology* on E is the final topology with respect to the g_α's, i.e., the coarsest topology on E such that the g_α's are continuous. The inductive limit topology on E is called *strict*, if the corestrictions $E_\alpha \to h_{\beta\alpha}(E_\alpha) \subseteq E_\beta$ are homeomorphisms.

Remark B.2.2. With the notation $\mathbb{R}^{(X)} := \{f : \mathbb{R} \to \mathbb{R} \,|\, f^{-1}(\mathbb{R}^\times) \subseteq X \text{ finite}\}$ for any subset $X \subseteq \mathbb{R}$, the vector space $\mathbb{R}^{(\mathbb{R})}$ is the union of the family $\{\mathbb{R}^{(F)} : F \subseteq \mathbb{R} \text{ finite}\}$. For each finite subset $F \subseteq \mathbb{R}$ the vector space $\mathbb{R}^{(F)}$ is isomorphic to $\mathbb{R}^n$ for $n = \#F \in \mathbb{N}_0$ and, by equipping each $\mathbb{R}^{(F)}$ with the natural Banach topology on $\mathbb{R}^n$, the inclusions $\mathbb{R}^{(F_1)} \to \mathbb{R}^{(F_2)}$ for finite sets $F_1 \subseteq F_2$ are homeomorphisms onto their images. Note that the strict LF-topology on $\mathbb{R}^{(\mathbb{R})}$ is not even Hausdorff (cf. Exercise II.6.12 of [Bou87]).

Issues of the above mentioned kind can be prevented by restricting to countable index sets. Indeed, the following statements, Proposition 19.4.1 of [Köt69] and Theorem II.6.6 of [SW99], give good reasons to study countable strict inductive limits.

Proposition B.2.3. *A countable strict inductive limit of topological vector spaces $\{E_n : n \in \mathbb{N}\}$ is a topological vector space and the inclusions of the E_n's are homeomorphisms onto their images.*

Theorem B.2.4. *A countable strict inductive limit of complete locally convex vector spaces is a complete locally convex vector space.*

Definition B.2.5. A countable strict inductive limit of Fréchet spaces is called a *strict LF-space*.

We can state Theorem III.2.2 of [SW99], the Open Mapping Theorem in its strict LF-version.

Theorem B.2.6. *Let $u : E \to F$ be a surjective continuous linear map between strict LF-spaces. Then u is open. In particular, continuous bijections between strict LF-spaces are isomorphisms of topological vector spaces.*

---

[1] In the literature, there are different definitions of "LF-space"; some authors include the strictness of the limit ([Trè67], [SW99]), some do not ([SN89]). We will explictly say "strict LF-space" to avoid any confusion.

[2] A topological space is Baire, if the union of any countable collection of closed subsets with empty interior has empty interior. Complete metrizable spaces are Baire.

Example B.2.7. Let M be a manifold and $(X_k)_{k\in\mathbb{N}}$ be an increasing exhaustion of M by compact sets. We can write $C_c^\infty(M,\mathbb{K}) = \bigcup_{k\in\mathbb{N}} C_k^\infty$, where $C_k^\infty := C_{X_k}^\infty(M,\mathbb{K})$ (cf. Definition 4.2.1), and equip each C_k^∞ with the subspace topology induced by $C^\infty(M,\mathbb{K})$. Since the latter space is Fréchet (cf. Definition B.1.8) and each $C_k^\infty \subseteq C^\infty(M,\mathbb{K})$ is closed, each C_k^∞ is Fréchet, too. For $i \leq j \in \mathbb{N}$, the natural inclusion $h_{ji} : C_i^\infty \to C_j^\infty$ is continuous, since the inclusion of compact subsets $\mathfrak{C}(X_i) \subseteq \mathfrak{C}(X_j)$ yields, for all $f \in C_i^\infty$, the identity

$$q_{D,m}^{(j)}(h_{ji}(f)) = \sup_{x\in D} |d^m\left(h_{ji}(f)\right)(x)| = \sup_{x\in D\cap X_i} |d^m(f)(x)| = q_{D\cap X_i, m}^{(i)}(f)$$

for the seminorms

$$q_{C,n}^{(i)} : C_i^\infty \to \mathbb{R}_+, \quad f \mapsto \sup_{x\in C} |d^n f(x)|,$$

of C_i^∞, where $C \in \mathfrak{C}(X_i)$, $n \in \mathbb{N}_0$, and

$$q_{D,m}^{(j)} : C_j^\infty \to \mathbb{R}_+, \quad f \mapsto \sup_{x\in D} |d^m f(x)|,$$

of C_j^∞, where $D \in \mathfrak{C}(X_j)$, $m \in \mathbb{N}_0$. Furthermore, for $i \leq j \in \mathbb{N}$, the inclusion $h_{ji} : C_i^\infty \to C_j^\infty$ has a closed image, so the fact that C_j^∞, and hence the image of h_{ji}, is Fréchet implies that h_{ji} is a homeomorphism onto its image by the Open Mapping Theorem. Thus $C_c^\infty(M,\mathbb{K})$ has a natural strict LF-topology such that the inclusions $C_k^\infty \to C_c^\infty(M,\mathbb{K})$ are continuous.

More generally, for finite-dimensional vector bundles $\pi : \mathbb{V} \to M$, there are natural strict LF-topologies on the spaces of compactly supported sections. In particular, the compactly supported vector bundle-valued p-forms on M and the compactly supported sections of a Lie algebra bundle form strict LF-spaces $\Omega_c^p(M,\mathbb{V})$ and $\Gamma_c(\mathfrak{K})$, respectively.

B.3. Non-Abelian Čech Cohomology

Čech cohomology is one of the important cohomology theories in algebraic topology, which is usually defined for any topological space X and a sheaf $\mathcal{F}$ of abelian groups on this space, to obtain abelian groups $\check{H}^n(X,\mathcal{F})$ for $n \in \mathbb{N}_0$. In particular, for a short exact sequence of sheaves of abelian groups

$$0 \to \mathcal{E} \overset{f}{\to} \mathcal{F} \overset{g}{\to} \mathcal{G} \to 0$$

there is an induced exact sequence of abelian groups as follows:

$$0 \to \check{H}^0(X,\mathcal{E}) \overset{[f]}{\to} \check{H}^0(X,\mathcal{F}) \overset{[g]}{\to} \check{H}^0(X,\mathcal{G}) \overset{d^1}{\to} \check{H}^1(X,\mathcal{E}) \overset{[f]}{\to} \check{H}^1(X,\mathcal{F}) \overset{[g]}{\to} \check{H}^1(X,\mathcal{G}) \overset{d^2}{\to} \check{H}^2(X,\mathcal{E}) \to \dots \tag{B.2}$$

We refer to Appendix A in Chapter II of [Wel80] for more details on how the spaces and induced maps in question are defined.

If M is a CW-complex or a manifold, then each singular cohomology group $H^n(M,A)$ is, for $n \in \mathbb{N}_0$, isomorphic to $\check{H}^n(M,\underline{A})$ (cf. Corollary E.6 of [Bre93]), where A is an abelian group and $\underline{A}$ the *constant sheaf* on M, i.e., $\underline{A}(U) = C(U,A)$ for open $U \subseteq M$.

There is a *non-abelian Čech cohomology* theory for manifolds and Lie groups respecting the smooth structures, such that there is, in low degrees, an interesting exact sequence similar to

(B.2) for extensions of Lie groups. Its restriction to commutative groups with discrete topology coincides with the usual Čech cohomology with a constant sheaf. This non-abelian Čech cohomology theory arises from the local data of (smooth and locally trivial) principal bundles and helps us to classify them. We now follow Chapter V of Neeb's manuscript [Nee04b]. Let M be a paracompact smooth manifold and G Lie group, both possibly infinite-dimensional.

Definition B.3.1. Let $\mathfrak{U} = (U_i)_{i \in I}$ be an open cover of M. We put $I_0 := I$ and define for $p \in \mathbb{N}$:

$$I_p := \left\{ (i_1, \ldots, i_{p+1}) \in I^{p+1} \big| U_{i_1} \cap \ldots \cap U_{i_{p+1}} \neq \emptyset \right\}.$$

For $\alpha = (i_1, \ldots, i_{p+1}) \in I_p$ we put $U_\alpha := U_{i_1} \cap \ldots \cap U_{i_{p+1}}$ and define the groups of p-*cochains*

$$\check{C}^p(\mathfrak{U}, \underline{G}) := \prod_{\alpha \in I^p} C^\infty(U_\alpha, G).$$

Then $\check{C}^0(\mathfrak{U}, \underline{G}) = \prod_{i \in I} C^\infty(U_i, G)$. Note that for discrete G, there is a natural isomorphism $C^\infty(V, G) \cong G$ for connected open $V \in \mathfrak{U}$. Define the maps

$$\mathrm{d}^0 : \check{C}^0(\mathfrak{U}, \underline{G}) \longrightarrow \check{C}^1(\mathfrak{U}, \underline{G}), \ (g)_{i \in I} \longmapsto (g_{ij})_{(i,j) \in I_1} \text{ by } g_{ij} := g_i g_j^{-1},$$

$$\mathrm{d}^1 : \check{C}^1(\mathfrak{U}, \underline{G}) \longrightarrow \check{C}^2(\mathfrak{U}, \underline{G}), \ (g_{ij})_{(i,j) \in I_1} \longmapsto (g_{ijk})_{(i,j,k) \in I_2} \text{ by } g_{ijk} := g_{ik}^{-1} g_{ij} g_{jk},$$

$$\mathrm{d}^2 : \check{C}^2(\mathfrak{U}, \underline{G}) \longrightarrow \check{C}^3(\mathfrak{U}, \underline{G}), \ (g_{ijk})_{(i,j,k) \in I_2} \longmapsto (g_{ijk\ell})_{(i,j,k,\ell) \in I_3} \text{ by } g_{ijk\ell} := g_{jk\ell} g_{ik\ell}^{-1} g_{ij\ell} g_{ijk}^{-1}.$$

Note that for abelian G, these maps are group morphisms, but not in general. For $p \in \{0, 1, 2\}$ we define p-*cocyles* to be the elements of the set

$$\check{Z}^p(\mathfrak{U}, \underline{G}) := \mathrm{d}^{p-1}(\mathbf{1}).$$

Note that $\mathrm{d}^p\left(\check{C}^p(\mathfrak{U}, \underline{G})\right) \subseteq \check{Z}^{p+1}(\mathfrak{U}, \underline{G})$ for $p \in \{0, 1\}$. A 0-cocycle is a collection of smooth maps that can be patched together to a smooth function on M, i.e., there is a bijection

$$C^\infty(M, G) \longrightarrow \check{H}^0(\mathfrak{U}, \underline{G}) := \check{Z}^0(\mathfrak{U}, \underline{G})$$
$$f \longmapsto \left(f_{|U_i}\right)_{i \in I}.$$

There is a group action

$$\check{C}^0(\mathfrak{U}, \underline{G}) \times \check{Z}^1(\mathfrak{U}, \underline{G}) \longrightarrow \check{Z}^1(\mathfrak{U}, \underline{G})$$
$$((g_i), (h_{jk})) \longmapsto \left(g_j h_{jk} g_k^{-1}\right)$$

and its orbits are called 1-*cohomology classes* and the set of 1-cohomology classes is denoted by

$$\check{H}^1(\mathfrak{U}, \underline{G}) := \check{Z}^1(\mathfrak{U}, \underline{G}) / \check{C}^0(\mathfrak{U}, \underline{G}).$$

If G is abelian, then all the $\check{C}^p(\mathfrak{U}, \underline{G})$'s are abelian and there is a group action

$$\check{C}^1(\mathfrak{U}, \underline{G}) \times \check{Z}^2(\mathfrak{U}, \underline{G}) \longrightarrow \check{Z}^2(\mathfrak{U}, \underline{G})$$
$$((g_{mn}), (h_{ijk})) \longmapsto \left(h_{ijk} g_{ik}^{-1} g_{ij} g_{jk}\right)$$

and its orbits are called 2-*cohomology classes* and the set of 2-cohomology classes is denoted by

$$\check{\mathrm{H}}^2(\mathfrak{U},\underline{G}) := \check{\mathrm{Z}}^2(\mathfrak{U},\underline{G})/\check{\mathrm{C}}^1(\mathfrak{U},\underline{G}).$$

For $p \in \{0,1,2\}$, if $\mathfrak{V}$ is a refinement of $\mathfrak{U}$, then there are induced maps $\check{\mathrm{C}}^p(\mathfrak{U},\underline{G}) \to \check{\mathrm{C}}^p(\mathfrak{V},\underline{G})$, leading to natural maps $\check{\mathrm{H}}^p(\mathfrak{U},\underline{G}) \to \check{\mathrm{H}}^p(\mathfrak{V},\underline{G})$, defining a direct system. We define

$$\check{\mathrm{H}}^p(M,\underline{G}) := \varinjlim_{\mathfrak{V}} \check{\mathrm{H}}^p(\mathfrak{V},\underline{G})$$

as the corresponding direct limit set. In particular, $\check{\mathrm{H}}^0(M,\underline{G}) = \mathrm{C}^\infty(M,G)$.

Remark B.3.2. There is a bijective correspondence of the equivalence classes of G-principal bundles over M with the 1-(Čech) cohomology classes with respect to M and G, symbolically expressed by $\mathrm{PBUN}(M,G) \cong \check{\mathrm{H}}^1(M,\underline{G})$:

Let $(P,M,\pi,G,R,\mathfrak{U},\Phi)$ be a principal bundle (cf. Definition 2.2.9) with transition map $g_{ji} \in \mathrm{C}^\infty\left(U_{(i,j)},G\right)$ for $(i,j) \in I_1$, so there is an equivalence as follows:

$$(\pi,\varphi_j)\circ(\pi,\varphi_i)^{-1}: U_{(i,j)}\times G \longrightarrow U_{(i,j)}\times G$$
$$(x,g) \longmapsto (x,g_{ji}(x)g).$$

The maps g_{ji} for $(i,j) \in I_1$ satisfy the condition $g_{kj}g_{ji} = g_{ki}$ for $(i,j,k) \in I_2$, thus $g = (g_{ij})_{(i,j)\in I_1}$ is a 1-cocyle.

Another trivialization of the same principal bundle on $\mathfrak{U}$ gives rise to a new family of smooth maps $\Psi = (\psi_i)_{i\in I} \in \prod_{i\in I}\mathrm{C}^\infty\left(\pi^{-1}(U_i),G\right)$ such that the corresponding 1-cocycle $g' = (g'_{ij})$ and the maps $h_j \in \mathrm{C}^\infty(U_j,G)$, defined by $(\pi,\psi_j)\circ(\pi,\varphi_j)^{-1}(x,g) = (x,h_j(x)g)$ for $j \in I$ and $(x,g) \in U_j\times G$, fulfill the relation $g'_{ji} = h_j g_{ji}h_i^{-1}$ for $(i,j) \in I_1$. So there is a map $\mathrm{PBUN}(M,G) \to \check{\mathrm{Z}}^1(\mathfrak{U},\underline{G})/\check{\mathrm{C}}^0(\mathfrak{U},\underline{G}) = \check{\mathrm{H}}^1(\mathfrak{U},\underline{G})$ which induces an injective function $F: \mathrm{PBUN}(M,G) \to \check{\mathrm{H}}^1(M,\underline{G})$ by transition to a maximal bundle atlas. The surjectivity of F can be shown as in the classical finite-dimensional case (cf. Theorem 5.3.2 of [Hus66]).

Remark B.3.3. There are some links to other cohomology theories for smoothly paracompact, so in particular finite-dimensional, manifolds.

1. If A is an abelian group considered as a discrete Lie group, then the $\check{\mathrm{H}}^p(M,\underline{A})$'s from Definition B.3.1 coincide with the singular cohomology groups $\mathrm{H}^p(M,A)$. Thus, by applying well-known facts on singular cohomology (cf., e.g., Section 3.1 of [Hat02]), we obtain:

   - If M is contractible, then $\check{\mathrm{H}}^p(M,\underline{A}) \cong \begin{cases} A & \text{if } p = 0 \\ 0 & \text{otherwise.} \end{cases}$

   - If M is path-connected, then $\check{\mathrm{H}}^1(M,\underline{A}) \cong \mathrm{Hom}(\pi_1(M),A)$.

   - $\check{\mathrm{H}}^1(\mathbb{S}^n,\underline{A}) \cong \mathrm{H}^1(\mathbb{S}^n,A) = \begin{cases} A & \text{if } n = 1 \\ 0 & \text{otherwise.} \end{cases}$

   - $\check{\mathrm{H}}^2(\mathbb{S}^n,\underline{A}) \cong \mathrm{H}^2(\mathbb{S}^n,A) = \begin{cases} A & \text{if } n = 2 \\ 0 & \text{otherwise.} \end{cases}$

2. If G is a group considered as a discrete Lie group and M is path-connected, then every G-principal bundle is flat and equivalent to some bundle P_χ, where $\chi \in \mathrm{Hom}(\pi_1(M), G)$ and P_χ is associated to the universal covering bundle $q_M : \tilde{M} \to M$ with respect to to the action

$$\pi_1(M) \times G \longrightarrow G$$
$$(\varphi, g) \longmapsto \chi(\varphi)g.$$

If $\chi, \xi \in \mathrm{Hom}(\pi_1(M), G)$ are in the same orbit of the group action

$$G \times \mathrm{Hom}(\pi_1(M), G) \longrightarrow \mathrm{Hom}(\pi_1(M), G)$$
$$(g, \chi) \longmapsto g.\chi \text{ by } (g.\chi)(d) := g\chi(d)g^{-1},$$

then P_χ, P_ξ are equivalent. So the set of orbits $\mathrm{H}^1(\pi_1(M), G) := \mathrm{Hom}(\pi_1(M), G)/G$ is in bijective correspondence with $\check{\mathrm{H}}^1(M, \underline{G})$ by Remark B.3.2.

One goal behind the use of non-abelian Čech theory is obtaining an exact sequence of cohomology sets in the category of pointed sets for every Lie group extension $q : H \to G$ with kernel N. The following statements are directly taken from Chapter V of [Nee04b] and can also be found in Chapitre I of [Fre57].

Lemma B.3.4. *The map* $\delta_0 : \check{\mathrm{H}}^0(M, \underline{G}) = \mathrm{C}^\infty(M, G) \to \check{\mathrm{H}}^1(M, \underline{N})$, $f \mapsto \left[d_0 \left(\hat{f} \right) \right]$ *is well-defined. Here,* $\hat{f} = \left(\hat{f}_i \right)_{i \in I} \subseteq \mathrm{C}^\infty(M, H)$ *is a local lift of* $f \in \mathrm{C}^\infty(M, G)$, *i.e., there is an open cover* $\mathfrak{U} = (U_i)_{i \in I}$ *such that* $q \circ \hat{f}_i = f_{|U_i}$ *for* $i \in I$. *If, in addition, N is abelian, then there is group action of* $\mathrm{C}^\infty(M, G)$ *on* $\check{\mathrm{H}}^1(M, \underline{N})$ *by automorphisms induced by conjugation of H on N. If N is central in H, then δ_0 is a group morphism.*

Theorem B.3.5. *For a Lie group extension* $q : H \to G$ *with kernel N, the induced sequence of pointed spaces*

$$1 \to \mathrm{C}^\infty(M, N) \xrightarrow{[\iota]} \mathrm{C}^\infty(M, H) \xrightarrow{[q]} \mathrm{C}^\infty(M, G) \xrightarrow{\delta_0} \check{\mathrm{H}}^1(M, \underline{N}) \xrightarrow{[\iota]} \check{\mathrm{H}}^1(M, \underline{H}) \xrightarrow{[q]} \check{\mathrm{H}}^1(M, \underline{G})$$

is exact.

Bibliography

[And92]   Nicolás Andruskiewitsch, *Some forms of Kac-Moody algebras*, J. Algebra **147** (1992), no. 2, 324–344.

[Bal00]   V. K. Balachandran, *Topological algebras*, North-Holland Mathematics Studies, vol. 185, North-Holland Publishing Co., Amsterdam, 2000, Reprint of the 1999 original.

[Bat00]   Punita Batra, *Invariants of real forms of affine Kac-Moody Lie algebras*, J. Algebra **223** (2000), no. 1, 208–236.

[Bau89]   Jean Bausch, *Étude et classification des automorphismes d'ordre fini et de première espèce des algèbres de Kac-Moody affines*, Algèbres de Kac-Moody affines, Inst. Élie Cartan, vol. 11, Univ. Nancy, Nancy, 1989, With appendices by Guy Rousseau, pp. 5–124.

[BM84]   Peter Breitenlohner and Dieter Maison, *Explicit and hidden symmetries of dimensionally reduced (super-) gravity theories*, Solutions of Einstein's equations: techniques and results (Retzbach, 1983), Lecture Notes in Phys., vol. 205, Springer, Berlin, 1984, pp. 276–310.

[BMR03]   Hechmi Ben Messaoud and Guy Rousseau, *Classification des formes réelles presque compactes des algèbres de Kac-Moody affines*, J. Algebra **267** (2003), no. 2, 443–513.

[BMR04]   _____, *Erratum to: "Classification of almost compact real forms of affine Kac-Moody algebras"*, J. Algebra **279** (2004), no. 2, 850–851.

[Bou87]   Nicolas Bourbaki, *Topological vector spaces. Chapters 1–5*, Elements of Mathematics (Berlin), Springer-Verlag, Berlin, 1987, Translated from the French by H. G. Eggleston and S. Madan.

[Bou89]   _____, *General topology. Chapters 5–10*, Elements of Mathematics (Berlin), Springer-Verlag, Berlin, 1989, Translated from the French, Reprint of the 1966 edition.

[BP87]   Stephen Berman and Arturo Pianzola, *Generators and relations for real forms of some Kac-Moody Lie algebras*, Comm. Algebra **15** (1987), no. 5, 935–959.

[BR89]   Jean Bausch and Guy Rousseau, *Involutions de première espèce des algèbres affines*, Algèbres de Kac-Moody affines, Inst. Élie Cartan, vol. 11, Univ. Nancy, Nancy, 1989, pp. 125–139.

Bibliography

[Bre93]   Glen E. Bredon, *Topology and geometry*, Graduate Texts in Mathematics, vol. 139, Springer-Verlag, New York, 1993.

[BtD85]   Theodor Bröcker and Tammo tom Dieck, *Representations of compact Lie groups*, Graduate Texts in Mathematics, vol. 98, Springer-Verlag, New York, 1985.

[BVBPBMR95]   Valérie Back-Valente, Nicole Bardy-Panse, Hechmi Ben Messaoud, and Guy Rousseau, *Formes presque-déployées des algèbres de Kac-Moody: classification et racines relatives*, J. Algebra **171** (1995), no. 1, 43–96.

[Car52]   Elie Cartan, *Œuvres complètes. Partie I. Groupes de Lie*, Gauthier-Villars, Paris, 1952.

[CE48]   Claude Chevalley and Samuel Eilenberg, *Cohomology theory of Lie groups and Lie algebras*, Trans. Amer. Math. Soc. **63** (1948), 85–124.

[Cor92a]   John F. Cornwell, *The conjugacy classes of the involutive automorphisms of affine Kac-Moody algebra $A_2^{(1)}$*, J. Phys. A **25** (1992), no. 14, 3955–3975.

[Cor92b]   ———, *General theory of the matrix formulation of the automorphisms of affine Kac-Moody algebras*, J. Phys. A **25** (1992), no. 8, 2311–2333.

[Cor92c]   ———, *Involutive automorphisms of the affine Kac-Moody algebra $A_1^{(1)}$*, J. Phys. A **25** (1992), no. 8, 2335–2358.

[Djo99]   Dragomir Ž. Djoković, *On real forms of complex semisimple Lie algebras*, Aequationes Math. **58** (1999), no. 1-2, 73–84, Dedicated to János Aczél on the occasion of his 75th birthday.

[Dyn52]   Eugene B. Dynkin, *Semisimple subalgebras of semisimple Lie algebras*, Mat. Sbornik N.S. **30(72)** (1952), 349–462 (3 plates).

[FLS74]   Moshé Flato, André Lichnerowicz, and Daniel Sternheimer, *Déformations 1 différentiables d'algèbres de Lie attachées à une variété symplectique ou de contact*, C. R. Acad. Sci. Paris Sér. A **279** (1974), 877–881.

[Fre57]   Jean Frenkel, *Cohomologie non abélienne et espaces fibrés*, Bull. Soc. Math. France **85** (1957), 135–220.

[Fuc95]   Jürgen Fuchs, *Affine Lie algebras and quantum groups*, Cambridge Monographs on Mathematical Physics, Cambridge University Press, Cambridge, 1995, An introduction, with applications in conformal field theory, Corrected reprint of the 1992 original.

[Gar80]   Howard Garland, *The arithmetic theory of loop groups*, Inst. Hautes Études Sci. Publ. Math. (1980), no. 52, 5–136.

[GHV72]   Werner Greub, Stephen Halperin, and Ray Vanstone, *Connections, curvature, and cohomology. Vol. I: De Rham cohomology of manifolds and vector bundles*, Academic Press, New York, 1972, Pure and Applied Mathematics, Vol. 47.

[Glö05] Helge Glöckner, *Infinite-dimensional lie groups*, lecture notes, online available from http://www.mathematik.tu-darmstadt.de, 2005.

[Gün10] Hasan Gündoğan, *The component group of the automorphism group of a simple Lie algebra and the splitting of the corresponding short exact sequence*, J. Lie Theory **20** (2010), no. 4, 709–737.

[Hat02] Allen Hatcher, *Algebraic topology*, Cambridge University Press, Cambridge, 2002.

[Hel78] Sigurdur Helgason, *Differential geometry, Lie groups, and symmetric spaces*, Pure and Applied Mathematics, vol. 80, Academic Press Inc. [Harcourt Brace Jovanovich Publishers], New York, 1978.

[HG09] Ernst Heintze and Christian Groß, *Finite order automorphisms and real forms of affine Kac-Moody algebras in the smooth and algebraic category*, ArXiv e-prints (2009), 80.

[HM06] Karl H. Hofmann and Sidney A. Morris, *The structure of compact groups*, augmented ed., de Gruyter Studies in Mathematics, vol. 25, Walter de Gruyter & Co., Berlin, 2006, A primer for the student—a handbook for the expert.

[HN11] Joachim Hilgert and Karl-Hermann Neeb, *An introduction to the structure and geometry of lie groups and lie algebras*, Springer-Verlag, Berlin, 2011.

[Hoc65] Gerhard P. Hochschild, *The structure of Lie groups*, Holden-Day Inc., San Francisco, 1965.

[Hus66] Dale Husemoller, *Fibre bundles*, McGraw-Hill Book Co., New York, 1966.

[Jam95] Ioan M. James (ed.), *Handbook of algebraic topology*, North-Holland, Amsterdam, 1995.

[JW10] Bas Janssens and Christoph Wockel, *Universal Central Extensions of Gauge Algebras and Groups*, ArXiv e-prints (2010), 8.

[JZ01] Quanqin Jin and Zhixue Zhang, *On automorphisms of affine Kac-Moody algebras*, Comm. Algebra **29** (2001), no. 7, 2827–2858.

[Kac68] Victor G. Kac, *Simple irreducible graded Lie algebras of finite growth*, Izv. Akad. Nauk SSSR Ser. Mat. **32** (1968), 1323–1367.

[Kac90] ———, *Infinite-dimensional Lie algebras*, third ed., Cambridge University Press, Cambridge, 1990.

[Kas84] Christian Kassel, *Kähler differentials and coverings of complex simple Lie algebras extended over a commutative algebra*, Proceedings of the Luminy conference on algebraic K-theory (Luminy, 1983), vol. 34, 1984, pp. 265–275.

[Kna96] Anthony W. Knapp, *Lie groups beyond an introduction*, Progress in Mathematics, vol. 140, Birkhäuser Boston Inc., Boston, MA, 1996.

Bibliography

[Kob86]   Zenji Kobayashi, *Automorphisms of finite order of the affine Lie algebra $A_l^{(1)}$*, Tsukuba J. Math. **10** (1986), no. 2, 269–283.

[Kob07]   Toshiyuki Kobayashi, *Visible actions on symmetric spaces*, Transform. Groups **12** (2007), no. 4, 671–694.

[Köt69]   Gottfried Köthe, *Topological vector spaces. I*, Translated from the German by D. J. H. Garling. Die Grundlehren der mathematischen Wissenschaften, Band 159, Springer-Verlag New York Inc., New York, 1969.

[KW92]   Victor G. Kac and John S. P. Wang, *On automorphisms of Kac-Moody algebras and groups*, Adv. Math. **92** (1992), no. 2, 129–195.

[Lam01]   T. Y. Lam, *A first course in noncommutative rings*, second ed., Graduate Texts in Mathematics, vol. 131, Springer-Verlag, New York, 2001.

[Lan02]   Serge Lang, *Algebra*, third ed., Graduate Texts in Mathematics, vol. 211, Springer-Verlag, New York, 2002.

[Lec79]   Pierre Lecomte, *Algèbres de lie d'ordre zéro sur une variété*, Ph.D. thesis, Université de Liège, Liège, 1979.

[Lec80]   ———, *Sur l'algèbre de Lie des sections d'un fibre en algèbres de Lie*, Ann. Inst. Fourier (Grenoble) **30** (1980), no. 4, 35–50.

[LeC94]   André LeClair, *Affine lie algebras in massive field theory and form factors from vertex operators*, Teoret. Mat. Fiz. **98** (1994), no. 3, 430–441.

[Lev88]   Fernando Levstein, *A classification of involutive automorphisms of an affine Kac-Moody Lie algebra*, J. Algebra **114** (1988), no. 2, 489–518.

[Loo69]   Ottmar Loos, *Symmetric spaces. I: General theory*, W. A. Benjamin, Inc., New York-Amsterdam, 1969.

[Mai02]   Peter Maier, *Central extensions of topological current algebras*, Geometry and analysis on finite- and infinite-dimensional Lie groups (Będlewo, 2000), Banach Center Publ., vol. 55, Polish Acad. Sci., Warsaw, 2002, pp. 61–76.

[Mas91]   William S. Massey, *A basic course in algebraic topology*, Graduate Texts in Mathematics, vol. 127, Springer-Verlag, New York, 1991.

[MHBRC94]   Daniel Moak, Konrad Heuvers, K. P. S. Bhaskara Rao, and Karen Collins, *An inversion relation of multinomial type*, Discrete Math. **131** (1994), no. 1-3, 195–204.

[Mic89]   Jouko Mickelsson, *Current algebras and groups*, Plenum Monographs in Nonlinear Physics, Plenum Press, New York, 1989.

[MN03]   Peter Maier and Karl-Hermann Neeb, *Central extensions of current groups*, Math. Ann. **326** (2003), no. 2, 367–415.

[Moo68]   Robert V. Moody, *A new class of Lie algebras*, J. Algebra **10** (1968), 211–230.

[MR93]   Alberto Medina and Philippe Revoy, *Algèbres de Lie orthogonales. Modules orthogonaux*, Comm. Algebra **21** (1993), no. 7, 2295–2315.

[MW09]   Christoph Müller and Christoph Wockel, *Equivalences of smooth and continuous principal bundles with infinite-dimensional structure group*, Adv. Geom. **9** (2009), no. 4, 605–626.

[Nar85]   R. Narasimhan, *Analysis on real and complex manifolds*, North-Holland Mathematical Library, vol. 35, North-Holland Publishing Co., Amsterdam, 1985, Reprint of the 1973 edition.

[Nee02]   Karl-Hermann Neeb, *Universal central extensions of Lie groups*, Acta Appl. Math. **73** (2002), no. 1-2, 175–219, The 2000 Twente Conference on Lie Groups (Enschede).

[Nee04a]   ———, *Current groups for non-compact manifolds and their central extensions*, Infinite dimensional groups and manifolds, IRMA Lect. Math. Theor. Phys., vol. 5, de Gruyter, Berlin, 2004, pp. 109–183.

[Nee04b]   ———, *Notes on non-abelian cohomology*, manuscript, 2004.

[NW09]   Karl-Hermann Neeb and Christoph Wockel, *Central extensions of groups of sections*, Ann. Global Anal. Geom. **36** (2009), no. 4, 381–418.

[Pro07]   Claudio Procesi, *Lie groups. an approach through invariants and representations*, Universitext, Springer, New York, 2007.

[PS86]   Andrew Pressley and Graeme Segal, *Loop groups*, Oxford Mathematical Monographs, The Clarendon Press Oxford University Press, New York, 1986, Oxford Science Publications.

[Rou88]   Guy Rousseau, *Formes réelles presque déployées des algèbres de Kac-Moody affines*, Harmonic analysis (Luxembourg, 1987), Lecture Notes in Math., vol. 1359, Springer, Berlin, 1988, pp. 252–264.

[Rou89a]   ———, *Almost split K-forms of Kac-Moody algebras*, Infinite-dimensional Lie algebras and groups (Luminy-Marseille, 1988), Adv. Ser. Math. Phys., vol. 7, World Sci. Publ., Teaneck, NJ, 1989, pp. 70–85.

[Rou89b]   ———, *Formes réelles presque-compactes des algèbres de Kac-Moody affines*, Algèbres de Kac-Moody affines, Inst. Élie Cartan, vol. 11, Univ. Nancy, Nancy, 1989, pp. 175–205.

[Ser66]   Jean-Pierre Serre, *Algèbres de Lie semi-simples complexes*, W. A. Benjamin, inc., New York-Amsterdam, 1966.

[SN89]   Stephen A. Saxon and P. P. Narayanaswami, *Metrizable [normable] (LF)-spaces and two classical problems in Fréchet [Banach] spaces*, Studia Math. **93** (1989), no. 1, 1–16.

Bibliography

[SP54]   M. E. Shanks and Lyle E. Pursell, *The Lie algebra of a smooth manifold*, Proc. Amer. Math. Soc. **5** (1954), 468–472.

[Ste51]   Norman Steenrod, *The Topology of Fibre Bundles*, Princeton Mathematical Series, vol. 14, Princeton University Press, Princeton, N. J., 1951.

[SW99]   Helmut H. Schaefer and Manfred P. Wolff, *Topological vector spaces*, second ed., Graduate Texts in Mathematics, vol. 3, Springer-Verlag, New York, 1999.

[Tan10]   Yoh Tanimoto, *Ground state representations of loop algebras*, ArXiv e-prints (2010), 22.

[Trè67]   François Trèves, *Topological vector spaces, distributions and kernels*, Academic Press, New York, 1967.

[Tur88]   Pawel Turkowski, *Low-dimensional real Lie algebras*, J. Math. Phys. **29** (1988), no. 10, 2139–2144.

[Vin94]   Èrnest B. Vinberg (ed.), *Lie groups and Lie algebras, III*, Encyclopaedia of Mathematical Sciences, vol. 41, Springer-Verlag, Berlin, 1994.

[Wag11]   Stefan Wagner, *A geometric approach to noncommutative principal bundles*, Ph.D. thesis, FAU Erlangen-Nürnberg, Erlangen, July 2011.

[Wel80]   Raymond O. Wells, Jr., *Differential analysis on complex manifolds*, second ed., Graduate Texts in Mathematics, vol. 65, Springer-Verlag, New York, 1980.

[Whi57]   Hassler Whitney, *Elementary structure of real algebraic varieties*, Ann. of Math. (2) **66** (1957), 545–556.

[Woc06]   Christoph Wockel, *Infinite-dimensional lie theory for gauge groups*, Ph.D. thesis, TU Darmstadt, Darmstadt, November 2006.

Index of Notations

Index of Terms

Lebenslauf

Januar 2011 - Oktober 2011	**Universität Erlangen-Nürnberg** Promotionsstudium am Department Mathematik Titel der Dissertation: *Classification and Structure Theory* *of Lie Algebras of Smooth Sections*
April 2008 - März 2011	Promotionsstipendiat der **Studienstiftung des deutschen Volkes**
April 2007 - März 2010	**Technische Universität Darmstadt** Promotionsstudium am Fachbereich Mathematik
Oktober 2005 - Mai 2007	**Technische Universität Darmstadt** Hauptstudium der Mathematik mit Nebenfach Philosophie Gesamturteil: sehr gut (1,06)
September 2004 - Juni 2005	**Université Catholique de Louvain** Auslandsstudium der Mathematik
Oktober 2002 - September 2004	**Technische Universität Darmstadt** Grundstudium der Mathematik mit Nebenfach Philosophie Gesamturteil: sehr gut (1,0)
September 2001 - Juni 2002	**Zivildienst** am Institut für Arbeitswissenschaft der Technischen Universität Darmstadt
August 1992 - Juni 2001	**Maria-Sibylla-Merian-Gymnasium** in Krefeld Allgemeine Hochschulreife Abiturnote: 1,0
31. Oktober 1981	Geburt in **Krefeld**